Lecture Notes in Mathematics

T0253828

For further volumes:
http://www.springer.com/series/304

Owen Dearricott • Fernando Galaz-García •
Lee Kennard • Catherine Searle • Gregor Weingart •
Wolfgang Ziller

Geometry of Manifolds with Non-negative Sectional Curvature

Editors: Rafael Herrera,
Luis Hernández-Lamoneda

 Springer

Owen Dearricott
Centro de Investigación en Matemáticas
Guanajuato, Mexico

Fernando Galaz-García
Westfälische Wilhelms-Universität
Münster, Nordrhein-Westfalen, Germany

Lee Kennard
University of California
Santa Barbara, CA, USA

Catherine Searle
Department of Mathematics
Oregon State University
Corvallis, OR, USA

Gregor Weingart
Instituto de Matemáticas - Cuernavaca
Universidad Nacional Autónoma de México
Cuernavaca, Morelos, Mexico

Wolfgang Ziller
Department of Mathematics
University of Pennsylvania
Philadelphia, PA, USA

ISBN 978-3-319-06372-0   ISBN 978-3-319-06373-7 (eBook)
DOI 10.1007/978-3-319-06373-7
Springer Cham Heidelberg New York Dordrecht London

Lecture Notes in Mathematics ISSN print edition: 0075-8434
                              ISSN electronic edition: 1617-9692

Library of Congress Control Number: 2014944318

Mathematics Subject Classification (2010): 53C21; 57S25; 53C23; 58A15; 58A20; 22EXX; 22FXX,
                 53CXX

Printed on acid-free paper

Springer is part of Springer Science+Business Media (www.springer.com)

Preface

The Third Mini-Meeting on Differential Geometry, "Recent Advances in the Geometry of Manifolds with Non-negative Sectional Curvature," was held from December 6 to 17, 2010, at the Center for Research in Mathematics (CIMAT), Guanajuato, Mexico.

The invited speakers included, in alphabetical order, Gil Bor, Owen Dearricott, Fernando Galaz-García, Luis Hernández-Lamoneda, Lee Kennard, Catherine Searle, Fabio Simas, Gregor Weingart, and Wolfgang Ziller (Distinguished Visiting Professor for the Mexican Academy of Sciences and the USA-Mexico Foundation for Science 2010). The lectures were organized into five advanced mini-courses (three lectures each) as well as two introductory mini-courses and two research reports. In addition, there were several question-and-answer sessions for graduate students, in which the lecturers participated actively.

This volume includes the lecture notes of the advanced mini-courses whose content we describe briefly. W. Ziller's notes give an introductory and up-to-date view of the subject with sufficient bibliographic references to allow further study. C. Searle's notes give a basic introduction to isometric group actions with a focus on applications to spaces with curvature bounded below. F. Galaz-García's notes deal with classification results of effective, isometric torus actions on Riemannian manifolds of maximal symmetry rank. O. Dearricott's notes deal with the construction of n-Sasakian manifolds and recall a good deal of preliminary material, such as isoparametric hypersurfaces in spheres, contact CR and CR submanifolds, and a very detailed introduction to Riemannian submersions. L. Kennard's notes are a condensed version of his research paper dealing with the Hopf conjecture with symmetry. Finally, G. Weingart's notes give an introduction to exterior differential systems.

We thank all the participants for making the meeting a successful and stimulating mathematical event. We also thank CIMAT's staff for the smooth running of the event. This meeting was the third edition of an annual event intended for researchers and graduate students, with the dual aim of combining a winter school and a research workshop. The meeting was supported by the Mexican Academy of Sciences

(AMC), the USA-Mexico Foundation for Science (FUMEC), and the Mexican Science and Technology Research Council (CONACyT). The organizers were Luis Hernández-Lamoneda (CIMAT, Mexico) and Rafael Herrera (CIMAT, Mexico).

Guanajuato, Mexico   Luis Hernández-Lamoneda
Guanajuato, Mexico   Rafael Herrera

Contents

Riemannian Manifolds with Positive Sectional Curvature

Wolfgang Ziller

1 History and Obstructions

It is fair to say that Riemannian geometry started with Gauss's famous "Disqui-sitiones generales" from 1827 in which one finds a rigorous discussion of what we now call the Gauss curvature of a surface. Much has been written about the importance and influence of this paper, see in particular the article [13] by Dombrowski for a careful discussion of its contents and influence during that time. Here we only make a few comments. Curvature of surfaces in 3-space had been studied previously by a number of authors and was defined as the product of the principal curvatures. But Gauss was the first to make the surprising discovery (his famous "Theorema Egregium") that this curvature only depends on the intrinsic metric and not on the embedding. Here one finds for example the formula for a metric in normal coordinates $ds^2 = dr^2 + f(r, \theta)^2 d\theta^2$, and Gauss showed that it has curvature $K = -f_{rr}/f$. He also proved a local version of what we nowadays call the Gauss–Bonnet theorem, which states that in a geodesic triangle Δ with angles α, β, γ the Gauss curvature measures the angle "defect":

$$\int_\Delta Kdvol = \alpha + \beta + \gamma - \pi$$

He also derived a similar formula for geodesic polygons, and Bonnet generalized it to the case where the sides are not necessarily geodesics.

These are notes from a series of lectures given in Guanajuato, Mexico in 2010. The author was supported by a grant from the National Science Foundation and by the Mexican National Academy of Sciences.

W. Ziller (✉)
University of Pennsylvania, Philadelphia, PA 19104, USA
e-mail: wziller@math.upenn.edu

O. Dearricott et al., *Geometry of Manifolds with Non-negative Sectional Curvature*,
Lecture Notes in Mathematics 2110, DOI 10.1007/978-3-319-06373-7_1,
© Springer International Publishing Switzerland 2014

Nowadays the Gauss–Bonnet theorem also goes under its global formulation for a compact surface:

$$\int_M Kdvol = 2\pi\chi(M)$$

where $\chi(M)$ is the Euler characteristic. This follows from the defect formula by using a triangulation, but it is actually not found in any of Gauss's papers. Of course no rigorous definition of a manifold or of the Euler characteristic existed at the time. Maybe the first time the above formulation can be found is in Blaschke's famous book "Vorlesungen ueber Differential Geometrie" from 1921 [7] (although it was already discussed in a paper by Boy in 1903 [8]).

In any case, the formula implies that a compact surface with positive curvature must be the 2-sphere, or the real projective plane. This is of course the beginning of the topic of these lectures on positive curvature.

The next big step was made by Riemann in his famous Habilitation from 1854, (8 months before Gauss's death). He started what we now aptly call Riemannian geometry by giving intrinsic definitions of what is now called sectional curvature (we will use sec for this notion instead of the more common one K). For each 2-plane $\sigma \subset T_p M$ one associates the sectional curvature $\sec(\sigma)$, which can be defined for example as the Gauss curvature of the two-dimensional surface spanned by going along geodesics in the direction of σ (this was in fact one of Riemann's definitions). Here one also finds for the first time an explicit formula for a space of constant curvature c:

$$ds^2 = \frac{dx_1^2 + \ldots dx_n^2}{1 + \frac{c}{4}(x_1^2 + \ldots x_n^2)}$$

including in particular the important case of the hyperbolic plane $c = -1$.

For our story, the next important development was Clifford's discovery in 1873 of the Clifford torus $\mathbb{S}^1(1) \times \mathbb{S}^1(1) \subset \mathbb{S}^3(\sqrt{2}) \subset \mathbb{R}^4$, which to his surprise has intrinsic curvature 0 (after all, something that locally looks like a plane would have to extend to infinity). This motivated Klein to formulate his famous Clifford–Klein space form problem, which in one formulation asks to classify surfaces of constant curvature. This has a painful history (one needs a good definition of completeness, a concept of an abstract manifold, and some understanding of covering space theory). In a beautiful paper by Hopf from 1926 [32] he gave us our present definition of completeness and solves the classification problem. It is amusing to note that Hopf, in a footnote, points out that the authors of previous papers on the subject, especially by Killing in 1891–1893, did not realize that the Moebius band has a flat metric.

For us the next development is of course the Bonnet–Myers theorem from 1941, which holds more generally for the positivity of an average of the sectional curvatures [39]:

Theorem (Bonnet–Myers). *If M has a complete metric with $\mathrm{Ric} \geq 1$ then the diameter is at most π, and the fundamental group is finite.*

This theorem also has an interesting history. Bonnet in 1855 only showed that the "extrinsic" diameter in 3-space has length at most π. The difficulty to obtain an intrinsic proof in higher dimensions was partially due to the fact that one needs a good formula for the second variation, which surprisingly took a long time to develop. Noteworthy are papers by Synge from 1925 [43] (who was the first one to show that a geodesic of length $> \pi$ cannot be shortest by a second variation argument), Hopf–Rinow from 1931 [33] (where they proved any two points can be joined by a minimal geodesic) Schoenberg from 1932 [47], Myers from 1935 [38] (here one finds for the first time the conclusion that $\pi_1(M)$ is finite) and Synge [44] from 1935 as well. There was a fierce competition between Myers and Synge for priorities ([38] and [44] appeared in the same issue of Duke Math. J. and in Myers paper one finds the mysterious footnote "Received by the Editors of the Annals of Mathematics, February 27, 1934, accepted by them, and later transferred to this journal"). Schoenbergs paper contains the formula for second variation that one now finds in books, and Synges papers the modern proof in the case of sectional curvature. In 1941 Myers used Synge's proof and summed over an orthonormal basis. Thus it would be fairer to call it the Bonnet–Synge–Myers theorem. Nevertheless, Myers paper created a lot of excitement at the time due to the importance of Ricci curvature in general relativity.

Important for our story is another paper by Synge from 1936 [45] where he proved:

Theorem (Synge). *If M is a compact manifold with positive sectional curvature, then $\pi_1(M)$ is 0 or $\mathbb{Z}_2$ if n is even, and M is orientable if n is odd.*

In particular, $\mathbb{RP}^n \times \mathbb{RP}^n$ does not admit a metric with positive curvature. I can also recommend reading Preissman's paper from 1936 [42] on negative curvature, still very readable for today's audience.

The surprising fact is that the above two theorems are the only known obstructions that deal with positive curvature only. There are a number of theorems that give obstructions to non-negative curvature. On the other hand, one expects that the class of manifolds admitting positive curvature is much smaller than the class admitting non-negative curvature (and this is born out in known examples). Since this is not the purpose of the present notes, we just summarize them:

- (Gromov) If M^n is a compact manifold with $\sec \geq 0$, then there exists a universal constant $c(n)$ such that $b_i(M^n, F) \leq c(n)$ for all i and any field of coefficients F. Furthermore, the fundamental group has a generating set with at most $c(n)$ elements.
- (Cheeger–Gromoll) If M^n is a compact manifold that admits a metric with non-negative sectional curvature, then there exists an abelian subgroup of $\pi_1(M^n)$ with finite index.
- (Lichnerowicz–Hitchin) The obstructions to positive scalar curvature imply that a compact spin manifold with $\hat{A}(M) \neq 0$ or $\alpha(M) \neq 0$ does not admit a metric

with non-negative sectional curvature, unless it is flat. In particular there exist exotic spheres, e.g., in dimension 9, which do not admit positive curvature.

- (Cheeger–Gromoll) If M^n is a non-compact manifold with a complete metric with sec ≥ 0, then there exists a totally geodesic compact submanifold S^k, called the soul, such that M^n is diffeomorphic to the normal bundle of S^k.

If we allow ourselves to add an upper as well as a lower bound on the sectional curvature it is convenient to introduce what is called the *pinching constant* which is defined as $\delta = \min \sec / \max \sec$. One then has the following recognition and finiteness theorems:

- (Berger–Klingenberg, Brendle–Schoen) If M^n is a compact manifold with $\delta \geq \frac{1}{4}$, then M is either diffeomorphic to a space form $\mathbb{S}^n / \Gamma$ or isometric to $\mathbb{CP}^n$, $\mathbb{HP}^n$ or $\mathrm{Ca}\mathbb{P}^2$ with their standard Fubini metric.
- (Cheeger) Given a positive constant ϵ, there are only finitely many diffeomorphism types of compact simply connected manifolds M^{2n} with $\delta \geq \epsilon$.
- (Fang–Rong, Petrunin–Tuschmann) Given a positive constant ϵ, there are only finitely many diffeomorphism types of compact manifolds M^{2n+1} with $\pi_1(M) = \pi_2(M) = 0$ and $\delta \geq \epsilon$.

Since our emphasis is on positive curvature, we will not discuss other results about non-negative curvature, except in passing. We finally mention some conjectures:

- (Hopf) There exists no metric with positive sectional curvature on $\mathbb{S}^2 \times \mathbb{S}^2$. It is natural to generalize this to a conjecture that there are no positively curved metrics on the product of two compact manifolds, or on a symmetric space of rank at least two.
- (Hopf) A compact manifold with sec ≥ 0 has non-negative Euler characteristic. An even dimensional manifold with positive curvature has positive Euler characteristic.
- (Bott–Grove–Halperin) A compact simply connected manifold M with sec ≥ 0 is *elliptic*, i.e., the sequence of Betti numbers of the loop space of M grows at most polynomially for every field of coefficients.

The latter conjecture, and its many consequences, were discussed in the literature for the first time in [24]. It is usually formulated for rational coefficients, where it is equivalent to the condition that only finitely many homotopy groups are not finite (called rationally elliptic). One can thus apply rational homotopy theory to obtain many consequences. For example, the conjecture implies, under the assumption of non-negative curvature, that $\sum b_i(M^n, F) \leq 2^n$ and that the Euler characteristic is non-negative (Hopf conjecture), and positive in even dimensions iff all odd Betti numbers vanish. The above more geometric formulation, which I will call topologically elliptic, is a natural generalization. If $n = 4$, rational homotopy theory implies that M, if compact, simply connected, and rationally elliptic, is homeomorphic to one of the known examples of non-negative curvature, i.e., $\mathbb{S}^4, \mathbb{CP}^2, \mathbb{S}^2 \times \mathbb{S}^2$ or $\mathbb{CP}^2 \# \pm \mathbb{CP}^2$. In [40] it was shown that a compact simply connected topologically elliptic 5-manifold is diffeomorphic to one of the known examples with non-negative curvature, i.e., one of $\mathbb{S}^5$, $SU(3)/SO(3)$, $\mathbb{S}^3 \times \mathbb{S}^2$ or the non-trivial $\mathbb{S}^3$ bundle over $\mathbb{S}^2$. In both cases, no curvature assumption is necessary.

Of course, one should also mention Hamilton's theorem which states that a 3 manifold with positive Ricci curvature is diffeomorphic to a space form $\mathbb{S}^3/\Gamma$. Thus in dimension 2 and 3, manifolds with positive curvature are classified.

We formulate some other natural conjectures:

- A compact simply connected 4 manifold with positive curvature is diffeomorphic to $\mathbb{S}^4$ or $\mathbb{CP}^2$.
- A compact simply connected 5 manifold with positive curvature is diffeomorphic to $\mathbb{S}^5$.
- (Klingenberg-Sakai) There are only finitely many diffeomorphism classes of positively curved manifolds in a given homotopy type.
- In a fixed even dimension, there are only finitely many diffeomorphism classes of positively curved manifolds, and all odd Betti numbers are 0.
- In a fixed odd dimension, there are only finitely many 2-connected manifolds with positive curvature.

The last two finiteness conjectures are probably too optimistic, but one should at least expect an upper bound on the Betti numbers, e.g., at most 2 in dimension 6.

2   Compact Examples of Positive Curvature

Homogeneous spaces that admit a homogeneous metric with positive curvature have been classified by Wallach in even dimensions [56] and by Bérard-Bergery in odd dimensions [4]. We now describe these examples, due to Berger, Wallach and Aloff–Wallach [2,5,56], as well as the biquotient examples [3,14,15]. In most cases we will also mention that they admit natural fibrations, a topic we will cover in Sect. 4.

1. The God given basic examples of positive curvature are the rank one symmetric spaces $\mathbb{S}^n$, $\mathbb{CP}^n$, $\mathbb{HP}^n$ or $\mathbb{Ca}\mathbb{P}^2$. (We do not know where in the literature it is first discussed that $\mathbb{Ca}\mathbb{P}^2$ carries a metric with positive curvature which is $1/4$ pinched.) They admit the well known homogeneous Hopf fibrations. Recall that a homogeneous fibration is of the form $K/H \to G/H \to G/K$ obtained from inclusions $H \subset K \subset G$.

$$\mathbb{S}^1 \to \mathbb{S}^{2n+1} \to \mathbb{CP}^n \text{ obtained from } SU(n) \subset U(n) \subset SU(n+1),$$

$$\mathbb{S}^3 \to \mathbb{S}^{4n+3} \to \mathbb{HP}^n \text{ obtained from } Sp(n) \subset Sp(n)Sp(1) \subset Sp(n+1),$$

$$\mathbb{S}^2 \to \mathbb{CP}^{2n+1} \to \mathbb{HP}^n \text{ obtained from } Sp(n)U(1) \subset Sp(n)Sp(1) \subset Sp(n+1).$$

$$\mathbb{S}^7 \to \mathbb{S}^{15} \to \mathbb{S}^8 \text{ coming from } Spin(7) \subset Spin(8) \subset Spin(9).$$

2. The homogeneous flag manifolds due to Wallach: $W^6 = \mathrm{SU}(3)/\mathrm{T}^2$, $W^{12} = \mathrm{Sp}(3)/\mathrm{Sp}(1)^3$ and $W^{24} = \mathrm{F}_4/\mathrm{Spin}(8)$. They are the total space of the following homogeneous fibrations:

$$\mathbb{S}^2 \to \mathrm{SU}(3)/\mathrm{T}^2 \to \mathbb{CP}^2,$$

$$\mathbb{S}^4 \to \mathrm{Sp}(3)/\mathrm{Sp}(1)^3 \to \mathbb{HP}^2,$$

$$\mathbb{S}^8 \to \mathrm{F}_4/\mathrm{Spin}(8) \to \mathrm{Ca}\mathbb{P}^2.$$

3. The Berger space $B^{13} = \mathrm{SU}(5)/\mathrm{Sp}(2) \cdot \mathrm{S}^1$, which admits a fibration

$$\mathbb{RP}^5 \to \mathrm{SU}(5)/\mathrm{Sp}(2) \cdot \mathrm{S}^1 \to \mathbb{CP}^4,$$

coming from the inclusions $\mathrm{Sp}(2) \cdot \mathrm{S}^1 \subset \mathrm{U}(4) \subset \mathrm{SU}(5)$. Here $\mathrm{Sp}(2) \subset \mathrm{SU}(4)$ is the usual embedding and S^1 is the center of $\mathrm{U}(4)$. Furthermore, the fiber is $\mathrm{U}(4)/\mathrm{Sp}(2) \cdot \mathrm{S}^1 = \mathrm{SU}(4)/\mathrm{Sp}(2) \cdot \mathbb{Z}_2 = \mathrm{SO}(6)/\mathrm{O}(5) = \mathbb{RP}^5$.

4. The Aloff–Wallach spaces $W^7_{p,q} = \mathrm{SU}(3)/\mathrm{diag}(z^p, z^q, \bar{z}^{p+q})$, $\gcd(p,q) = 1$. By interchanging coordinates we can assume $p \geq q \geq 0$. They have positive curvature, unless $(p,q) = (1,0)$. They also admit interesting fibrations

$$\mathbb{S}^3/\mathbb{Z}_q \to W_{p,q} \to \mathbb{CP}^2,$$

coming from the inclusions $\mathrm{diag}(z^p, z^q, \bar{z}^{p+q}) \subset \mathrm{U}(2) \subset \mathrm{SU}(3)$. Hence, as long as $q > 0$, the fiber is the lens space $\mathrm{U}(2)/\mathrm{diag}(z^p, \bar{z}^{p+q}) = \mathrm{SU}(2)/\mathrm{diag}(z^p, \bar{z}^{p+q})$ with $z^q = 1$. In the special case of $p = q = 1$, we obtain a principal $\mathrm{SO}(3)$ bundle.

Another fibration is of the form

$$\mathbb{S}^1 \to W_{p,q} \to \mathrm{SU}(3)/\mathrm{T}^2,$$

coming from the inclusions $\mathrm{diag}(z^p, z^q, \bar{z}^{p+q}) \subset \mathrm{T}^2 \subset \mathrm{SU}(3)$.

5. The Berger space: $B^7 = \mathrm{SO}(5)/\mathrm{SO}(3)$. To describe the embedding $\mathrm{SO}(3) \subset \mathrm{SO}(5)$, we recall that $\mathrm{SO}(3)$ acts orthogonally via conjugation on the set of 3×3 symmetric traceless matrices. This space is special since this embedding of $\mathrm{SO}(3)$ is maximal in $\mathrm{SO}(5)$ and hence does not admit a homogeneous fibration. On the other hand, in [22] it was shown that the manifold is diffeomorphic to an $\mathbb{S}^3$ bundle over $\mathbb{S}^4$. It is also what is called isotropy irreducible, i.e., the isotropy action of $\mathrm{SO}(3)$ on the tangent space is irreducible. This implies that there is only one $\mathrm{SO}(5)$ invariant metric up to scaling.

Thus all of these examples in (2)–(5) are the total space of a fibration. This property will be interesting to us in Sect. 4.

A natural generalization of homogeneous spaces are so called biquotients, discussed for the first time in [23]. For this, let G/H be a homogeneous space and $K \subset G$ a subgroup. Then K acts on G/H on the left, and in some cases the action is free, in which case the manifold K\G/H is a biquotient. An equivalent formulation is as follows: Take a subgroup $U \subset G \times G$ and let U act one the left and right $(u_1, u_2) * g = u_1 g u_2^{-1}$. The action is free, if for any $(u_1, u_2) \in U$ the element u_1 is not conjugate to u_2 unless $u_1 = u_2$ lies in the center of G. We denote the quotient by $G//U$. The biinvariant metric on G (or $G \times G$) induces a metric on $G//U$ with non-negative sectional curvature. In some cases, this can be deformed (via a Cheeger deformation) into one with positive curvature. We now describe these biquotient examples, due to Eschenburg and Bazaikin, explicitly.

6. There is an analogue of the six-dimensional flag manifold which is a biquotient of SU(3) under an action of $T^2 = \{(z, w) \mid z, w \in \mathbb{C}, |z| = |w| = 1\}$. It is given by:

$$E^6 = \mathrm{SU}(3)//T^2 = \mathrm{diag}(z, w, zw)\backslash \mathrm{SU}(3)/\mathrm{diag}(1, 1, z^2 w^2)^{-1}.$$

The action by T^2 is clearly free. In order to show that this manifold is not diffeomorphic to the homogeneous flag W^6, one needs to compute the cohomology with integer coefficients. The cohomology groups are the same for both manifolds, but the ring structure is different [15]. The examples W^6 and E^6, which have $b_2 = 2$, as well as $\mathbb{S}^6$ and $\mathbb{CP}^3$, are the only known examples of positive curvature in dimension 6. It is thus a natural question whether positive curvature in dimension 6 implies that the Betti numbers satisfy $b_1 = b_3 = b_5 = 0$ and $b_2 = b_4 \leq 2$.

The inhomogeneous flag also admits a fibration of a (different) sphere bundle similar to the flag manifold:

$$\mathbb{S}^2 \to \mathrm{SU}(3)//T^2 \to \mathbb{CP}^2$$

7. We now describe the seven-dimensional family of Eschenburg spaces $E_{k,l}$, which can be considered as a generalization of the Aloff Wallach spaces. Let $k := (k_1, k_2, k_3)$ and $l := (l_1, l_2, l_3) \in \mathbb{Z}^3$ be two triples of integers with $\sum k_i = \sum l_i$. We can then define a two-sided action of $S^1 = \{z \in \mathbb{C} \mid |z| = 1\}$ on SU(3) whose quotient we denote by $E_{k,l}$:

$$E_{k,l} = \mathrm{SU}(3)//S^1 = \mathrm{diag}(z^{k_1}, z^{k_2}, z^{k_3})\backslash \mathrm{SU}(3)/\mathrm{diag}(z^{l_1}, z^{l_2}, z^{l_3})^{-1}.$$

The action is free if and only if $\mathrm{diag}(z^{k_1}, z^{k_2}, z^{k_3})$ is not conjugate to $\mathrm{diag}(z^{l_1}, z^{l_2}, z^{l_3})$, i.e.,

$$\gcd(k_1 - l_i, k_2 - l_j) = 1, \quad \text{for all } i \neq j, i, j \in \{1, 2, 3\}.$$

Eschenburg showed that $E_{k,l}$ has positive curvature if

$$k_i \notin [\min(l_1, l_2, l_3), \max(l_1, l_2, l_3)].$$

Among the biquotients $E_{k,l}$ there are two interesting subfamilies. $E_p = E_{k,l}$ with $k = (1, 1, p)$ and $l = (1, 1, p + 2)$ has positive curvature when $p > 0$. It admits a large group acting by isometries. Indeed, $G = SU(2) \times SU(2)$ acting on $SU(3)$ on the left and on the right, acts by isometries in the Eschenburg metric and commutes with the S^1 action. Thus it acts by isometries on E_p and one easily sees that E_p/G is one-dimensional, i.e., E_p is cohomogeneity one. A second family consists of the cohomogeneity two Eschenburg spaces $E_{a,b,c} = E_{k,l}$ with $k = (a, b, c)$ and $l = (1, 1, a + b + c)$. Here $c = -(a + b)$ is the subfamily of Aloff–Wallach spaces. The action is free iff a, b, c are pairwise relatively prime and the Eschenburg metric has positive curvature iff, up to permutations, $a \geq b \geq c > 0$ or $a \geq b > 0, c < -a$. For these spaces $G = U(2)$ acts by isometries on the right and $E_{a,b,c}/G$ is two-dimensional. For a general Eschenburg space $G = T^3$ acts by isometries and $E_{k,l}/G$ is four-dimensional. In [28] it was shown that these groups G are indeed the id component of the full isometry group of a positively curved Eschenburg space, unless it is an Aloff–Wallach space.

There are again natural fibrations. In the case of $E_{a,b,a+b}$ with $a \geq b > 0$, we have the circle fibrations:

$$\mathbb{S}^1 \to E_{a,b,a+b} \to SU(3)//T^2,$$

and the lens space fibrations:

$$\mathbb{S}^3/\mathbb{Z}_{a+b} \to E_{a,b,a+b} \to \mathbb{CP}^2$$

which, in the case of $a = b = 1$, gives a second $SO(3)$ principal bundle over $\mathbb{CP}^2$.

The cohomogeneity two Eschenburg spaces admit orbifold fibrations, which will also be of interest to us in Sect. 4.

$$F \to E_{a,b,c} \to \mathbb{CP}^2[a + b, a + c, b + c],$$

where the fiber F is $\mathbb{RP}^3$ if all a, b, c's are odd, and $F = \mathbb{S}^3$ otherwise. Here the base is a two-dimensional weighted complex projective space. A general Eschenburg space is the total space of an orbifold circle bundle [18].

8. We finally have the 13-dimensional Bazaikin spaces B_q, which can be considered as a generalization of the Berger space B^{13}. Let $q = (q_1, \ldots, q_5)$ be a 5-tuple of integers with sum q_0 and define

$$B_q = \text{diag}(z^{q_1}, \ldots, z^{q_5}) \backslash SU(5)/\text{diag}(z^{-q_0}, A)^{-1},$$

where $A \in Sp(2) \subset SU(4) \subset SU(5)$. Here we follow the treatment in Ziller (1998, Homogeneous spaces, biquotients, and manifolds with positive curvature. Lecture Notes , unpublished) of Bazaikin's work [3] (see also [16]). First, one easily shows that the action of $Sp(2) \cdot S^1$ is free if and only if

$$\text{all } q_i\text{'s are odd} \quad \text{and} \quad \gcd(q_{\sigma(1)} + q_{\sigma(2)}, q_{\sigma(3)} + q_{\sigma(4)}) = 2,$$

for all permutations $\sigma \in S_5$. On SU(5) we choose an Eschenburg metric by scaling the biinvariant metric on SU(5) in the direction of U(4) $\subset$ SU(5). The right action of Sp(2) $\cdot$ S^1 is then by isometries. Repeating the same arguments as in the previous case, one shows that the induced metric on SU(5)//Sp(2) $\cdot$ S^1 satisfies

$$\sec > 0 \quad \textit{if and only if} \quad q_i + q_j > 0 \text{ (or } < 0) \textit{ for all } i < j.$$

The special case of $q = (1, 1, 1, 1, 1)$ is the homogeneous Berger space. One again has a one parameter subfamily that is cohomogeneity one, given by $B_p = B_{(1,1,1,1,2p-1)}$ since U(4) acting on the left induces an isometric action on the quotient. It has positive curvature when $p \geq 1$.

There is another equivalent description of the Bazaikin spaces given by

$$B_q = \text{diag}(z^{q_1}, \dots, z^{q_6}) \backslash SU(6)/Sp(3)$$

with $\sum q_i = 0$.

For these manifolds one has natural fibrations obtained from both descriptions, given by

$$\mathbb{S}^1 \to SU(6)/Sp(3) \to B_q,$$

and

$$\mathbb{S}^5 \to SU(6)/Sp(3) \to \mathbb{S}^9.$$

But B_q is not the total space of a fibration, unless it is homogeneous. On the other hand, if we allow orbifold fibrations they all admit one:

$$\mathbb{RP}^5 \to B_q \to \mathbb{CP}^4[q_0 + q_1, q_0 + q_2, \dots, q_0 + q_6].$$

Unlike in the homogeneous case, there is no general classification of positively curved biquotients, except in the following cases. We call a metric on $G//H$ torus invariant if it is induced by a left invariant metric on G which is also right invariant under the action of a maximal torus. The main theorem in [15] states that an even dimensional biquotient $G//H$ with G simple and which admits a positively curved torus invariant metric is diffeomorphic to a rank one symmetric space or the biquotient $SU(3)//T^2$. In the odd dimensional case he shows that $G//H$ with a positively curved torus invariant metric and G of rank 2 is either diffeomorphic to a homogeneous space or a positively curved Eschenburg space.

There are only two more examples which are not homogeneous or biquotients. One is a seven-dimensional exotic sphere due to Petersen–Wilhelm [41,53] (although the rather delicate calculations have not yet been verified). The method is via deforming a natural metric of non-negative curvature on a biquotient description of the exotic sphere to positive curvature. Deforming non-negative curvature to positive curvature is an important problem, and not yet well understood.

The second example is due to Grove–Verdiani–Ziller [30], and independently Dearricott [11], and will be discussed in Sect. 4. It arises as the total space of an orbifold fibration.

It is also interesting to examine the topology of the known examples. In [37] it was shown that there exist pairs of positively curved Aloff Wallach spaces which are homeomorphic but not diffeomorphic. This turns out to happen more frequently for the Eschenburg spaces [10]. For such positively curved pairs M, M', one knows that $M = M' \# \Sigma^7$, for some exotic sphere Σ^7. It is not hard to check that among the examples in [10], every exotic 7-sphere can occur as a factor Σ^7, whereas this does not seem to be the case for Aloff Wallach spaces. On the other hand, [19] provides evidence that positively curved Bazaikin spaces are homeomorphically distinct.

3 Positive Curvature with Symmetry

As we saw in Sect. 1, not much is known as far as general obstructions to positive curvature is concerned. A very successful program was suggested by Grove, motivated by the Hsiang–Kleiner theorem below [34], that one should examine positive curvature under the additional assumption of a large symmetry group.

Theorem (Hsiang–Kleiner). *If M is a compact simply connected 4-manifold on which a circle acts by isometries, then M is homeomorphic to $\mathbb{S}^4$ or $\mathbb{CP}^2$.*

Topological results on circle actions [17] imply that they are diffeomorphic, and a recent result by Grove–Wilking [26] shows that the S^1 action is linear. Thus a counter example to the Hopf conjecture would have to have a finite isometry group.

In higher dimensions, one obtains obstructions assuming that a torus of large dimension acts [25].

Theorem (Grove–Searle). *If M^n is a compact simply connected manifold with positive curvature on which a torus T^s acts by isometries, then $s \leq n/2$ in even dimensions, and $s \leq (n + 1)/2$ in odd dimensions. Equality holds iff M is diffeomorphic to $\mathbb{S}^n$ or $\mathbb{CP}^n$.*

See the article in this volume where it is shown that the torus action is linear as well. Great strides were made by Wilking [54, 55] who showed

Theorem (Wilking). *Let M^n be a compact simply connected manifold with positive curvature:*

(a) *If $n \neq 7$ and T^s acts by isometries with $s \geq \frac{n+1}{4}$, then M is homotopy equivalent to a rank one symmetric space.*

(b) *If $0 < \dim M/G < \sqrt{n/18} - 1$, then M is homotopy equivalent to a rank one symmetric space.*

(c) *If* $\dim \mathrm{Isom}(M) \geq 2n - 6$ *then* M *is either homotopy equivalent to a rank one symmetric space or isometric to a homogeneous space with positive curvature.*

One of the main new tools is the so called connectedness Lemma, which turns out to be very powerful.

Lemma (Connectedness Lemma). *If* M^n *has positive curvature, and* N *is a totally geodesic submanifold of codimension* $k \leq (n + 1)/2$, *then the inclusion* $N \hookrightarrow M$ *is* $n - 2k + 1$ *connected.*

The proof is surprisingly simple and similar to the proof of Synge's theorem. One shows that in the loop space $\Omega_N(M)$ of curves starting and ending at N every critical point, i.e., geodesic starting and ending perpendicular to N, has index at least $n - 2k + 1$. Indeed, due to the assumption on the codimension, there are $n - 2k + 1$ parallel Jacobi fields starting and ending perpendicular to N. This implies that the inclusion $N \rightarrow \Omega_N(M)$ is $n - 2k$ connected, and hence $N \rightarrow M$ is $n - 2k + 1$ connected.

An important consequence is a certain kind of periodicity in cohomology:

Lemma (Periodicity Theorem). *If* M^n *has positive curvature, and* N *is a totally geodesic submanifold of codimension* k, *then there exists cohomology class* $e \in H^k(M, \mathbb{Z})$ *such that* $\cup e : H^i(M, \mathbb{Z}) \rightarrow H^{i+k}(M, \mathbb{Z})$ *is an isomorphism for* $k \leq i \leq n - 2k$.

Along the way to proving these results, he obtains a number of fundamental obstructions on the structure of the possible isotropy groups of the action. We mention a few striking examples.

Theorem (Wilking). *Let* M^n *be a compact simply connected manifold with positive curvature on which* G *acts by isometries with principal isotropy group* H.

(a) *If* H *is non-trivial, then* $\partial(M/G)$ *is non-empty.*
(b) *Every irreducible subrepresentation of* G/H *is a subrepresentation of* K/H *where* K *is an isotropy group and* K/H *is a sphere.*
(c) *If* $\dim(M/G) = k$, *then* $\partial(M/G)$ *has at most* $k + 1$ *faces, and in the case of equality* M/G *is homeomorphic to a simplex.*

Part (a) is powerful since the distance to the boundary is a strictly concave function. In general one can use Alexandrov geometry on the quotient as an important tool, see the article by Fernando Galaz-Garcia in this volume. Part (b) has strong implications for the pair (G, H), with a very short list of possibilities when the rank of H is bigger than 1.

Recently, Kennard proved two theorems concerning the Hopf conjectures with symmetry [35, 36], see also his article in this volume.

Theorem (Kennard). *Let* M^n *be a compact simply connected manifold with positive sectional curvature.*

(a) *If* $n = 4k$ *and* T^r *acts effectively and isometrically with* $r \geq 2 \log_2(n)$, *then* $\chi(M) > 0$.

(b) *Suppose M^n has the rational cohomology of a compact, simply connected irreducible symmetric space N. If T^r acts isometrically with $r \geq 2\log_2 n + 7$, then N is either a rank one symmetric space, or a rank p Grassmannian $SO(p + q) = SO(p)SO(q)$ with $p = 2$ or $p = 3$.*

The last two exceptions are due to the fact that, up to high degrees, the rational cohomology is that of a complex or quaternionic projective space. The main new tool is to use the action of the Steenrod algebra to improve periodicity theorems. For example, the analogue of the connectedness Lemma is

Theorem (Kennard). *Let M^n be a compact simply connected manifold.*

(a) *If M^n has positive curvature and contains a pair of totally geodesic, transversely intersecting submanifolds of codimensions k_1, k_2 such that $2k_1 + 2k_2 \leq n$, then $H^*(M; Q)$ is $\gcd(4, k_1, k_2)$-periodic.*
(b) *If $H^*(M; \mathbb{Z})$ is k-periodic with $3k \leq n$, then $H^*(M; Q)$ is $\gcd(4, k)$-periodic.*

Here the cohomology is called k-periodic if there exists cohomology class $e \in H^k(M, \mathbb{Z})$ such that $\cup e : H^i(M, \mathbb{Z}) \to H^{i+k}(M, \mathbb{Z})$ is an isomorphism for $0 < i < n - k$, surjective for $i = 0$ and injective for $i = n - k$. In particular, $H^{ik}(M, \mathbb{Z}) \simeq \mathbb{Z}$ for $0 \leq i \leq n - 2k - 1$.

One conclusion one may draw from Wilking's results is that positive curvature with a large isometry group can only be expected in low dimensions (apart from the rank one symmetric spaces). This is born out in the classification of cohomogeneity one manifolds with positive curvature. We first need to describe the structure of such manifolds.

A simply connected compact cohomogeneity one manifold is the union of two homogeneous disc bundles see egg. [1, 27]. Given compact Lie groups H, K^-, K^+ and G with inclusions $H \subset K^\pm \subset G$ satisfying $K^\pm/H = \mathbb{S}^{\ell\pm}$, the transitive action of $K^\pm$ on $\mathbb{S}^{\ell\pm}$ extends to a linear action on the disc $\mathbb{D}^{\ell\pm+1}$. We can thus define

$$M = G \times_{K^-} \mathbb{D}^{\ell_-+1} \cup G \times_{K^+} \mathbb{D}^{\ell_++1}$$

glued along the boundary $\partial(G \times_{K^\pm} \mathbb{D}^{\ell\pm+1}) = G \times_{K^\pm} K^\pm/H = G/H$ via the identity. G acts on M on each half via left action in the first component. This action has principal isotropy group H and singular isotropy groups $K^\pm$. One possible description of a cohomogeneity one manifold is thus simply in terms of the Lie groups $H \subset \{K^-, K^+\} \subset G$.

The simplest example is $\{e\} \subset \{S^1, S^1\} \subset S^1$ which is the manifold $\mathbb{S}^2$ with $G = S^1$ fixing north and south pole (and thus $K^\pm = G$) and principal isotropy trivial. The isotropy groups $\{e\} \subset \{S^1 \times \{e\}, \{e\} \times S^1\} \subset S^1 \times S^1$ describe the 3-sphere $\mathbb{S}^3 \subset \mathbb{C} \oplus \mathbb{C}$ on which $S^1 \times S^1$ acts in each coordinate. More subtle is the example $S(O(1)O(1)O(1)) \subset \{S(O(2)O(1)), S(O(1)O(2))\} \subset SO(3)$. This is the 4-sphere, thought of as the unit sphere in the set of 3×3 symmetric traceless matrices, on which $SO(3)$ acts via conjugation.

The first new family of cohomogeneity one manifolds we denote by $P_{(p_-,q_-),(p_+,q_+)}$, and is given by the group diagram

$$H = \{\pm(1,1), \pm(i,i), \pm(j,j), \pm(k,k)\}$$
$$\subset \{(e^{ip_-t}, e^{iq_-t}) \cdot H \, , \, (e^{ip_+t}, e^{iq_+t}) \cdot H\} \subset S^3 \times S^3.$$

where $\gcd(p_-,q_-) = \gcd(p_+,q_+) = 1$ and all four integers are congruent to 1 mod 4.

The second family $Q_{(p_-,q_-),(p_+,q_+)}$ is given by the group diagram

$$H = \{(\pm 1, \pm 1), (\pm i, \pm i)\} \subset \{(e^{ip_-t}, e^{iq_-t}) \cdot H \, , \, (e^{ip_+t}, e^{iq_+t}) \cdot H\} \subset S^3 \times S^3,$$

where $\gcd(p_-,q_-) = \gcd(p_+,q_+) = 1$, q_+ is even, and p_-, q_-, p_+ are congruent to 1 mod 4.

Special among these are the manifolds $P_k = P_{(1,1),(1+2k,1-2k)}$, $Q_k = Q_{(1,1),(k,k+1)}$ with $k \geq 1$, and the exceptional manifold $R^7 = Q_{(-3,1),(1,2)}$. In terms of these descriptions, we can state the classification, see [29, 50].

Theorem (Verdiani, n Even, Grove–Wilking–Ziller, n Odd). *A simply connected cohomogeneity one manifold M^n with an invariant metric of positive sectional curvature is equivariantly diffeomorphic to one of the following:*

- *An isometric action on a rank one symmetric space.*
- *One of E_p^7, B_p^{13} or B^7.*
- *One of the 7-manifolds $P_k = P_{(1,1),(1+2k,1-2k)}$, $Q_k = Q_{(1,1),(k,k+1)}$ with $k \geq 1$, or the exceptional manifold $R^7 = Q_{(-3,1),(1,2)}$*

with one of the actions described above.

Here P_k, Q_k, R should be considered as candidates for positive curvature. Recently the exceptional manifold R was excluded [52]:

Theorem (Verdiani–Ziller). *Let M be one of the 7-manifolds $Q_{(p_-,q_-),(p_+,q_+)}$ with its cohomogeneity one action by $G = S^3 \times S^3$ and assume that M is not of type $Q_k, k \geq 0$. Then there exists no analytic metric with non-negative sectional curvature invariant under G, although there exists a smooth one.*

In particular, it cannot carry an invariant metric with positive curvature.

Among the candidates P_k, Q_k the first in each sequence admit an invariant metric with positive curvature since $P_1 = \mathbb{S}^7$ and $Q_1 = W_{1,1}^7$. The first success in the Grove program to find a new example with positive curvature is the following [11, 30]:

Theorem (Grove–Verdiani–Ziller, Dearricott). *The cohomogeneity one manifold P_2 carries an invariant metric with positive curvature.*

As for the topology of this manifold one has the following classification [21]:

Theorem (Goette). *The cohomogeneity one manifold* P_k *is diffeomorphic to* $E_k \# \Sigma^{\frac{k(k+1)}{2}}$, *where* E_k *is the* $\mathbb{S}^3$ *principal bundle over* $\mathbb{S}^4$ *with Euler class* k, *and* Σ *is the Gromoll–Meyer generator in the group of exotic 7-spheres.*

In particular, it follows that P_2 is homeomorphic but not diffeomorphic to $T_1 \mathbb{S}^4$, and thus indeed a new example of positive curvature.

4 Fibrations with Positive Curvature

As we saw in Sect. 2, many of the examples of positive curvature are the total space of a fibration. It is thus natural to ask under what condition the total space admits positive curvature, if the base and fiber do. One should certainly expect conditions, since the bundle could be trivial.

Weinstein examined this question in the context of Riemannian submersions with totally geodesic fibers [57]. He called a bundle fat if $\sec(X, U) > 0$ for all 2-planes spanned by a vector U tangent to the fibers and X orthogonal to the fibers. For simplicity, we restrict ourselves for the moment to principle bundles. Let $\pi \colon P \to B$ be a G-principle bundle. Given a metric on the base $\langle ., . \rangle_B$, a principal connection $\theta \colon TP \to \mathfrak{g}$, and a fixed biinvariant metric Q on G, one defines a Kaluza Klein metric on P as:

$$g_t(X, Y) = tQ(\theta(X), \theta(Y)) + g(\pi_*(X), \pi_*(Y)).$$

Here one has the additional freedom to modify t, in fact $t \to 0$ usually increases the curvature.

The projection π is then a Riemannian submersion with totally geodesic fibers isometric to (G, tQ). Weinstein observed that the fatness condition (for any t) is equivalent to requiring that the curvature Ω of θ has the property that $\Omega_u = Q(\Omega, u)$ is a symplectic 2-form on the horizontal space, i.e., the vector space orthogonal to the fibers, for every $u \in \mathfrak{g}$. If $G = \mathbb{S}^1$, this is equivalent to the base being symplectic. Fatness of the principal connection is already a strong condition on the principal bundle which one can express in terms of the characteristic classes of the bundle.

Weinstein made the following observation. Assume that G and B^{2n} are compact and connected. For each $y \in \mathfrak{g}$, we have a polynomial $q_y : \mathfrak{g} \to \mathbb{R}$ given by

$$q_y(\alpha) = \int_G \langle Ad_g(y), \alpha \rangle^{2n} dg$$

which one easily checks is Ad_G-invariant (in α and y). By Chern–Weyl theory, there exists a closed $2n$-form ω_y on B^{2n} such that $\pi^* \omega_y = q_y(\Omega)$ and $[\omega_y] \in H^{2n}(B, \mathbb{R})$ represents a characteristic class of the bundle. If the bundle is fat, $\Omega_y^{2n} \neq 0$ and hence $\langle Ad_g(y), \Omega \rangle^{2n}$ is nowhere zero and (if G is connected) has constant sign when g varies along G. This implies that the integral $q_y(\Omega)$ is nonzero, and hence

ω_y is a volume form of B^{2m}, in particular B^{2m} is orientable. Fatness of θ thus implies that the characteristic number $\int_B \omega_y$ is nonzero. In [20] this characteristic number is called the Weinstein invariant associated to $y \in \mathfrak{t}$ and was computed explicitly in terms of Chern and Pontrjagin numbers. For each adjoint orbit in $\mathfrak{t}$ one obtains obstructions to fatness in terms of Chern and Pontryagin numbers. The above discussion easily generalizes to fiber bundles associated to principle bundles to obtain obstructions. We call the metric on the fiber bundle a connection metric if the fibers are totally geodesic. Some sample obstructions are:

Theorem. (a) ([12]) *The only $\mathbb{S}^3$ bundle over $\mathbb{S}^4$ that admits a fat connection metric is the Hopf bundle $\mathbb{S}^3 \to \mathbb{S}^7 \to \mathbb{S}^4$.*
(b) ([20]) *The only two $\mathbb{S}^3$ sphere bundles over $\mathbb{CP}^2$ that may admit a fat connection metric are the complex sphere bundles with $c_1^2 = 9$ and $c_2 = 1$ or 2. In particular, $T_1\mathbb{CP}^2 \to \mathbb{CP}^2$ does not have a fat connection metric.*
(c) ([20]) *If a sphere bundle over B^{2n} admits a fat connection metric, then the Pontryagin numbers satisfy $\det(p_{j-i+1})_{1 \le i,j \le n} \ne 0$.*

On the other hand, many of the examples in Sect. 2 are fat fiber bundles, and the new example of positive curvature in Sect. 3 is a fat bundle as well.

If one wants to achieve positive curvature on the total space, we need to assume, in addition to the base having positive curvature, that $G = S^1$, SU(2) or SO(3). In [9] a necessary and sufficient condition for positive curvature of such metrics was given. The proof carries over immediately to the category of orbifold principal bundles, which includes the case where the G action on P has only finite isotropy groups.

Theorem 4.1 (Chaves–Derdzinski–Rigas). *A connection metric g_t on an orbifold G-principal bundle with $\dim G \le 3$ has positive curvature, for t sufficiently small, if and only if*

$$(\nabla_x \Omega_u)(x, y)^2 < |i_x \Omega_u|^2 k_B(x, y),$$

for all linearly independent horizontal vectors x, y and $0 \ne u \in \mathfrak{g}$.

Here $k_B(x, y) = g(R_B(x, y)y, x)$ is the unnormalized sectional curvature and $i_x \Omega_u \ne 0$ is precisely the above fatness condition.

All the known examples of principle bundles with positive curvature satisfy this condition (called hyperfatness in [58]). The new example in Sect. 3 also is such an $G = S^3$ principle bundle, if one allows the action of G to be almost free (the base is an orbifold homeomorphic to $\mathbb{S}^4$).

The condition is of course trivially satisfied if $\nabla\Omega = 0$. This is equivalent to the metric on B being quaternionic Kähler. Thus a quaternionic Kähler manifold with positive sectional curvature gives rise to a positively curved metric on the S^3 or SO(3) principal bundle defined canonically by its structure. Unfortunately, if $\dim > 4$, a positively curved quaternionic Kähler metric is isometric to $\mathbb{HP}^n$ (Berger) and thus the construction only gives rise to the metric on $\mathbb{S}^{4n+3}(1)$. And in dimension 4, the only smooth quaternionic Kähler metric (with positive scalar curvature) is isometric to $\mathbb{S}^4(1)$ and $\mathbb{CP}^2$ giving rise to $\mathbb{S}^7(1)$ and the positively curved Aloff Wallach space $W_{1,1}$. Thus in the smooth category, it gives nothing new.

But if the quotient is a quaternionic Kähler orbifold, there are many examples. In the case of the candidate P_k, the subgroup $S^3 \times \{e\} \subset S^3 \times S^3$ acts almost freely and the quotient is an orbifold homeomorphic to $\mathbb{S}^4$. Similarly, for Q_k, the subgroup $S^3 \times \{e\} \subset S^3 \times S^3$ acts almost freely with quotient an orbifold homeomorphic to $\mathbb{CP}^2$. On these two orbifolds, Hitchin [31] constructed a quaternionic Kähler metric with positive scalar curvature, and in [29] it was shown that the total space of the canonically defined principal S^3 resp. SO(3) bundle is equivariantly diffeomorphic to P_k resp. Q_k. In [59] it was shown that the Hitchin metrics have a large open set on which the curvature is positive, but not quite everywhere. Nevertheless, in [29] and [11] this Hitchin metric was the starting point. In [29] a connection metric was constructed on P_2 by defining the metric piecewise via low degree polynomials, both for the metric on the base, and the principle connection. The metric is indeed very close to the Hitchin metric. In [11] the quaternionic Kähler Hitchin metric on $\mathbb{S}^4$ was deformed, but the principle connection stayed the same. The metric on $\mathbb{S}^4$ was then approximated by polynomials in order to show the new metric has positive curvature.

In both cases, the proof that the polynomial metric has positive curvature, was carried out by using a method due to Thorpe (and a small modification of it in [11]).

We finish by describing this Thorpe method since it is not so well known, but very powerful (see [48,49] and also [46]). In fact, one of the problems of finding new examples is that after constructing a metric, showing that it has positive curvature is difficult. Even the linear algebra problem for a curvature tensor on a vector space is highly non-trivial! Here is where the Thorpe method helps.

Let V be a vector space with an inner product and R a 3–1 tensor which satisfies the usual symmetry properties of a curvature tensor. We can regard R as a linear map

$$\hat{R} : \Lambda^2 V \to \Lambda^2 V,$$

which, with respect to the natural induced inner product on $\Lambda^2 V$, becomes a symmetric endomorphism. The sectional curvature is then given by:

$$\sec(v, w) = \langle \hat{R}(v \wedge w), v \wedge w \rangle$$

if v, w is an orthonormal basis of the 2-plane they span.

If $\hat{R}$ is positive definite, the sectional curvature is clearly positive as well. But this condition is exceedingly strong since a manifold with $\hat{R} > 0$ is covered by a sphere [6]. But one can modify the curvature operator by using a 4-form $\eta \in \Lambda^4(V)$. It induces another symmetric endomorphism $\hat{\eta} : \Lambda^2 V \to \Lambda^2 V$ via $\langle \hat{\eta}(x \wedge y), z \wedge w \rangle = \eta(x, y, z, w)$. We can then consider the modified curvature operator $\hat{R}_\eta = \hat{R} + \hat{\eta}$. It satisfies all symmetries of a curvature tensor, except for the Bianchi identity. Clearly $\hat{R}$ and $\hat{R}_\eta$ have the same sectional curvature since

$$\langle \hat{R}_\eta(v \wedge w), v \wedge w \rangle = \langle \hat{R}(v \wedge w), v \wedge w \rangle + \eta(v, w, v, w) = \sec(v, w)$$

If we can thus find a 4-form η with $\hat{R}_\eta > 0$, the sectional curvature is positive. Thorpe showed [49] that in dimension 4, one can always find a 4-form such that the smallest eigenvalue of $\hat{R}_\eta$ is also the minimum of the sectional curvature, and similarly a possibly different 4-form such that the largest eigenvalue of $\widehat{R_\eta}$ is the maximum of the sectional curvature. Indeed, if an eigenvector ω to the largest eigenvalue of $\hat{R}$ is decomposable, the eigenvalue is clearly a sectional curvature. If it is not decomposable, then $\omega \wedge \omega \neq 0$ and one easily sees that $\widehat{R_\eta}$ with $\eta = \omega \wedge \omega$ has a larger eigenvalue.

This is not the case anymore in dimension bigger than 4 [60]. Nevertheless this can be an efficient method to estimate the sectional curvature of a metric. In fact, Püttmann [46] used this to compute the maximum and minimum of the sectional curvature of most positively curved homogeneous spaces, which are not spheres. It is peculiar to note though that this method does not seem to work to determine which homogeneous metrics on $\mathbb{S}^7$ have positive curvature, see [51].

References

1. A.V. Alekseevsy, D.V. Alekseevsy, G-manifolds with one dimensional orbit space. Adv. Sov. Math. **8**, 1–31 (1992)
2. S. Aloff, N. Wallach, An infinite family of 7–manifolds admitting positively curved Riemannian structures. Bull. Am. Math. Soc. **81**, 93–97 (1975)
3. Y. Bazaikin, On a family of 13-dimensional closed Riemannian manifolds of positive curvature. Siberian Math. J. **37**, 1068–1085 (1996)
4. L. Bérard Bergery, Les variétés riemanniennes homogènes simplement connexes de dimension impaire à courbure strictement positive. J. Math. pure et appl. **55**, 47–68 (1976)
5. M. Berger, Les variétés riemanniennes homogènes normales simplement connexes à courbure strictement positive. Ann. Scuola Norm. Sup. Pisa **15**, 179–246 (1961)
6. C. Böhm, B. Wilking, Manifolds with positive curvature operators are space forms. Ann. Math. **167**, 1079Ű-1097 (2008)
7. W. Blaschke, *Vorlesungen ber Differentialgeometrie* (Springer, Berlin, 1921)
8. W. Boy, Uber die Curvatura integra und die Topologie geschlossener Flächen. Math. Ann. **57**, 151–184 (1903)
9. L. Chaves, A. Derdzinski, A. Rigas, A condition for positivity of curvature. Bol. Soc. Brasil. Mat. **23**, 153–165 (1992)
10. T. Chinburg, C. Escher, W. Ziller, Topological properties of Eschenburg spaces and 3-Sasakian manifolds. Math. Ann. **339**, 3–20 (2007)
11. O. Dearricott, A 7-manifold with positive curvature. Duke Math. J. **158**, 307–346 (2011)
12. A. Derdzinski, A. Rigas, Unflat connections in 3-sphere bundles over $\mathbb{S}^4$. Trans. Am. Math. Soc. **265**, 485–493 (1981)
13. P. Dombrowski, *150 years after Gauss' Disquisitiones generales circa superficies curvas*, Astérisque 62 (Société Mathématique de France, Paris, 1979)
14. J.H. Eschenburg, New examples of manifolds with strictly positive curvature. Invent. Math. **66**, 469–480 (1982)
15. J.H. Eschenburg, Freie isometrische Aktionen auf kompakten Lie-Gruppen mit positiv gekrümmten Orbiträumen. Schriftenr. Math. Inst. Univ. Münster **32**, 1–177 (1984)
16. J. Eschenburg, A. Kollross, K. Shankar, Free isometric circle actions on compact symmetric spaces. Geom. Dedicata **103**, 35–44 (2003)

17. R. Fintushel, Circle actions on Simply Connected 4-manifolds. Trans. Am. Math. Soc. **230**, 147–171 (1977)
18. L.A. Florit, W. Ziller, Orbifold fibrations of Eschenburg spaces. Geom. Ded. **127**, 159–175 (2007)
19. L.A. Florit, W. Ziller, On the topology of positively curved Bazaikin spaces. J. Eur. Math. Soc. **11**, 189Ű-205 (2009)
20. L.A. Florit, W. Ziller, Topological obstructions to fatness. Geom. Topol. **15**, 891–925 (2011)
21. S. Goette, Adiabatic limits of Seifert fibrations, Dedekind sums, and the diffeomorphism type of certain 7-manifolds. J. Eur. Math. Soc. (2014, in press)
22. S. Goette, N. Kitchloo, K. Shankar, Diffeomorphism type of the Berger space SO(5)/SO(3). Am. Math. J. **126**, 395–416 (2004)
23. D. Gromoll, W. Meyer, An exotic sphere with nonnegative sectional curvature. Ann. Math. (2) **100**, 401–406 (1974)
24. K. Grove, S. Halperin, Contributions of rational homotopy theory to global problems in geometry. Publ. Math. I.H.E.S. **56**, 171–177 (1982)
25. K. Grove, C. Searle, Positively curved manifolds with maximal symmetry rank. J. Pure Appl. Algebra **91**, 137–142 (1994)
26. K. Grove, B. Wilking, A knot characterization and 1-connected nonnegatively curved 4-manifolds with circle symmetry (2011, preprint)
27. K. Grove, W. Ziller, Curvature and symmetry of Milnor spheres. Ann. Math. **152**, 331–367 (2000)
28. K. Grove, R. Shankar, W. Ziller, Symmetries of Eschenburg spaces and the Chern problem. Special Issue in honor of S.S. Chern. Asian J. Math. **10**, 647–662 (2006)
29. K. Grove, B. Wilking, W. Ziller, Positively curved cohomogeneity one manifolds and 3-Sasakian geometry. J. Differ. Geom. **78**, 33–111 (2008)
30. K. Grove, L. Verdiani, W. Ziller, An exotic $T_1\mathbb{S}^4$ with positive curvature. Geom. Funct. Anal. **21**, 499–524 (2011)
31. N. Hitchin, A new family of Einstein metrics, in *Manifolds and Geometry* (Pisa, 1993). Sympos. Math., XXXVI (Cambridge University Press, Cambridge, 1996), pp. 190–222
32. N. Hopf, Zum Clifford Kleinschen Raumproblem. Math. Ann. **95**, 313–339 (1926)
33. N. Hopf, W. Rinow, Über den Begriff der vollständigen differential-geometrischen Fläche. Comm. Math. Helvetici **3**, 209–Ű225 (1931)
34. W.Y. Hsiang, B. Kleiner, On the topology of positively curved 4-manifolds with symmetry. J. Differ. Geom. **29**, 615–621 (1989)
35. L. Kennard, On the Hopf conjecture with symmetry. Geom. Topol. **161**, 563–593 (2013)
36. L. Kennard, Positively curved Riemannian metrics with logarithmic symmetry rank bounds. Comment. Math. Helv. (in press)
37. M. Kreck, S. Stolz, Some non diffeomorphic homeomorphic homogeneous 7-manifolds with positive sectional curvature. J. Differ. Geom. **33**, 465–486 (1991)
38. S.B. Myers, Riemannian manifolds in the large. Duke Math. J. **1**, 42–43 (1935)
39. S.B. Myers, Riemannian manifolds with positive mean curvature. Duke Math. J. **8**, 401–404 (1941)
40. G.P. Paternain, J. Petean, Minimal entropy and collapsing with curvature bounded from below. Invent. Math. **151**, 415–450 (2003)
41. P. Petersen, F. Wilhelm, An exotic sphere with positive sectional curvature (2008, preprint)
42. A. Preissmann, Quelques propriétés globales des espaces de Riemann. Comm. Math. Helv. **15**, 175–216 (1943)
43. J.L. Synge, The first and second variations of the length integral in Riemann space. Proc. Lond. Math. Soc. **25**, 247–264 (1925)
44. J.L. Synge, On the neighborhood of a geodesic in Riemannian space. Duke Math. J. **1**, 527–537 (1935)
45. J.L. Synge, On the connectivity of spaces of positive curvature. Q. J. Math. **7**, 316–320 (1936)
46. T. Püttmann, Optimal pinching constants of odd dimensional homogeneous spaces. Invent. math. **138**, 631–684 (1999)

47. I.J. Schoenberg, Some applications of the calculus of variations to Riemannian geometry. Ann. Math. **33**, 485–495 (1932)
48. J.A. Thorpe, On the curvature tensor of a positively curved 4-manifold, in *Mathematical Congress*, Dalhousie Univ., Halifax, N.S. Canad. Math. Congr., Montreal, Que. (1972), pp. 156–159
49. J.A. Thorpe, The zeros of nonnegative curvature operators. J. Differ. Geom. **5**, 113–125 (1971). Erratum J. Diff. Geom. **11**, 315 (1976)
50. L. Verdiani, Cohomogeneity one manifolds of even dimension with strictly positive sectional curvature. J. Differ. Geom. **68**, 31–72 (2004)
51. L. Verdiani, W. Ziller, Positively curved homogeneous metrics on spheres. Math. Zeitschrift **261**, 473–488 (2009)
52. L. Verdiani, W. Ziller, Concavity and rigidity in non-negative curvature. J. Differ. Geom. **97** (2014)
53. F. Wilhelm, An exotic sphere with positive curvature almost everywhere. J. Geom. Anal. **11**, 519–560 (2001)
54. B. Wilking, Manifolds with positive sectional curvature almost everywhere. Invent. Math. **148**, 117–141 (2002)
55. B. Wilking, Positively curved manifolds with symmetry. Ann. Math. **163**, 607–668 (2006)
56. N. Wallach, Compact homogeneous Riemannian manifolds with strictly positive curvature. Ann. Math. **96**, 277–295 (1972)
57. A. Weinstein, Fat bundles and symplectic manifolds. Adv. Math. **37**, 239–250 (1980)
58. W. Ziller, Examples of Riemannian manifolds with nonnegative sectional curvature, in *Metric and Comparison Geometry*. Surveys in Differential Geometry, vol. 11, ed. by K. Grove, J. Cheeger, International Press, (2007), pp. 63–102
59. W. Ziller, On the geometry of cohomogeneity one manifolds with positive curvature, in *Riemannian Topology and Geometric Structures on Manifolds*, in honor of Charles P. Boyer's 65th birthday. Progress in Mathematics, Birkhaueser, vol. 271 (2009), pp. 233–262
60. S. Zoltek, Nonnegative curvature operators: some nontrivial examples. J. Differ. Geom. **14**, 303–315 (1979)

An Introduction to Isometric Group Actions with Applications to Spaces with Curvature Bounded from Below

Catherine Searle

2000 *Mathematics Subject Classification*. Primary: 53C20; Secondary: 57S25, 51M25

1 Lecture 1: An Introduction to Isometric Groups Actions

Positively curved manifolds remain unclassified and in fact, the known examples of positively curved manifolds are few. In dimensions greater than or equal to 24, all known examples are diffeomorphic to the CROSSes, that is, the compact rank one symmetric spaces, which are $S^n, \mathbb{C}P^k, \mathbb{H}P^l$ and CaP^2. One observes immediately that all of these examples are highly symmetric: that is, they all admit transitive or nearly transitive isometric actions by a compact Lie group. In lower dimensions, until recently, all known examples were homogeneous spaces (cf. [2, 4, 40]) and bi-quotients, that is, quotients of compact Lie groups G, equipped with a biinvariant metric, by a free isometric two sided action of a subgroup $H \subset G \times G$ [3, 10]. In the last few years, new methods for constructing examples with positive curvature have been proposed in [33] on the Gromoll-Meyer exotic 7-sphere and in [22] and [9] on a 7-manifold homeomorphic but not diffeomorphic to $T^1 S^4$.

The author was supported in part by CONACYT Project #SEP-106923.

C. Searle (✉)
Department of Mathematics, Oregon State University, 368 Kidder Hall, Corvallis, OR 97331, USA
e-mail: searleca@math.oregonstate.edu

O. Dearricott et al., *Geometry of Manifolds with Non-negative Sectional Curvature*, Lecture Notes in Mathematics 2110, DOI 10.1007/978-3-319-06373-7_2, © Springer International Publishing Switzerland 2014

To further complicate matters, there are only two classifical results that give us topological obstructions:

Theorem 1.1 (Bonnet-Myers Theorem). *Let M be a manifold of strictly positive sectional curvature. Then M is compact and $\pi_1(M) < \infty$.*

Theorem 1.2 (Synge's Theorem). *Let M be a Riemannian manifold of strictly positive sectional curvature. Then the following are true:*

(1) *If M is even-dimensional, $\pi_1(M) = 0$ if M is orientable and $\pi_1(M) = \mathbb{Z}_2$ if M is non-orientable.*
(2) *If M is odd-dimensional, then it is orientable.*

In the 1990s Wu-Yi Hsiang and Karsten Grove promoted the idea of attacking the classification of positive curved manifolds by considering "large" isometric group actions, where "large" can be interpreted as one likes. With this in mind, we'll begin with the basics of group actions, eventually restricting our attention to isometric ones. In the second and third talk, we'll see some results that have been obtained for closed, simply-connected manifolds of both positive and negative curvature.

Now, it is a well-known fact that the isometry group of a connected compact manifold is a compact Lie group (cf. [27]) and therefore we will restrict our attention to compact Lie groups. Before we start, we need to introduce some general notation and terminology.

Let $\phi : G \times M \to M$ be a (left) group action defined as $\phi(g, x) = gx$. We may similarly define a (right) group action.

An action will be called *effective* if the only element that fixes M pointwise is the identity element. An action is called *almost effective* if a finite group fixes M pointwise. An action defines a natural map

$$\Theta : G \to \mathrm{Diff}(M)$$

$$g \mapsto \Theta_g$$

and we can rephrase the above definitions as follows. G acts *effectively* if $\ker \Theta$ is trivial and *almost effectively* if $\ker \Theta$ is finite.

For most purposes, it suffices to consider effective actions because if $N = \ker \Theta$ is non-trivial, then there exists a canonically induced effective action of G/N on M.

As some trivial examples of G-actions we have the following:

1. G, a compact Lie group, can act on itself by conjugation:

$$G \times G \to G$$

$$(g, h) \mapsto ghg^{-1}$$

and $\ker \Theta = Z(G)$, the center of G.

2. G acts on itself by left translation

$$G \times G \rightarrow G$$
$$(g, h) \mapsto gh$$

and $\ker \Theta = \{e\}$.

3. G acts on the coset space G/H by left translation.

$$G \times G/H \rightarrow G/H$$
$$(g', gH) \mapsto g'gH$$

and $\ker \Theta = \cap_{g \in G} gHg^{-1}$.

Given a subset $A \subset M$, the stabilizer of A under the G-action is

$$G_A = \{g \in G : g(A) = A\}.$$

In the particular case where $A = \{p\}$, G_p is called the *isotropy group* of the point p. We call $G(p) = \{gp : g \in G\}$ the *orbit* of p.

Lemma 1.3. *In each orbit, all isotropy groups are equal up to conjugacy, that is, for some $g \in G$,*

$$G_p = g^{-1} G_q g,$$

where $q = gp$.

The proof is left as an exercise for the reader.

We call an action *free* if $G_x = \{e\}$ for all $x \in M$, *semi-free* if G_x is G or $\{e\}$ for all $x \in M$ and *almost free* if G_x is finite for all $x \in M$. We illustrate these definitions with the following examples:

1. The Hopf action is an example of a free action:

$$S^1 \times S^{2n+1} \rightarrow S^{2n+1}$$
$$e^{2\pi i\theta}, (z_1, \ldots, z_n) \mapsto (e^{2\pi i\theta} z_1, \ldots, e^{2\pi i\theta} z_n).$$

2. The "generalized" Hopf action on S^3 is given by:

$$S^1 \times S^3 \rightarrow S^3$$
$$e^{2\pi i\theta}, (z_1, z_2) \mapsto (e^{2\pi ik\theta} z_1, e^{2\pi il\theta} z_2).$$

This action is effective provided $(k, l) = 1$ and almost free, since $G_{(z_1,0)} = \mathbb{Z}_k$, $G_{(0,z_2)} = \mathbb{Z}_l$ and $G_{(z_1,z_2)} = \{e\}$, otherwise. This action gives us a Seifert fibration of S^3 over S^2.

3. Finally, there is a semi-free action of the circle on S^2.

$$SO(2) \times S^2 \to S^2$$

$$\begin{pmatrix} \cos(\theta) & \sin(\theta) & 0 \\ -\sin(\theta) & \cos(\theta) & 0 \\ 0 & 0 & 1 \end{pmatrix}, \begin{pmatrix} x_1 \\ x_2 \\ x_3 \end{pmatrix} \mapsto \begin{pmatrix} \cos(\theta)x_1 + \sin(\theta)x_2 \\ -\sin(\theta)x_1 + \cos(\theta)x_2 \\ x_3 \end{pmatrix}.$$

Here $G_{(0,0,\pm 1)} = SO(2)$ and $G_{(x_1,x_2,x_3)} = \{e\}$, otherwise.

This last example is an action of *cohomogeneity one*, where a *cohomogeneity k* action is defined as an action for which the principal orbit is of codimension k.

Now a smooth map $f : M \to N$, where both M and N admit a G-action is called *equivariant* if the map commutes with the G-action, that is

$$f(g(x)) = g(f(x)) \text{ for all } g \in G, \text{ for all } x \in M.$$

Observe that the inverse, f^{-1} will also be equivariant:

$$f^{-1}(g(y)) = f^{-1}(g(f(x)) = f^{-1}(f(g(x)) = g(x) = g(f^{-1}(y)).$$

Example 1.4. If G acts on M then there is an induced action on TM, the tangent bundle over M. The induced action is given by:

$$G \times TM \to TM$$
$$(g, v_p) \mapsto Dg_p V_p$$

The exponential map $\exp : TM \to M$ and the canonical projection $\pi : TM \to M$ are both G-equivariant maps since $\exp_{gp}(gv) = g(\exp_p v)$ and $\pi(gv) = g\pi(v)$. When $\exp_p : T_pM \to M$ is a diffeomorphism, we obtain a G-equivalence.

A G-equivariant diffeomorphism $f : M \to N$ is called an *equivalence* of M and N. In this case, $G_p = G_{f(p)}$. When two manifolds are not equivalent, but still related by an equivariant map, one can show that $G_p \subset G_{f(p)}$. That is, suppose that $g \in G_p$ then $f(p) = f(g(p)) = g(f(p))$ and therefore $g \in G_{f(p)}$.

There is also a notion of *weak equivalence*, where the map differs by an automorphism of G, that is,

$$f(g(x)) = \alpha(g)(f(x)), \text{ for all } g \in G, \text{ for all } x \in M$$

where $\alpha \in \text{Aut}(G)$.

We present an example to illustrate.

Example 1.5. Let $G = \mathbb{Z}_5 = <\gamma>$ be generated by the fifth roots of unity. Let $\phi_1 : G \times S^1 \to S^1$ send $(\gamma, z) \mapsto \gamma z$ and let $\phi_2 : G \times S^1 \to S^1$ send $(\gamma, z) \mapsto \gamma^2 z$. Let $f : S^1 \to S^1$ be the identity. Then f is a weak equivalence since if we let $z = re^{2\pi i \theta}$, then

$$re^{2\pi i(\theta+1/5)} = \gamma z = f(\gamma(z)) \neq \gamma f(z) = \gamma^2 z = re^{2\pi i(\theta+2/5)}.$$

Now we would like to consider the orbit spaces of G-actions. As mentioned previously, $G(x) = \{gx : g \in G\}$ is called the *orbit* of $x \in M$ under the action of G. Observe that if $gx = hy$ for some $g, h \in G, x, y \in M$ then $G(x) = G(y)$ and thus we see that orbits of G are either *disjoint* or *equal*.

Let $M/G = M^*$ denote the set whose elements are the orbits $G(x) = x^*$ of G on M. Then $x^* = y^*$ if and only if $x, y \in G(x)$. Let $\pi : M \to M/G$ denote the natural map taking x into its orbit $x^* = G(x)$. M/G is endowed with the quotient topology, that is, $U \subset M/G$ is open if and only if $\pi^{-1}(U)$ is open in M, and is called the orbit space of M with respect to the G-action.

Example 1.6. Consider the action of $\mathbb{R}$ on T^2 producing an irrational flow. The quotient space has the trivial (indiscrete) topology and is therefore not very interesting . For actions of compact groups, however, we will see that the orbit space exhibits lots of nice properties as in the following theorem (cf. [5]).

Theorem 1.7. *If $G \times M \to M$ with G compact then the following hold:*

(1) *M/G is Hausdorff.*
(2) *$\pi : M \to M/G$ is closed.*
(3) *$\pi : M \to M/G$ is proper.*
(4) *M is compact if and only if M/G is compact.*
(5) *M is locally compact if and only if M/G is locally compact.*

Moreover, in some special cases M/G will be a manifold (although not necessarily smooth): for example, when the action of G is free or of low cohomogeneity, namely if it is transitive (in which case $M = G/H$) or of cohomogeneity one or two. If M is compact and simply connected and G has connected orbits, then if the action is of cohomogeneity three M/G is also a manifold (see Sect. 2, Theorem 2.13).

However, in the more general case where we consider M a Riemannian manifold with a lower curvature bound, then M/G is an Alexandrov space with a lower curvature bound. Recall that an *Alexandrov space* is a finite-dimensional length space with a lower curvature bound defined in the comparison sense. The space of Alexandrov spaces properly contains Riemannian manifolds with a lower curvature bound. They are not manifolds in general and have "singular" points. For example, the spherical suspension of $\mathbb{R}P^2$ is not a manifold (the only spherical suspension that is a manifold is the spherical suspension of a sphere) but it is an Alexandrov space with two singular points, namely the two cone points.

We have the following theorem:

Theorem 1.8. *If G acts freely on M then M/G has the structure of a smooth manifold such that $\pi : M \to M/G$ is a submersion.*

The proof is left to the reader. The following corollary is immediate since H acts freely on G by left multiplication.

Corollary 1.9. *If H is a closed Lie subgroup of H then G/H is a smooth submanifold such that $\pi : G \to G/H$ is a submersion.*

Corollary 1.10. *Suppose $G \times M \to M$. Then $G(p) = G/G_p \subset M$ is an embedded submanifold.*

We now proceed to define the concept of a slice. This will allow us to find a canonical decomposition of an ϵ-tubular neighborhood of an orbit and gives us more information about orbit types as well as valuable information about the action itself.

Recall by Corollary 1.10 that $G(p)$ is an embedded submanifold of M. Thus, there is an $\epsilon > 0$ such that for any $p \in G(q)$ the orthogonal ball $B_\epsilon^\perp = \{v \in T_pG(q))^\perp : |v| < \epsilon\}$ is normal to $T_pG(q)$. The image of $B_\epsilon^\perp$ via the exponential map is an ϵ-neighborhood S of $G(q)$ in M.

We define S as follows:

Definition 1.11. Given $p \in G(q)$, the embedded submanifold S_p normal to the orbit $G(q)$ given by $\exp_p v, v \in B_\epsilon^\perp$ is the *slice* through p.

Let $S = \cup_{x \in G(p)} S_x = \cup_{g \in G} g \cdot S_x$, then the following are true:

(1) If $S_p \cap S_q \neq \emptyset$ then $p = q$.
(2) $g \cdot S_p = S_{gp}$.
(3) $S = \cup_{g \in G} S_{gx}$.
(4) If $p' \in S_p$ then $G_{p'}$ is conjugate to a subgroup of G_p.
(5) (Slice Representation) $G_p \times S_p \to S_p$ and S_p is G_p-equivalent to the orthogonal action of G_p on T_pS_p via the exponential map.
(6) The neighborhood S can be represented as a product if the action of G_x is free, otherwise we have the following G-equivariant map:

$$\Psi : G \times_{G_p} S_p \to S$$
$$[g, p'] \mapsto gp',$$

where $G \times_{G_p} S_p = (G \times S)/G_p$ is given by the diagonal action $\tilde{g} * (g, p') = (g\tilde{g}, \tilde{g}^{-1} p')$. G also acts on $G \times_{G_p} S_p$ via a left action:

$$(\tilde{g}, [g, p']) \mapsto [g\tilde{g}, p'].$$

(7) If G_p acts trivially on S_p then $G \times_{G_p} S_p = G/G_p \times S_p$.
(8) If every isotropy group of the G action on M is conjugate to some fixed subgroup $H \subset G$ (that is, we only have principal orbits), then M/G is a manifold.

For [6] we need to verify the following:

- Ψ is well-defined:

$$\Psi([g\tilde{g}, \tilde{g}^{-1}p']) = g\tilde{g}\tilde{g}^{-1}p' = gp' = \Psi([g, p']).$$

- Ψ is a G-equivariant map:

$$\Psi(\tilde{g}[g, p']) = \Psi([\tilde{g}g, p']) = \tilde{g}gp' = \tilde{g}\Psi([g, p']).$$

- Ψ is one-to-one:
  Suppose $g_1 p_1 = g_2 p_2$, then $\Psi([g_1, p_1]) = \Psi([g_2, p_2])$. But $p_2 = g_2^{-1}g_1 p_1$ and therefore

$$[g_2, p_2] = [g_2 g_2^{-1} g_1, (g_2^{-1}g_1)^{-1}p_2] = [g_1, (g_2^{-1}g_1)^{-1}(g_2^{-1}g_1 p_1)] = [g_1, p_1].$$

- Ψ is onto.
  Given $q \in S$, then $q = \exp_{gp}(gv) = g\exp_p v = \Psi([g, \exp_p v])$ which implies that $S = G(S_p)$.
- Ψ is differentiable by definition and its inverse is differentiable.

We now define the different possible types of orbits.

Definition 1.12. We call an orbit *principal* if its isotropy subgroup G_x is minimal with respect to the following partial order. $(H) \subseteq (K)$ if and only if H is conjugate to some subgroup of K. The ordering is partial because there may be elements that we cannot compare.

An orbit is called *exceptional* if its isotropy subgroup is not minimal but has the same dimension as the minimal (principal) isotropy subgroup. An orbit is called *singular* if its isotropy subgroup has strictly larger dimension than that of the principal isotropy subgroup.

We have the following theorem (cf. [5]):

Theorem 1.13 (Principal Orbit Theorem). *Let G be a compact group acting on M. Then the following are true.*

(1) *There exists a unique minimal isotropy type.*
(2) *$M_r = \{p \in M : p \in G(q)$ a principal orbit $\}$ is open and dense. $M_r/G = M_r^*$ is open dense and convex (and therefore connected) in M^*.*
(3) *There are only finitely many isotropy types.*

We now give some examples of G-actions with different orbit types.

Example 1.14. Let $SO(n) \times S^n \rightarrow S^n$ as follows.

$$\begin{pmatrix} A & \\ & 1 \end{pmatrix} \begin{pmatrix} x_1 \\ \vdots \\ x_{n+1} \end{pmatrix} \mapsto \begin{pmatrix} a_{11}x_1 + \ldots + a_{1n} \\ \vdots \\ a_{n1}x_1 + \ldots + a_{nn}x_n \\ x_{n+1} \end{pmatrix}$$

The G-action is not transitive, but on the subsphere $S^{n-1} = \{(x_1, \ldots, x_n, 0) : \sum x_i^2 = 1\}$ it is. The principal orbits have isotropy $SO(n-1)$. The points $(0, \ldots, 0, \pm 1)$ are distinct singular orbits with isotropy $SO(n)$. This is an example of a cohomogeneity one actions and it is easy to see that $S^n/SO(n) = I$ an interval of length π.

Example 1.15. Let $SO(n) \times S^n \rightarrow S^n$ as above. The action clearly commutes with the antipodal map on S^n and induces an action on $\mathbb{R}P^n$ with principal orbits $SO(n)/SO(n-1) = S^{n-1}$ and one exceptional orbit $SO(n)/O(n-1) = \mathbb{R}P^{n-1}$ and a singular orbit $SO(n)/SO(n) = \{p\}$.

Now we have the following result about the action of G_x on its slice (cf. [5]).

Theorem 1.16. *Let $G \times M \rightarrow M$, G compact, $G(x)$ the orbit of x and G_x its isotropy subgroup. Let S be the slice and S_x the slice at x. Then the following are true.*

(1) *$G_x(p)$, $p \in S_x$ is a principal, exceptional or singular orbit for the G_x action on S_x according as $G(p)$ is principal, exceptional or singular.*
(2) *$G(x)$ is principal if and only if G_x acts trivially on S_x, that is $G \times_{G_x} S_x \simeq G/G_x \times S_x$.*

Observe that in the previous example, the principal isotropy was $SO(n-1)$ and it acts trivially on the slice to its orbit at any point of the orbit. The singular orbit $SO(n)/SO(n) = \{p\}$ has an induced action on its slice by $SO(n)$ and the orbits of the slice are principal with isotropy $SO(n-1)$. That is, the isotropy group at the point p acts transitively on the normal S^{n-1} with isotropy $SO(n-1)$. The exceptional orbit is $SO(n)/O(n-1)$ and the $O(n-1)$ isotropy subgroup acts on its normal S^0 with isotropy $SO(n-1)$. This action is also transitive.

Here the ϵ-neighborhood around a principal orbit $SO(n)/SO(n-1)$ is $S^{n-1} \times S_x$, where $S_x \simeq [-\epsilon, \epsilon]$. The ϵ-tubular neighborhood around the singular orbit is $(SO(n) \times S_x)/SO(n)$ and around the exceptional orbit it is $(SO(n) \times S_x)/O(n-1)$.

We will finish this lecture with the following non-trivial example.

Example 1.17. Let $SO(3) \times \mathbb{R}^5 \rightarrow \mathbb{R}^5$. We may represent $\mathbb{R}^5$ as the set of real 3×3 symmetric matrices of trace 0. Since the action is isometric, we may also consider the action on $S^4 \subset \mathbb{R}^5$. Then the action of $SO(3)$ on $\mathbb{R}^5$ is given by the following. Let $A \in SO(3)$, $X \in \mathbb{R}^5$.

$$(A, X) \mapsto AXA^{-1}$$

It is a well-known fact of linear algebra that $Y \in G(X)$ if and only if the eigenvalues of Y and X are the same counting multiplicities. The orbit space consists of representatives

$$\begin{pmatrix} \lambda_1 & & \\ & \lambda_2 & \\ & & \lambda_3 \end{pmatrix},$$

where $\lambda_1 \geq \lambda_2 \geq \lambda_3$ and $\lambda_1 + \lambda_2 + \lambda_3 = 0$. If we restrict out attention to S^4, we have the additional condition that $\lambda_1^2 + \lambda_2^2 + \lambda_3^2 = 1$.

The quotient space $\mathbb{R}^5/SO(3)$ is a triangular wedge of $\mathbb{R}^2$ starting at the origin and forming an angle of $\pi/3$. The quotient space $S^4/SO(3)$ is the arc of length $\pi/3$ corresponding to the intersection of the unit circle in $\mathbb{R}^2$ with $\mathbb{R}^5/SO(3)$. Here we have four distinct orbit types for the cohomogeneity two action on $\mathbb{R}^5$ and three distinct orbit types for the cohomogeneity one action on S^4.

The principal orbits correspond to elements in the quotient space with $\lambda_1 > \lambda_2 > \lambda_3$. It is clear for points in the principal orbit that the isotropy subgroup is $S(O(1) \times O(1) \times O(1))$. The points in the quotient space that correspond to those for which $\lambda_1 = \lambda_2 \geq \lambda_3$ form a line of singular orbits with isotropy $S(O(2)O(1))$ and those for which $\lambda_1 \geq \lambda_2 = \lambda_3$ form a line of singular orbits with isotropy $S(O(1)O(2))$. These two lines intersect in the point $\lambda_1 = \lambda_2 = \lambda_3 = 0$ and the isotropy of this point is the entire group, $SO(3)$. One may easily calculate the angle of the triangular wedge by calculating the Euclidean distance between the two endpoints of the quotient space $S^4/SO(3)$, which, without loss of generality, can be taken to be $(\frac{1}{\sqrt{6}}, \frac{1}{\sqrt{6}}, -\frac{2}{\sqrt{6}})$ and $(\frac{2}{\sqrt{6}}, \frac{1}{\sqrt{6}}, \frac{1}{\sqrt{6}})$, and converting this distance into the corresponding arc-length of the arc via the formula

$$\theta = 2\arcsin(\frac{d}{2}).$$

2 Lecture 2: The Hopf Conjecture and Torus Actions of Low Cohomogeneity

The main goal of this lecture is to give a detailed proof of the Hsiang–Kleiner theorem [24], which we now state.

Theorem 2.1 ([24]). *Let $S^1 \times M^4 \to M^4$ be an isometric, effective action with M^4 a compact, simply-connected positively curved Riemannian manifold. Then M^4 is homeomorphic to S^4 or $\mathbb{C}P^2$.*

In order to give a proof we first need to establish some facts about fixed point sets of isometries. We have the following theorem (cf. [25]).

Theorem 2.2. *Let M be a Riemannian manifold and G any set of isometries of M. Let $F = \text{Fix}(M; S^1) = \{x \in M : g(x) = x \forall x \in M, \forall g \in G\}$. Then each connected component of F is a closed totally geodesic submanifold of M.*

We will prove this theorem, but first let us recall what totally geodesic means.

Definition 2.3. We say that $N \subset M$ is a *totally geodesic* submanifold if a geodesic in N is also a geodesic in M.

Trivially, any geodesic $\gamma(t) \subset M$ is a totally geodesic submanifold of M, since γ gives us an embedded submanifold in M. In the sphere, S^2, the great circles are totally geodesic but no other closed embedded curve is. In S^3 we can easily find totally geodesic S^2 submanifolds, but there are many more S^2 submanifolds that are not totally geodesic. For example, most embedded submanifolds of $\mathbb{R}^3$ are not totally geodesic, such as T^2, S^2, the catenoid and more, whereas any straight line or flat, embedded two-plane will be.

In general, if N is a totally geodesic submanifold of M, then

$$\sec(N) = \sec(M)$$

by the Gauss equation and therefore if we have a lower curvature bound for M, we also have the same lower bound for N.

We will now prove Theorem 2.2.

Proof. Suppose that $\text{Fix}(M; G)$ does not consist only of isolated fixed points. Let $F \subset \text{Fix}(M; G)$ be a connected component of positive dimension. Let $p \in F$ and let U be a normal neighborhood of p which we assume to be convex. That is, we assume $\exp_p : U_o \subset T_p M \to U$ is a diffeomorphism onto U. Let $V \subset T_p M$ be the subspace of vectors left fixed by all elements of G. Recall that a G-action on N induces a G-action on the tangent space and for directions corresponding to fixed points the G action will be trivial. Then it is easy to see that $U \cap F = \exp_x(U_o \cap V)$. This tells us that a neighborhood, $U \cap F$, of $p \in F$ is a submanifold, $\exp_p(U_o \cap V)$. Thus $\text{Fix}(M; G)$ consists of submanifolds of M. It is clear that each connected component will be closed.

Now it remains to show that F is totally geodesic. Suppose that p and q are contained in the same connected component of $\text{Fix}(M; G)$ and that they are sufficiently close so that we can find a unique geodesic joining them. Let $\gamma : [0, 1] \to M$ be this geodesic and such that $\gamma(0) = p$ and $\gamma(1) = q$. Since isometries map geodesics to geodesics, $g\gamma(t)$ is another geodesic joining p to q for any $g \in G$. By uniqueness, $g\gamma(t) = \gamma(t)$ for all $g \in G$ and for all $t \in [0, 1]$ and therefore $\gamma(t) \subset F$.   □

As an immediate corollary we obtain the following (cf. [25]).

Corollary 2.4. *Assume M is a complete Riemannian manifold and G is a connected group acting isometrically on M. Let p and q belong to different connected components of $\text{Fix}(M; G)$. Then q is a conjugate point of p.*

Note that in non-positive curvature this means that if we have a fixed point set it has a unique connected component since for non-positively curved manifolds there are no conjugate points. For non-negative sectional curvature this means that the connected components of $\text{Fix}(M; G)$ are as far away from each other as they can be.

The following theorem allows us to characterize the fixed point set components of isometric circle actions (cf. [25])

Theorem 2.5. *Let (M, g) be a Riemannian manifold and suppose that $S^1 \times M \to M$ acts isometrically and effectively. Let $Fix(M; S^1) = \cup_i N_i$ be a decomposition of the fixed point set into its connected components. Then the following are true.*

(1) *Each N_i is a closed totally geodesic submanifold of even codimension.*
(2) *If M is orientable then each N_i is orientable.*

Proof. By the previous theorem we know that the N_i are closed and totally geodesic. Let $x_i \in N_i$. By the slice representation we know that at each x_i, $G \times S_{x_i} \to S_{x_i}$ and $S_{x_i} \cong T_{x_i} M$. Further we know that the action is G-equivalent to a linear action on $T_{x_i} M$. Now, G acts trivially on $T_{x_i} N$ and non-trivially on $T_{x_i}^\perp$. This implies that the S^1 action, for a suitable choice of basis looks like

$$
\begin{pmatrix}
0 & & & & & \\
 & \ddots & & & & \\
 & & 0 & & & \\
 & & & \text{rot}(\theta) & & \\
 & & & & \ddots & \\
 & & & & & \text{rot}(\theta)
\end{pmatrix}.
$$

Since the rotation matrices are 2×2 real matrices it follows that

$$
\dim(N_i) = \dim(T_{x_i} N_i) = \dim(M) - \dim(T_{x_i} N_i^\perp) = \dim(M) - 2k,
$$

where k is the number of rotation matrices. Thus N_i is of even codimension. Now, for the second part it is possible to put a complex structure on $T_{x_i} N_i^\perp$ and since M is orientable and $T_{x_i} = T_{x_i} N_i + T_{x_i} N_i^\perp$ it follows that N_i is orientable. $\qquad\square$

We now present some illustrative examples.

Example 2.6. Let $S^1 \times S^{2n} \to S^{2n}$ as follows:

$$
(\theta, x_1, \dots, x_{2n+1}) \mapsto
\begin{pmatrix}
\text{rot}(\theta) & & & \\
 & \ddots & & \\
 & & \text{rot}(\theta) & \\
 & & & 1
\end{pmatrix}
\begin{pmatrix}
x_1 \\
\vdots \\
x_{2n} \\
x_{2n+1}
\end{pmatrix}.
$$

Clearly $\{(0, \ldots, 0, \pm 1)\} \cong S^0$ are the only fixed points in S^{2n} under this action. Isolated fixed points are trivially totally geodesic submanifolds.

Example 2.7. One can easily generalize this example to obtain any S^{2k} as a fixed point set as follows.

Let $S^1 \times S^{2n} \to S^{2n}$ as follows:

$$(\theta, x_1, \ldots, x_{2n+1}) \mapsto \begin{pmatrix} \text{rot}(\theta) & & & & \\ & \ddots & & & \\ & & \text{rot}(\theta) & & \\ & & & 1 & \\ & & & & \ddots \\ & & & & & 1 \end{pmatrix} \begin{pmatrix} x_1 \\ \vdots \\ x_{2n-2k} \\ x_{2n-2k+1} \\ \vdots \\ x_{2n+1} \end{pmatrix}.$$

Clearly $\{(0, \ldots, 0, x_{2n-2k+1}, \ldots, x_{2n+1}) : \sum_{i=2n-2k+1}^{2n+1} x_i^2 = 1\} \cong S^{2k}$ is fixed by the circle action.

Example 2.8. Let $S^1 \times \mathbb{C}P^2 \to \mathbb{C}P^2$ as follows. Consider $\mathbb{C}P^2 = \{[z_1, z_2, z_3] : (z_1, z_2, z_3)(e^{2\pi i \theta} z_1, e^{2\pi i \theta} z_2, e^{2\pi i \theta} z_3), \theta \in [0, 2\pi)\}$. The circle action is then given by

$$e^{2\pi i \phi}, [z_1, z_2, z_3] \mapsto [e^{2\pi i \phi} z_1, e^{-2\pi i \phi} z_2, z_3]$$

and the circle action fixes $\{[z_1, z_2, 0]\} \cong S^2$ and an isolated point, $\{[0, 0, z_3]\}$.

Note here that for homogeneous spaces $\text{Fix}(M; S^1)$ will consist of only homogeneous submanifolds with the same sectional curvature. For the case of strictly positive sectional curvature this alone gives us a lot of information as will become evident after stating the following topological result (cf. [25]).

Theorem 2.9. *Let M be a compact Riemannian manifold. Let $S^1 \times M \to M$ act isometrically. Let $\text{Fix}(M; S^1) = \cup_i N_i$. Then the following are true.*

(1) $\sum_k (-1)^k \dim(H_k(M; K)) = \sum_i \sum_k (-1)^k \dim(H_k(N_i; K))$ *and*
(2) $\sum_k \dim(H_k(M; K)) \geq \sum_i \sum_k \dim(H_k(N_i; K))$,

for any coefficient field K.

We will prove (1) and refer the reader to [25] for the proof of (2).

Proof. Let A_i be the closure of an ϵ-neighborhood around N_i. Take $\epsilon > 0$ to be small enough so that every point of A_i can be joined to the nearest point of N_i by a unique geodesic of length less than or equal to ϵ and such that $A_i \cap A_j = \emptyset, i \neq j$. Thus A_i is a fiber bundle over N_i whose fibers are closed solid balls of radius ϵ. Set $A = \cup_i A_i$. Let B be the closure of the open set $M \smallsetminus A$. Then $A \cap B$ is the boundary of A.

Recall that a sequence is exact if

$$\cdots \to U_k \xrightarrow{\phi_k} V_k \xrightarrow{\psi_k} W_k \to \cdots$$

satisfies $\ker \psi_k = \operatorname{im} \phi_k$.

Now observe that if

$$\cdots \to U_k \to V_k \to W_k \to U_{k-1} \to V_{k-1} \to W_{k-1} \to \cdots$$

is an exact sequence of vector spaces then

$$\sum_k (-1)^k \dim(U_k) - \sum_k (-1)^k \dim(V_k) + \sum_k (-1)^k \dim(W_k) = 0.$$

We apply the formula to the exact relative sequences in homology of (M, B) and $(A, A \cap B)$, that is

$$\cdots \to H_k(M, B) \to H_{k-1}(B) \to H_{k-1}(M) \to H_{k-1}(M, B) \to \cdots$$

and

$$\cdots \to H_k(A, A \cap B) \to H_{k-1}((A \cap B) \to H_{k-1}(A) \to H_{k-1}(A, A \cap B) \cdots$$

and we obtain

$$\chi(B) - \chi(M) + \chi(M, B) = 0$$

and

$$\chi(A \cap B) - \chi(A) + \chi(A, A \cap B) = 0.$$

The excision axiom tells us that $H_i(X, C) \cong H_i(X \smallsetminus U, C \smallsetminus U)$, where $U \subset C \subset X$. It then follows by the excision axiom that the pairs (M, B) and $(A, A \cap B)$ have the same relative homology. In particular, this tells us that $\chi(M, B) = \chi(A, A \cap B)$ and therefore

$$\chi(M) = \chi(A) - \chi(B) - \chi(A \cap B).$$

Since S^1 fixes no points in B or in $A \cap B$, the Lefschetz theorem tells us that $\chi(B) = \chi(A \cap B) = 0$ and therefore $\chi(M) = \chi(A)$. Moreover, since $\chi(A_i) = \chi(N_i)$ this tells us that $\chi(M) = \sum_i \chi(N_i)$. □

Note that in the case where the fixed point set is zero-dimensional, the Euler characteristic of the manifold tells us exactly how many isolated fixed points there will be. In the case of an even-dimensional sphere, this tells us that if we have an isometric circle action, there must be a non-empty fixed point set and the fixed point set must have Euler characteristic equal to two.

The following theorem tells us when we have fixed points of an isometry (cf. [25]).

Theorem 2.10. *Let* (M, g) *be a compact Riemannian manifold of positive sectional curvature. Let* f *be an isometry of* M.

(1) *If* $n = \dim(M)$ *is even and* f *is orientation preserving then* f *has a fixed point.*
(2) *If* $n = \dim(M)$ *is odd and* f *is orientation reversing then* f *has a fixed point.*

As a corollary we obtain Synge's theorem.

Corollary 2.11. *Let* (M, g) *be a compact Riemannian manifold of positive sectional curvature.*

(1) *If* $n = \dim(M)$ *is even and* M *is orientable then* $\pi_1(M) = 0$.
(2) *If* $n = \dim(M)$ *is odd then* M *is orientable.*

Proof. We begin by proving part (1). Let $\tilde{M}$ be the universal cover of M. Every deck transformation of $\tilde{M}$ is an orientation preserving isometry without fixed points. This is a contradiction by Theorem 2.10 unless $\pi_1(M) = 0$ to begin with.

We now prove part (2). If M is not orientable, let $\tilde{M}$ be its orientable double cover. Then the nontrivial deck transformation of $\tilde{M}$ is orientation reversing and therefore it must have a fixed point by Theorem 2.10. This gives us a contradiction once again and therefore M must be orientable.   □

We now need to gather a few facts about the orbit space of a Riemannian manifold under an isometric group action.

Let M/G be the quotient space of M where $G \times M \to M$ acts by isometries with G compact and M a Riemannian manifold with sectional curvature bounded below by some positive constant $\alpha \in \mathbb{R}$. Then, M/G is an Alexandrov space with curvature bounded below by α. The metric on M/G is the orbital distance metric, that is, $\text{dist}_{M/G}(x^*, y^*) = \text{dist}_M(G(x), G(y))$. The space of directions Σ_{x^*}, for any point $x^* \in M/G$ will be S^k/G_x, where S^k is the unit normal sphere to any point of the orbit $G(x)$ and G_x is the isotropy subgroup of x. This tells us that the directions in M/G are geodesic directions. We now define the q-extent of a metric space as the maximum average distance between q points and we have the following geometrical result for points $x^* \in M/G$.

Extent Lemma 2.12 ([15, 17]). *Let* $\overline{p}_0, \ldots, \overline{p}_q$ *be* $q + 1$ *distinct points in* $X = M/G$. *If* $\text{curv } X \geq 0$, *then*

$$\frac{1}{q+1} \sum_{i=0}^{q} \text{xt}_q(\Sigma_{\overline{p}_i} X) \geq \pi/3.$$

We remark that in the case of strictly positive curvature, the inequality is also strict.

We now recall the following topological result from Bredon [5].

Theorem 2.13. *Let G be a compact Lie group acting by cohomogeneity three on M, a compact, simply-connected smooth manifold. If all orbits are connected, then M^* is a simply-connected topological 3-manifold with or without boundary.*

It follows from the resolution of the Poincaré conjecture (cf. [30–32]) that M^* is homeomorphic to one of S^3, D^3, $S^2 \times I$ or, more generally, to S^3 with a finite number of disjoint open 3-balls removed.

We also recall the following general result of Bredon [5] about the fundamental group of the orbit space:

Theorem 2.14. *Let G be a compact Lie group acting on X, a topological space. If either G is connected or G has a nonempty fixed point set, then the orbit projection $\pi : X \to X/G$ induces an onto map on fundamental groups.*

Finally, we recall the Soul theorem for Alexandrov spaces (cf. [7]).

Soul Theorem 2.15. *Let $X = M/G$. If curv $X \geq 0$ and $\partial X \neq \emptyset$, then there exists a totally convex compact subset $S \subset X$ with $\partial S = \emptyset$, which is a strong deformation retract of X. If curv $M/G > 0$, then $S = \bar{s}$ is a point, and ∂X is homeomorphic to $\Sigma_{\bar{s}} X \simeq S_s^{\perp}/G_s$.*

We now have all the tools necessary to prove Theorem 2.1.

Proof. By the classification results of Freedman for 4-manifolds (cf. [11]), in order to obtain the result it is sufficient to show that $\chi(M) \leq 3$. By Theorem 2.10, we know that $\mathrm{Fix}(M^4; S^1)$ is non-empty and since its components must be of even codimension they are either isolated points or orientable, positively curved 2-manifolds. In the latter case, it follows directly from the Gauss-Bonnet theorem that they must be S^2.

There are two cases to consider: case one, where $\dim(\mathrm{Fix}(M^4; S^1)) = 0$ and case two, where there is a component of the fixed point set of dimension two.

We begin with case one. We may apply the Extent Lemma 2.12 to see that four isolated points may not occur and therefore there are at most three.

We now consider case two. Let $N^2 = S^2 \subset \mathrm{Fix}(M^4; S^1)$. Consider $X^3 = M^4/S^1$. N^2 will correspond to the boundary of X^3 and applying the Soul theorem for Alexandrov spaces we see that there is a unique point, p^*, at maximal distance from N^2 in X^3. Using the convexity of the distance function from the boundary, we also see easily that all points in $X^3 \setminus \{N^2 \cup \{p\}\}$ are regular points and therefore correspond to principal orbits in M^4. The orbit in M^4 corresponding to $p^* \in X^3$ will either be S^1, $S^1/\mathbb{Z}_k$ or $\{p\}$. Note that in the latter case, we have a fixed point. This shows that $\mathrm{Fix}(M^4; S^1)$ consists of either S^2 or $S^2 \cup \{p\}$. In particular, once again $\chi(M^4) \leq 3$ as desired. $\square$

In case two, we can actually say more about the manifold. In particular, we can decompose M^4 as the union of disc bundles over N^2 and the orbit of p. This result generalizes in a variety of ways (cf. [17, 18, 35, 37]).

3 Lecture 3: Cohomogeneity One Alexandrov Spaces

The goal of this lecture is to prove a structure result for cohomogeneity one Alexandrov spaces. Alexandrov spaces play an important role in Riemannian geometry. They are a natural synthetic generalization of Riemannian manifolds with a lower curvature bound and the natural process of taking Gromov-Hausdorff limits is closed in Alexandrov geometry.

In order to better motivate why one might want to study Alexandrov spaces of cohomogeneity one or with symmetries in general, we'll first take a short detour and consider Gromov-Hausdorff convergence.

The *Hausdorff distance* between two subsets $A, B \subset X$, where X is a metric space, is defined as

$$d_H(A, B) = \inf\{r : B \subset B(A, r) \text{ and } A \subset B(B, r)\}.$$

Hausdorff distance does not give a metric on subsets of a metric space X. Consider any dense (proper) subset $B \subset X$, then $d_H(B, X) = 0$ since $B(B, \epsilon) \supset X$ for all $\epsilon > 0$ but $B \neq X$.

Example 3.1. Let $A = \mathbb{Q}, B = \mathbb{R} \smallsetminus \mathbb{Q}, X = \mathbb{R}$ and as we saw above, $d_H(A, B) = 0$.

Gromov modified this distance as follows.

Definition 3.2. The *Gromov-Hausdorff distance* between X and Y is

$$d_{GH}(X, Y) = \inf\{d_H(i(X), j(Y) : i : X \to Z, j : Y \to Z\},$$

where i, j are isometric embeddings of X and Y into a metric space Z.

Observe that the Gromov-Hausdorff distance is not a metric, but rather a pseudo metric for the same reasons as detailed above. However, the set $\mathfrak{U} :=$ {isometry classes of compact metric spaces} together with d_{GH} does form a complete metric space.

We observe as well that $i(X) = X'$ and $j(Y) = Y'$ are metric spaces considered with the restriction of the metric of the ambient space, as opposed to the induced intrinsic metric. For example, consider $X = S^2(1)$ with the standard round metric. Then Z cannot be $\mathbb{R}^3$ because $S^2(1) \subset \mathbb{R}^3$ has the restricted metric and X and X' are path-isometric but not isometric.

Naturally one says that a sequence $\{X_i\}_{i=1}^{\infty}$ of (compact) metric spaces converges in the Gromov-Hausdorff sense to a (compact) metric space X if $d_{GH}(X_i, X)$ converges to 0. Note that we have the following nice property for Gromov-Hausdorff distance. We first define an ϵ-net to be a subset S of a metric space X such that for any point $x \in X$ there exists a point $p \in S$ such that $\text{dist}_{GH}(X, Y) < \epsilon$. Then, if Y is an ϵ-net in a metric space X, $\text{dist}_{GH}(X, Y) \leq \epsilon$. Indeed, we can take $Z = X' = X$ and $Y' = Y$.

Now, how does this distance differ from Hausdorff distance? It is clear that $\text{dist}_{GH}(X, Y) \le \text{dist}_H(X, Y)$. For example, let $X = \{p\}$ and $Y = D^2(1)$. Then $\text{dist}_{GH}(X, Y) = 1/2$, since we can always isometrically embed X as the center of the disk, whereas if $p \in \partial D^2(1)$, then $\text{dist}_H(X, Y) = 1$.

Observe that Gromov-Hausdorff distance does not detect small scale behaviour. For example, let X be a circle with very thin spikes such that $X \smallsetminus \{$ spikes $\} \subset S^1(1)$. Then we can find an ϵ-net Y of X such that Y is also an ϵ-net of $S^1(1)$ and therefore $\text{dist}_{GH}(X, S^1) < \epsilon$ whereas $\text{dist}_H(X, S^1) = l$, where l is the length of the spikes.

Gromov-Hausdorff convergence does not preserve dimension. As an example, one can take a sequence of three-balls progressively shrinking along one diameter. The limit will be an interval.

Gromov-Hausdorff convergence does not behave well with respect to fundamental group. We can take a sequence of tori, which are spheres with progressively smaller handles. The limit of this sequence will be a sphere.

The Gromov-Hausdorff limit of a sequence of manifolds with a lower curvature bound will be an Alexandrov space. There is also a notion of convergence for noncompact spaces which is called *pointed Gromov-Hausdorff convergence*. That is, given a sequence $\{(X_i, p_i)\}$ of locally compact complete length metric spaces with marked points. The sequence converges to (Y, p) if for any $R > 0$ the closed R-balls around $p_i \in X_i$ converge to the R-ball around $p \in Y$ in the usual Gromov-Hausdorff sense.

The following general problem is still unsolved.

General Problem . Given $n \in \mathbb{Z}, n \ge 2, k, K \in \mathbb{R}, k \le K$, let $\mathfrak{M}_k(n)$ (respectively $\mathfrak{M}_k^K(n)$) be the class of complete, connected, pointed Riemannian n-manifolds with sectional curvatures greater than or equal to k (respectively $k \le \text{sec}(M) \le K$, $M \in \mathfrak{M}_k^K(n)$). Describe the metric spaces in the Gromov-Hausdorff closures of $\mathfrak{M}_k(n)$ and $\mathfrak{M}_k^K(n)$, denoted by $\overline{\mathfrak{M}_k(n)}$ and $\overline{\mathfrak{M}_k^K(n)}$, respectively.

As has been mentioned, the spaces in $\overline{\mathfrak{M}_k(n)}$ are Alexandrov spaces with curvature bounded below by k and dimension $\le n$. Fukaya showed that all spaces in $\overline{\mathfrak{M}_k(n)}$ look like quotient spaces $N/O(n)$, where N is a Riemannian manifold and $O(n)$ acts on N by isometries (cf. [12]).

However, not all quotients $N/O(n)$ are limits of Riemannian manifolds with bounded curvature. In order to see this we need the following three results from Petersen, Wilhelm and Zhu (cf. [34]).

Theorem 3.3. *Let M be a convex Riemannian manifold with $k \le \text{sec}(M) \le K$. Let $G \times M \to M$ isometrically with closed orbits. Then M/G with the natural quotient metric can be obtained as a limit of a sequence of Riemannian manifolds with a uniform lower curvature bound and fixed dimension.*

A good example here is $\Sigma_{\sin}\mathbb{R}P^n$, which can be seen as the quotient of S^{n+1} by a finite group as follows. Let A be the antipodal map on S^n and since $\Sigma_{\sin}S^n = S^{n+1}$ we may simply "suspend" the antipodal action on S^n to obtain $\Sigma_{\sin}\mathbb{R}P^n$. Without Theorem 3.3, it is not obvious that $\Sigma_{\sin}\mathbb{R}P^n$ is the limit of a sequence of Riemannian manifolds with a lower curvature bound.

Theorem 3.4. *Let N be a connected Riemannian manifold with dimension ≥ 2, $\sec \geq 1$, $\pi_1(N) = \pi_2(N) = 0$. If $\sum_{\sin} N \in \mathfrak{M}_k^K(m)$ for some k, K, m then N is diffeomorphic to a standard sphere.*

In particular, $N = \mathbb{H}P^n$, $n \geq 2$ satisfies the hypothesis of the theorem but since it is not a standard sphere, this tells us that $\sum_{\sin} \mathbb{H}P^n$, $n \geq 2$ is not in $\overline{\mathfrak{M}_k^K}(n)$. Note further that $\sum_{\sin} \mathbb{H}P^n = \sum_{\sin} S^{4n+3}/SU(2) = (\sum_{\sin} S^{4n+3})/SU(2) = S^{4n+4}/SU(2)$. Hence not all quotient spaces are limits of Riemannian manifolds with an upper and a lower sectional curvature bound.

Another question we can ask is whether an Alexandrov space of curvature $\geq k$ is in $\overline{\mathfrak{M}_k}(n)$ for some n. This is not true for $k = 1$ as the following theorem shows.

Theorem 3.5. *Let N be a connected Riemannian manifold of dimension $n \geq 2$ and sectional curvature ≥ 1. If $\sum_{\sin} N$ is in $\overline{\mathfrak{M}_k}(m)$ for some $k > 1/4$ and $m \geq 2$ then:*

(1) *N is homeomorphic to the sphere,*
(2) *N has the homotopy type of $\mathbb{C}P^{\frac{n}{2}}$ or*
(3) *N has the cohomology algebra of $\mathbb{H}P^{\frac{n}{4}}$.*

If $n \neq 4, 8$ and if N is homeomorphic to the sphere then N is diffeomorphic to the sphere.

Consider the following example.

Example 3.6. Let $N = CaP^2$ r $\mathbb{R}P^n$. Then $\sum_{\sin} N \notin \overline{\mathfrak{M}_k}(n)$ when $k > 1/4$. By Theorem 3.3 $\sum_{\sin} \mathbb{R}P^n \in \overline{\mathfrak{M}_k}(n)$ for some k and some m since $\sum_{\sin} \mathbb{R}P^n = S^{n+1}/\mathbb{Z}_2$. However, it is not known whether or not $\sum_{\sin} CaP^2 \in \overline{\mathfrak{M}_k}(m)$ for some k, m, because we do not know how to write it as a quotient $N/O(n)$. Recall that $CaP^2 = F^4/Spin(9)$. Previously we could write $\sum_{\sin} N$ as $N'/O(n)$ since $N = M/G$ and $M = S^m$. Note that $\sum_{\sin} F^4$ is a topological space but is not even an Alexandrov space, since the space of directions at the two cone points is not an Alexandrov space of curvature ≥ 1. However, $(\sum_{\sin} F^4)/Spin(9) = \sum_{\sin} CaP^2$ is an Alexandrov space . This gives us an example of an Alexandrov space that arises as an isometric quotient of a topological length space that is not Alexandrov.

Also, while $\sum_{\sin} \mathbb{H}P^{\frac{n}{4}} = S^{n+4}/SU(2) \in \overline{\mathfrak{M}_k}(n+4)$ and $\sum_{\sin} \mathbb{C}P^{\frac{n}{2}} \in \overline{\mathfrak{M}_k}(n+2)$ for some (possibly different) k, we don't know whether k can be chosen to be strictly larger than $1/4$ since we only know that N in such a case would have the homotopy type of $\mathbb{H}P^{\frac{n}{4}}$ or the cohomology algebra of $\mathbb{C}P^{\frac{n}{2}}$.

Now, as was mentioned in the first lecture, the isometry group of a Riemannian manifold is a Lie group (cf. [27]). In 1994, Fukaya and Yamaguchi [13] showed that *Isom(A)* is also a Lie group for X, a length space which is locally compact and has curvature bounded away from $-\infty$. Therefore the isometry group of an Alexandrov space is a Lie group.

Now, the fact that *Isom(X)* is a Lie group is extremely powerful for obvious reasons. If X is a homogeneous Alexandrov space then it is in fact a Riemannian manifold.

Just as in the Riemannian case it makes sense to try to understand what cohomogeneity one Alexandrov spaces are. Cohomogeneity one manifolds have been studied extensively (see, e.g., [1, 19–21, 23, 26, 28, 29, 36, 38, 39]). The concept of cohomogeneity one was introduced by Mostert in 1957 [26] and he (along with Neumann [28]) classified the three-dimensional cohomogeneity one manifolds (without curvature restrictions).

We'll first discuss the structure of cohomogeneity one manifolds and some properties of the same. From here on, all spaces to be considered are compact. A compact Riemannian manifold admits a cohomogeneity one action if there exists a compact Lie group, G, acting isometrically on M^n such that $\dim(M^n/G) = 1$ or equivalently if $\dim(G/H) = n - 1$, where G/H is a principal orbit. Now, in this particular case, the orbit space must also be compact and can therefore only be S^1 or I. If $M/G = S^1$ then it follows that all orbits are principal and we have a fibration over S^1 with fiber G/H. The structure group of the fibration is $N(H)/H$. This particular case is not very interesting and we will now restrict our attention to the case where $M/G = I$. Here the situation changes some: the principal orbits are dense in M and will correspond to the inverse images of the interior points of the interval whereas the inverse images of the two endpoints will correspond to singular or exceptional orbits. Additionally the principal orbits will fiber over the singular orbits with fiber a sphere. In fact, a cohomogeneity one G action on a closed manifold with orbit space an interval determines a group diagram

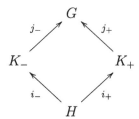

where $i_\pm$ and $j_\pm$ are the inclusion maps, $K_\pm$ are the isotropy groups of the singular orbits at the endpoints of the interval, and H is the principal isotropy group of the action. Further, $K_\pm/H \cong S^{k_\pm}$.

In dimensions ≤ 4 a complete classification of such manifolds is due to Mostert and Neumann [26, 28] and Parker and Hoelscher [23, 29]. It is easy to write down the manifolds in dimension three. Note first that the only groups acting effectively are T^2 and $SO(3)$ and we see that for the T^2 action we obtain S^3, $\mathbb{R}P^3$, $L_{p.q}$, $S^2 \times S^1$, $S^2 \tilde{\times} S^1$, Kl $\times S^1$, A, $\mathbb{R}P^2 \times S^1$ and T^3, where the last is the only example of a cohomogeneity one action by T^2 with quotient space S^1. For the $SO(3)$ action we obtain S^3, $\mathbb{R}P^3$, $\mathbb{R}P^3 \# \mathbb{R}P^3$, $S^2 \times S^1$, $S^2 \tilde{\times} S^1$ and $\mathbb{R}P^2 \times S^1$, where the last three are examples of a cohomogeneity one $SO(3)$ action with quotient space S^1.

In dimension 4 we have more groups to deal with: namely, if we only consider effective actions, we have T^3, $SO(3)T^1$ and $SO(3) \times SO(3)$. Here there are more than 60 manifolds that admit a cohomogeneity one action. Parker classified these manifolds in [29] and Hoelscher found a few missing examples in his thesis [23].

In dimensions 5–7, Hoelscher classified simply-connected cohomogeneity one manifolds.

There are many results about cohomogeneity one manifolds that are of a purely group theoretic nature. This is the case for theorems about the basic decomposition of a cohomogeneity one manifold. Since the fibres over the singular or exceptional orbits are spheres, by the slice theorem we know that the tubular neighborhoods of $G/K_{\pm}$ have the form $G \times_{K_{pm}} D_{\pm}$ and therefore M decomposes as a union of disc bundles, that is, $M = G \times_{K_-} D_- \cup_{G/H} G \times_{K_+} D_+$, where $S^{k\pm} = \partial D_{\pm} = K_{\pm}/H$. Thus we have a description of M in terms of its group diagram. Conversely, given a group diagram of compact Lie groups (G, H, K_+, K_-) where $K_{\pm}/H \cong S^{k\pm}$, then by the classification of transitive actions on spheres (see [6]) the $K_{\pm}$ actions on $S^{k\pm}$ are linear and therefore extend to an action on the corresponding discs D_{pm} bounded by $S^{k\pm}$ for each $K_{\pm}$ and one can construct a cohomogeneity one manifold from the group diagram. This implies that a cohomogeneity one manifold M with $M/G = I$ is in one-to-one correspondence with its group diagram.

In particular, this reduces the problem of classifying cohomogeneity one manifolds whose quotient space is an interval to the problem of classifying the possible group diagrams.

Normally we classify these actions up to "weak equivalence", but for two diagrams $(G_1, H_1, K_{1_+}, K_{1_-})$ and $(G_2, H_2, K_{2_+}, K_{2_-})$, if $G_1 = G_2 = G$ we ask for G-equivariant equivalence.

The following operations on the group diagram result in a G-equivariantly diffeomorphic manifold:

(1) Switching K_- and K_+.
(2) Conjugating each group by the same element of G.
(3) Replacing K_- with aKa^{-1} for $a \in N(H)$.

Conversely, two group diagrams are G-equivariantly equivalent via some combination of these operations. Now the idea is to see what results can be extended to Alexandrov spaces. We have the following structure result from [14].

Theorem 3.7. *Let X be a closed Alexandrov space with an effective isometric action of G by cohomogeneity one with principal isotropy group H.*

(1) *If the orbit space of the action is an interval, then X is the union of two fiber bundles over the two singular orbits whose fibers are cones over positively curved homogeneous spaces, that is,*

$$X = G \times_{K_-} C(K_-/H) \cup_{G/H} G \times_{K_+} C(K_+/H).$$

The group diagram of the action is given by (G, H, K_-, K_+), where $K_{\pm}/H$ are positively curved homogeneous spaces. Conversely, a group diagram (G, H, K_-, K_+), where $K_{\pm}/H$ are positively curved homogeneous spaces, determines a cohomogeneity one Alexandrov space.

(2) *If the orbit space of the action is a circle, then X is equivariantly homeomorphic to a G/H-bundle over a circle with structure group $N(H)/H$. In particular, X is a manifold.*

As mentioned before, a homogeneous Alexandrov space is a manifold, so it is interesting and natural to try to understand at what level of symmetry do Alexandrov spaces appear which are no longer manifolds. Since one- and two-dimensional Alexandrov spaces are topological manifolds [8], it follows from work of Mostert [26] that any closed one- or two-dimensional cohomogeneity one Alexandrov space is equivariantly homeomorphic to a closed manifold with the same action.

In contrast, there is one three-dimensional Alexandrov space of cohomogeneity one that is not a manifold as we see in the following theorem from [14].

Theorem 3.8. *Let X^3 be a closed, three-dimensional Alexandrov space with an effective isometric cohomogeneity one G action. Then G is SO(3) or T^2 and the only non-manifold we obtain is $\Sigma(\mathbb{R}P^2)$.*

The following corollary from [14] is immediate.

Corollary 3.9. *A closed Alexandrov space X^n of dimension $n \geq 4$ with an effective isometric T^{n-1} action is equivariantly homeomorphic to the product of T^{n-3} with one of T^3, S^3, $L_{p,q}$, $S^2 \times S^1$, $S^2 \tilde{\times} S^1$, Kl $\times S^1$, $\mathbb{R}P^2 \times S^1$ or A.*

We will now give a brief idea of how we obtain these results. First, the group considerations pass from the Riemannian case to the Alexandrov case pretty much directly. Further, just as for manifolds, the quotient space of an Alexandrov space by an isometric group action (with closed orbits) is once again an Alexandrov space. So, for a closed, cohomogeneity one Alexandrov space X, X/G is an Alexandrov space of dimension one and hence a topological manifold. Since X is closed, X/G is either S^1 or I.

The proof of the first theorem relies on the slice representation which can be applied in this case since G is compact and an Alexandrov space is a topological space. We can show then that at non-principal orbits where $\dim(G/K_\pm) > 0$, the space of directions for any $x \in \Sigma_x$ splits as S_x, the unit tangent space to G/G_x, and the set $v(S_x)) = \{v \in \Sigma_x : \text{dist}(v, w) = \text{diam}\,\Sigma_x/2 \text{ for all } w \in S_x\}$, which is a compact, totally geodesic Alexandrov subspace of Σ_x with curvature bounded below by 1, and finally, the space of directions Σ_x is isometric to the join $S_x * v(S_x)$. Further, either $v(S_x)$ is connected or it contains exactly two points at distance π.

Recall that Σ_x is a compact Alexandrov space with curvature ≥ 1. This tells us that our "normal space of directions" to any orbit is of the same type and since the action on the normal space is either homogeneous or trivial, it gives us the structure result. Note that if the orbit is exceptional or of codimension one then the cone bundle has to be a disc bundle.

To prove the result on splitting we note that the compactness of $v(S_x)$ follows from the continuity of the distance function. The rest follows from the Join Lemma (cf. [15, 16]) once we show that S_x is isomorphic to the unit round sphere and is totally geodesic in Σ_X.

Lemma 3.10 (Join Lemma). *Let X be an n-dimensional Alexandrov space with curvature bounded below by 1. If X contains the unit round sphere S_1^m isometrically, then $E = \{ x \in X : \mathrm{dist}(x, S_1^m) = \pi/2 \}$ is an isometrically embedded $(n-m-1)$-dimensional Alexandrov space with curvature bounded below by 1, and X is isometric to $S_1^m * E$ with the standard join metric.*

The fact that S_x is isomorphic to the unit round sphere follows from the fact that the tangent cone at x splits isometrically as $\mathbb{R}^n \times U$, where $\mathbb{R}^n$ correspond to the tangent space to $G(x) \cong G/G_x$ and U is a cone.

Using this structure theorem we can classify Alexandrov spaces of cohomogeneity one through dimension four.

References

1. A.V. Alekseevskii, D.V. Alekseevskii, G-manifolds with one dimensional orbit space. Adv. Sov. Math. **8**, 1–31 (1992)
2. S. Aloff, N. Wallach, An infinite family of 7-manifolds admitting positively curved Riemannian structures. Bull. Am. Math. Soc. **81**, 93–97 (1975)
3. Y.V. Bazaikin, On a certain family of closed 13-dimensional Riemannian manifolds of positive curvature. Sib. Math. J. **37**(6), 1219–1237 (1996)
4. M. Berger, Les variétés riemanniennes homogenes normales simplement connexes a courbure strictment positive. Ann. Scuola Norm. Sup. Pisa **15**, 191–240 (1961)
5. G. Bredon, *Introduction to Compact Transformation Groups*, vol. 46 (Academic, New York, 1972)
6. A.L. Besse, *Manifolds all of Whose Geodesics are Closed*. Ergebnisse Series, vol. 93 (Springer, Berlin, 1978)
7. Y. Burago, M. Gromov, G. Perelman, A.D. Alexandrov's spaces with curvatures bounded below (Russian). Uspekhi Mat. Nauk, **47**(2)(284), 3–51, 222 (1992) [Translation in Russian Math. Surv. **47**(2), 1–58 (1992)]
8. D. Burago, Y. Burago, S. Ivanov, *A Course in Metric Geometry*. Graduate Studies in Mathematics, vol. 33 (American Mathematical Society, Providence, 2001)
9. O. Dearricott, A 7-manifold with positive curvature (2010, preprint)
10. J.-H. Eschenburg, New examples of manifolds with strictly positive curvature, Invent. Math. **66**(3), 469–480 (1986)
11. M. Freedman, The topology of four-dimensional manifolds. J. Differ. Geom. **17**(3), 357–453 (1982)
12. K. Fukaya, A boundary of the set of Riemannian manifolds with bounded curvatures and diameters. J. Differ. Geom. **28**, 1–21 (1988)
13. K. Fukaya, T. Yamaguchi, Isometry groups of singular spaces. Math. Z. **216**, 31–44 (1994)
14. F. Galaz-García, C. Searle, Cohomogeneity one Alexandrov spaces. Transform. Groups, **16**(1), 91–107 (2011)
15. K. Grove, S. Markvorsen, New extremal problems for the Riemannian recognition problem via Alexandrov geometry. J. Am. Math. Soc. **8**, 1–28 (1995)
16. K. Grove, P. Petersen, A radius sphere theorem. Invent. Math. **112**, 577–583 (1993)
17. K. Grove, C. Searle, Positively curved manifolds with maximal symmetry rank. J. Pure Appl. Algebra **91**, 137–142 (1994)
18. K. Grove, C. Searle, Differential topological restrictions curvature and symmetry. J. Differ. Geom. **47**, 530–559 (1997)

19. K. Grove, W. Ziller, Curvature and symmetry of Milnor spheres. Ann. Math. **152**, 331–367 (2000)
20. K. Grove, W. Ziller, Cohomogeneity one manifolds with positive Ricci curvature. Invent. Math. **149**, 619–646 (2002)
21. K. Grove, B. Wilking, W. Ziller, Positively curved cohomogeneity one manifolds and 3-Sasakian geometry. J. Differ. Geom. **78**(1), 33–111 (2008)
22. K. Grove, L. Verdiani, W. Ziller, An exotic $T^1 S^4$ with positive curvature (2010, preprint)
23. C. Hoelscher, Classification of cohomogeneity one manifolds in low dimensions (2009). ArXiv:0712.1327v1
24. W.-Y. Hsiang, B. Kleiner, On the topology of positively curved 4-manifolds with symmetry. J. Differ. Geom. **29**, 615–621 (1989)
25. S. Kobayashi, *Transformation Groups in Differential Geometry*. Classics in Mathematics (Springer, Berlin, 1995)
26. P.S. Mostert, On a compact Lie group acting on a manifold. Ann. Math. **65**(2), 447–455 (1957)
27. S.B. Myers, N.E. Steenrod, The group of isometries of a Riemannian manifold. Ann. Math. **40**, 400–416 (1939)
28. W.D. Neumann, 3-Dimensional G-manifolds with 2-dimensional orbits, in *Proceedings of Conference on Transformation Groups*, New Orleans, 1968, pp. 220–222
29. J. Parker, 4-Dimensional G-manifolds with 3-dimensional orbit. Pac. J. Math. **125**(1), 187–204 (1986)
30. G. Perelman, The entropy formula for the Ricci Flow and its geometric applications, 11 Nov 2002. arXiv:math.DG/0211159
31. G. Perelman, Ricci Flow with surgery on three-manifolds, 10 Mar 2003. arXiv:math.DG/0303109
32. G. Perelman, Finite extinction time for the solutions to the Ricci flow on certain three-manifolds, 17 Jul 2003. arXiv:math.DG/0307245.
33. P. Petersen, F. Wilhelm, An exotic sphere with positive sectional curvature (2010, preprint)
34. P. Petersen, F. Wilhelm, S. Zhu, Spaces on and beyond the boundary of existence. J. Geom. Anal. **5**(3), 419–426 (1995)
35. X. Rong, Positively curved manifolds with almost maximal symmetry rank. Geometriae Dedicata **95**, 157–182 (2002)
36. C. Searle, Cohomogeneity and positive curvature in low dimensions. Mathematische Zeitschrift **214**(3), 491–498 (1993). Corrigendum, Math. Z. **226**, 165–167 (1997)
37. C. Searle, D. Yang, On the topology of non-negatively curved simply-connected 4-manifolds with continuous symmetry. Duke Math. J. **74**(2), 547–556 (1994)
38. L. Verdiani, Cohomogeneity one Riemannian manifolds of even dimension with strictly positive sectional curvature, I. Math. Z **241**, 329–339 (2002)
39. L. Verdiani, Cohomogeneity one manifolds of even dimension with strictly positive sectional curvature. J. Differ. Geom. **68**(1), 31–72 (2004)
40. N. Wallach, Compact homogeneous Riemannian manifolds with strictly positive curvature. Ann. Math. **96**, 277–295 (1972)

A Note on Maximal Symmetry Rank, Quasipositive Curvature, and Low Dimensional Manifolds

Fernando Galaz-Garcia

2000 *Mathematics Subject Classification.* 53C20.

1 Introduction and Main Results

The topological classification of compact Riemannian n-manifolds with positive or nonnegative (sectional) curvature is a fundamental question in Riemannian geometry. The classification in dimension 2 is well-known and follows from the Gauss-Bonnet theorem. In dimension 3, the classification follows from Hamilton's work [22]; in particular, a compact, simply connected 3-manifold of positive curvature must be diffeomorphic to the 3-sphere. In dimension $n \geq 4$, in contrast, a complete solution to the classification problem remains elusive to this day, as evidenced by the relative scarcity of examples and techniques for the construction of compact manifolds with positive or nonnegative curvature. Given these difficulties, it has been helpful to first consider the classification of the most symmetric spaces in these classes, that is, those with a "large" group of isometries. This approach, proposed by Grove [16], allows for flexibility in deciding which isometry groups are to be considered "large". The classification of simply connected positively curved homogeneous spaces (cf. [1–3,46,48]), for example, may be framed in this program,

The author "Fernando Galaz-Garcia" is part of SFB 878: *Groups, Geometry & Actions*, at the University of Münster.

F. Galaz-Garcia (✉)

Mathematisches Institut, Westfälische Wilhelms-Universität Münster, Münster, Germany

e-mail: f.galaz-garcia@uni-muenster.de

O. Dearricott et al., *Geometry of Manifolds with Non-negative Sectional Curvature*,
Lecture Notes in Mathematics 2110, DOI 10.1007/978-3-319-06373-7__3,
© Springer International Publishing Switzerland 2014

which has led to other classification results and to new examples of positively and nonnegatively curved Riemannian manifolds (cf. [6, 19–21, 44, 45, 52, 53]).

Let (M, g) be a (compact) Riemannian manifold and let $\text{Isom}(M, g)$ be its isometry group, which is a (compact) Lie group (cf. [27, 29]). There are several possible measures for the size of $\text{Isom}(M, g)$, e.g., the *cohomogeneity*, defined as the dimension of the orbit space of the action of $\text{Isom}(M, g)$ on (M, g), the *symmetry degree*, defined as the dimension of $\text{Isom}(M, g)$, or the *symmetry rank*, defined as the rank of $\text{Isom}(M, g)$ and denoted by $\text{symrank}(M, g)$. In this note we will focus our attention on this last invariant in the cases when (M, g) has positive curvature and when (M, g) has *quasipositive curvature*, i.e., (M, g) has nonnegative curvature and a point with strictly positive curvature.

The following three problems arise naturally in the study of Riemannian manifolds and their symmetry rank:

(a) *Maximal symmetry rank:* Given a class $\mathcal{M}^n$ of Riemannian n-manifolds, find an optimal upper bound K for the symmetry rank of the elements in $\mathcal{M}^n$.
(b) *Topological classification:* Classify, up to diffeomorphism, all manifolds in $\mathcal{M}^n$ with symmetry rank $k \leq K$.
(c) *Equivariant classification:* Let $M \in \mathcal{M}^n$ with $\text{symrank}(M) = k$. Classify, up to equivariant diffeomorphism, all possible (effective) isometric actions of T^k on M and realize these actions via appropriate Riemannian metrics on M.

These problems have received particular attention when $\mathcal{M}^n$ is the class of compact, positively curved n-manifolds or the class of compact, simply connected n-manifolds of nonnegative curvature (cf. [12–14, 17, 23, 26, 41, 42, 50, 51]). In a curvature-free setting, analogs of problems (a), (b) and (c) for compact, simply connected smooth n-manifolds, $3 \leq n \leq 6$, have also been extensively studied (cf. [10, 11, 18, 28, 30–36, 40]).

The maximal symmetry rank problem and the topological classification of compact, positively curved manifolds of maximal symmetry rank were first considered by Grove and Searle [17]:

Theorem 1.1 (Grove and Searle [17]). *Let (M^n, g) be a compact Riemannian n-manifold of positive curvature. Then the following hold:*

(1) $\text{symrank}(M^n, g) \leq \lfloor (n + 1)/2 \rfloor$.
(2) *If* $\text{symrank}(M^n, g) = \lfloor (n + 1)/2 \rfloor$, *then* M^n *is diffeomorphic to a sphere, a lens space or to a real or complex projective space.*

Let (M, g_0) be isometric to any of the manifolds listed in Theorem 1.1(2), equipped with its standard Riemannian metric g_0. As pointed out in [17], (M, g_0) has maximal symmetry rank. We will refer to the isometric torus actions on (M, g_0) as *linear* torus actions. Our first result is the equivariant classification of torus actions of maximal rank on compact, positively curved manifolds of maximal symmetry rank:

Theorem 1.2. *Any effective, isometric torus action of maximal rank on a compact, positively curved Riemannian manifold of maximal symmetry rank is equivariantly diffeomorphic to a linear action.*

It is natural to ask to what extent the conclusions of Theorem 1.1 hold under weaker curvature conditions, e.g., nonnegative curvature. In this case, an upper bound on the symmetry rank smaller than the dimension of the manifold, as in Theorem 1.1, cannot be achieved in full generality, since the n-dimensional flat torus has maximal symmetry rank n. Under the additional hypothesis of simple connectivity, it has been conjectured (cf. [12]) that if (M^n, g) is a compact, simply connected nonnegatively curved Riemannian n-manifold, then symrank$(M^n, g) \leq \lfloor 2n/3 \rfloor$ and that, if $n = 3k$ and (M^n, g) has maximal symmetry rank, then M^n must be diffeomorphic to the product of k copies of the 3-sphere $\mathbb{S}^3$. This conjectural bound on the symmetry rank has been verified in dimensions at most 9; the topological classification of compact, simply connected Riemannian manifolds of nonnegative curvature and maximal symmetry rank has also been completed in dimensions at most 6, verifying the diffeomorphism conjecture in dimensions 3 and 6 (cf. [12]).

In addition to nonnegatively curved (Riemannian) manifolds, one may consider manifolds with *almost positive curvature*, i.e., nonnegatively curved manifolds with positive curvature on an open and dense set, or manifolds with *quasipositive curvature*, i.e., nonnegatively curved manifolds with a point at which all tangent 2-planes have positive curvature. These two families may be considered as intermediate classes between positively and nonnegatively curved manifolds, and may be used as test cases to determine to what extent the collections of positively and nonnegatively curved manifolds differ from each other. In the noncompact case, it follows from Perelman's proof of the Soul Conjecture [37] that a complete, noncompact manifold with quasipositive curvature must be diffeomorphic to $\mathbb{R}^n$; in particular, it admits positive curvature. In the compact case, $\mathbb{R}P^2 \times \mathbb{R}P^3$ admits a metric with quasipositive curvature (cf. [49]) and cannot support a metric of positive curvature. In contrast to this, in the simply connected case there are no known obstructions distinguishing compact manifolds with positive, almost positive, quasipositive or nonnegative curvature .

Although there are many examples of manifolds with quasipositive or almost positive curvature (cf. [7–9, 15, 24, 38, 43, 47, 49]), including an exotic 7-sphere, the topological classification of these spaces remains open and one may consider problems (a), (b) and (c) for these classes of Riemannian manifolds. Problem (a) was solved by Wilking [54], who showed that the bound for the symmetry rank in Theorem 1.1(1) also holds for Riemannian manifolds with quasipositive curvature:

Theorem 1.3 (Wilking [54]). *If (M^n, g) is an n-dimensional Riemannian manifold of quasipositive curvature, then* symrank$(M^n, g) \leq \lfloor (n + 1)/2 \rfloor$.

Our second result is the topological classification of compact, simply-connected 4- and 5-manifolds of quasipositive curvature and maximal symmetry rank:

Theorem 1.4. *Let M^n be a compact, simply connected Riemannian n-manifold with quasipositive curvature and maximal symmetry rank. Then the following hold:*

(1) *If $n = 4$, then M^4 is diffeomorphic to $\mathbb{S}^4$ or $\mathbb{C}P^2$.*
(2) *If $n = 5$, then M^5 is diffeomorphic to $\mathbb{S}^5$.*

It follows from work of Orlik and Raymond [34], in dimension 4, and of Oh [32], in dimension 5, that smooth, effective T^2 actions on $\mathbb{S}^4$ and $\mathbb{C}P^2$, and smooth, effective T^3 actions on $\mathbb{S}^5$, are equivalent to linear actions. Therefore, the isometric torus actions in Theorem 1.4 must be equivalent to linear actions; any such action can be realized via a standard metric that is (trivially) quasipositively curved, whence the equivariant classification follows.

Recall that the known examples of simply connected 4- and 5-manifolds of nonnegative curvature that are not known to admit positively curved metrics are $\mathbb{S}^2 \times \mathbb{S}^2$, $\mathbb{C}P^2 \# \pm \mathbb{C}P^2$, $\mathbb{S}^2 \times \mathbb{S}^3$, the non-trivial bundle $\mathbb{S}^2 \tilde{\times} \mathbb{S}^3$ and the Wu manifold $SU(3)/SO(3)$. Out of these manifolds, only $\mathbb{S}^2 \times \mathbb{S}^2$ and $\mathbb{C}P^2 \# \pm \mathbb{C}P^2$ admit metrics with nonnegative curvature and maximal symmetry rank 2, and the bundles $\mathbb{S}^2 \times \mathbb{S}^3$ and $\mathbb{S}^2 \tilde{\times} \mathbb{S}^3$ are the only ones admitting metrics of nonnegative curvature and maximal symmetry rank 3; the Wu manifold $SU(3)/SO(3)$, equipped with its standard nonnegatively curved homogeneous metric, has symmetry rank 2 (cf. [13, 14]). On the other hand, the trivial sphere bundle $\mathbb{S}^3 \times \mathbb{S}^2$ carries an almost positively curved metric with symmetry rank 1 (cf. [49]) and it is not known if the remaining 4- and 5-manifolds listed in this paragraph admit metrics of quasipositive curvature. Theorem 1.4 implies that any such metric would have symmetry rank at most 1, in dimension 4, and at most 2, in dimension 5.

We conclude these remarks by recalling the so-called *deformation conjecture* (cf. [15, 49]), which states that if (M, g) is a complete Riemannian manifold of quasipositive curvature, then M admits a metric with positive curvature. As pointed out above, this conjecture is true if M is noncompact, false if M is compact and not simply connected, and remains open if M is compact and simply connected (see [39] for the construction of a metric with positive curvature on the Gromoll-Meyer sphere, an exotic 7-sphere with quasipositive curvature). Theorem 1.4 may be seen as supporting this conjecture when (M, g) is compact, simply connected and has maximal symmetry rank.

The contents of this note are organized as follows. In Sect. 2 we collect some background material and recall the proof of Theorem 1.3. This result was not available in the literature; for the sake of reference, we have included Wilking's proof as conveyed to us by M. Kerin. In Sect. 3 we prove Theorem 1.2 and in Sect. 4 we prove Theorem 1.4. The proofs follow easily from restrictions on the structure of the manifolds and their orbit spaces imposed by the curvature hypotheses and the rank of the actions. As the reader may have already noticed, we have strived to give extensive references to the literature.

2 Preliminaries

2.1 Basic Setup and Notation

Let $G \times M \to M$, $m \mapsto g(m)$, be a smooth action of a compact Lie group G on a smooth manifold M. The orbit $G(p)$ through a point $p \in M$ is diffeomorphic to the quotient G/G_p, where $G_p = \{g \in G : g(p) = p\}$ is the *isotropy* subgroup of G at p. If G_p acts trivially on the normal space to the orbit at p, then G/G_p is called a *principal orbit*. The set of principal orbits is open and dense in M. Since the isotropy groups of principal orbits are all conjugate in G, all principal orbits have the same dimension. The isotropy group of principal orbits is the *principal isotropy subgroup*. If G/G_p has the same dimension as a principal orbit and G_p acts non-trivially on the normal space at p, then G/G_p is called an *exceptional orbit*. An orbit that is neither principal nor exceptional is called a *singular orbit*. When $G_p = G$, the point p is called a *fixed point* of the action. Recall that the *ineffective kernel* of the action is $K := \{g \in G : g(m) = m, \text{ for all } m \in M\}$. The action is *effective* if the ineffective kernel is trivial. The group $\tilde{G} = G/K$ always acts effectively on M.

Given a subset $X \subset M$, we will denote its projection under the orbit map $\pi : M \to M/G$ by X^*. Following this convention, we will denote the orbit space M/G by M^*.

Recall that a finite dimensional length space (X, dist) is an *Alexandrov space* if it has curvature bounded from below in the triangle comparison sense (cf. [5]). When (M, g) is a complete, connected Riemannian manifold and G is a compact Lie group acting on (M, g) by isometries, the orbit space M^* can be made into a metric space (M^*, dist) by defining the distance between orbits p^* and q^* in M^* as the distance between the orbits $G(p)$ and $G(q)$ as subsets of (M, g). If, in addition, (M, g) has sectional curvature bounded below by k, then the orbit space (M^*, dist) equipped with this so-called *orbital metric* is an Alexandrov space with curvature bounded below by k. The *space of directions* of a general Alexandrov space at a point x is, by definition, the completion of the space of geodesic directions at x. In the case of an orbit space $M^* = M/G$, the space of directions $\Sigma_{p^*} M^*$ at a point $p^* \in M^*$ consists of geodesic directions and is isometric to

$$\mathbb{S}_p^{\perp} / G_p,$$

where $\mathbb{S}_p^{\perp}$ is the unit normal sphere to the orbit $G(p)$ at $p \in M$.

2.2 Proof of Theorem 1.3 (Wilking [54])

Let (M^n, g) be an n-dimensional Riemannian manifold of quasipositive curvature with an (effective) isometric T^k action. It suffices to show that if $k > (n + 1)/2$, then (M^n, g) cannot have quasipositive curvature. Throughout the proof we will let $\Omega_p = T^k(p)$ be a principal orbit of the T^k action for some $p \in M$. Given $q \in \Omega_p$, we let $T_q(\Omega_p)^\perp$ be the orthogonal complement of $T_q(\Omega_p)$ in $T_p M$. Recall that the second fundamental form at $q \in \Omega_p$ is given by

$$\alpha : T_q(\Omega_p) \times T_q(\Omega_p) \to T_q(\Omega_p)^\perp.$$

Let $u \in T_q(\Omega_p)$, $|u| = 1$, and let $u^\perp$ be its orthogonal complement in $T_q(\Omega_p)$. Note that

$$\dim u^\perp = \dim \Omega_p - 1 > \frac{n-1}{2},$$

$$\dim T_q(\Omega_p)^\perp = \dim M - \dim \Omega_p < \frac{n-1}{2}.$$

Consider $\alpha(u, \cdot) : u^\perp \to T_q(\Omega_p)^\perp$. For dimension reasons there exists a unit vector $w \in u^\perp$ such that $\alpha(u, w) = 0$.

We will now show that there exists a unit vector $v \in T_q(\Omega_p)$ such that $\alpha(v, v) = 0$. Suppose that there is no such vector in $T_q\Omega_p$. Then there exists $u \in T_q(\Omega_p)$, $|u| = 1$, such that $|\alpha(u, u)| > 0$ is minimal. By the preceding paragraph, there exists $w \in u^\perp$, $|w| = 1$, such that $\alpha(u, w) = 0$. Consider the function

$$f(t) := |\alpha((\cos t)u + (\sin t)w, (\cos t)u + (\sin t)w)|^2$$
$$= |(\cos^2 t)\alpha(u, u) + (\sin^2 t)\alpha(w, w)|^2.$$

Since $f(0) = |\alpha(u, u)|^2$ is minimal, $f'(0) = 0$ and

$$0 \leq f''(0) = 4(\langle \alpha(u, u), \alpha(w, w) \rangle - |\alpha(u, u)|^2).$$

In particular, since $|\alpha(u, u)|^2 > 0$, we have that

$$\langle \alpha(u, u), \alpha(w, w) \rangle > 0.$$

It then follows from the Gauss formula that

$$\sec_{\Omega_p}(u, w) = \sec_M(u, w) + \langle \alpha(u, u), \alpha(w, w) \rangle > 0.$$

This yields a contradiction, since Ω_p is a torus equipped with a left-invariant metric, hence flat.

It follows from the preceding paragraph that there exist orthogonal unit vectors $u, v \in T_q(\Omega_p)$ such that $\alpha(u, u) = \alpha(u, v) = 0$. Then, by the Gauss formula,

$$\sec_{\Omega_p}(u, v) = \sec_M(u, v).$$

Since the choice of principal orbit Ω_p and $q \in \Omega_p$ was arbitrary, it follows that there is an open and dense set of points $p \in M$ with a tangent plane $\Pi_p \subset T_p(\Omega_p)$ such that

$$0 = \sec_{\Omega_p}(\Pi_p) = \sec_M(\Pi_p).$$

Therefore, (M, g) cannot have quasipositive curvature. □

3 Proof of Theorem 1.2

Let (M^n, g) be a compact, positively curved Riemannian n-manifold of maximal symmetry rank with an (effective) isometric action of a torus T^k of maximal rank. We first recall some basic facts from [17, 18]. There exists a circle subgroup $T^1 \leq T^k$ with fixed point set a totally geodesic codimension 2 submanifold F^{n-2} of M^n such that $F^{n-2} = \partial(M^n/T^1)$. The orbit space M^n/T^1 is a positively curved Alexandrov space homeomorphic to the cone over an orbit p^* at maximal distance from $\partial(M^n/T^1)$. The isotropy subgroup of p^* is either T^1, $\mathbb{Z}_k$, $k \geq 2$, or 1, the trivial subgroup of T^1. Since F^{n-2} is diffeomorphic to the space of directions of M^n/T^1 at p^*, F^{n-2} is diffeomorphic to a sphere, if p^* is a principal orbit; a lens space or an even-dimensional real projective space, if p^* is an exceptional orbit; or to a complex projective space, if p^* is a fixed point. Moreover, there exists an invariant disc bundle decomposition

$$M^n = D(F^{n-2}) \cup_E D(G(p)),$$

where $D(F^{n-2})$ is a tubular neighborhood of F^{n-2}, $D(G(p))$ is a tubular neighborhood of the orbit $G(p)$ corresponding to the vertex p^* of the orbit space, and E is the common boundary $\partial D(F^{n-2}) = \partial D(G(p))$. The manifold M^n is diffeomorphic to a sphere if p^* is a principal orbit; a lens space or a real projective space, if p^* is an exceptional orbit; or to a complex projective space, if p^* is a fixed point.

We will now prove the theorem in the case where M^n is diffeomorphic to an n-sphere $\mathbb{S}^n$. We proceed by induction on the dimension n. For $n = 2$, it is well known that any smooth T^1 action on $\mathbb{S}^2$ is equivalent to a linear action. Fix $n > 2$ and let $(\mathbb{S}^n, g)$ be a positively curved n-sphere of maximal symmetry rank, so that there exists an effective isometric T^k action on $(\mathbb{S}^n, g)$ with $k = \lfloor (n+1)/2 \rfloor$. As recalled in the preceding paragraph, there is a circle subgroup $T^1 \leq T^k$ with fixed point set a totally geodesic sphere $\mathbb{S}^{n-2} \subset \mathbb{S}^n$ of codimension 2. The invariant decomposition of $\mathbb{S}^n$ into a union of disc bundles induced by the T^1 action is given by

$$\mathbb{S}^n \simeq D(\mathbb{S}^{n-2}) \cup_{\partial D(\mathbb{S}^1)} D(\mathbb{S}^1)$$
$$\simeq (\mathbb{S}^{n-2} \times D^2) \cup_{\mathbb{S}^{n-2} \times \mathbb{S}^1} (D^{n-1} \times \mathbb{S}^1),$$

where $D(\mathbb{S}^{n-2})$ is a tubular neighborhood of the fixed point set $\mathbb{S}^{n-2}$ of T^1 and $D(\mathbb{S}^1)$ is a tubular neighborhood of the orbit $T^1(p) \simeq \mathbb{S}^1$ whose projection p^* is the vertex point of the orbit space $\mathbb{S}^n/T^1$. As in [17], the T^1 action on

$$D(\mathbb{S}^{n-2}) \simeq \mathbb{S}^{n-2} \times D^2$$

is equivalent to the T^1 action on $\mathbb{S}^{n-2} \times_{T^1} D^2$, the associated disc bundle to the T^1 action on the (trivial) normal bundle of the fixed point set $\mathbb{S}^{n-2}$.

We may write $T^k = T^1 \oplus T^{k-1}$, where T^{k-1} is the orthogonal complement of T^1 in T^k. Observe now that T^{k-1} acts effectively and isometrically on $\mathbb{S}^{n-2} \subset \mathbb{S}^n$. By induction, the action of T^{k-1} on $\mathbb{S}^{n-2}$ is linear. It follows that the T^k action on $\mathbb{S}^{n-2} \times D^2$ is given by the product of a linear T^{k-1} action on $\mathbb{S}^{n-2}$ and a linear T^1 action on D^2. Consequently, on the boundary $\mathbb{S}^{n-2} \times \mathbb{S}^1$ the T^k action is the product of a linear T^{k-1} action on $\mathbb{S}^{n-2}$ and a linear T^1 action on $\mathbb{S}^1$. On $D^{n-1} \times \mathbb{S}^1$, the other half of the disc bundle decomposition of $\mathbb{S}^n$, the T^k action is given by the product of a linear T^{k-1} action on D^{n-1} and a linear action T^1 action on $\mathbb{S}^1$. Observe that the linear T^{k-1} action on D^{n-1} is the cone over the linear action of T^{k-1} on the $\mathbb{S}^{n-2}$ factor of the boundary $D(\mathbb{S}^1) \simeq \mathbb{S}^{n-2} \times \mathbb{S}^1$. Hence, the T^k action on $(\mathbb{S}^n, g)$ is equivariantly diffeomorphic to the linear T^k action on $\mathbb{S}^n = \mathbb{S}^1 * \mathbb{S}^{n-2} \subset \mathbb{R}^2 \times \mathbb{R}^{n-1}$ given by letting T^1 act orthogonally on $\mathbb{R}^2$ and T^{k-1} act orthogonally on $\mathbb{R}^{n-1}$.

When M^n is diffeomorphic to a lens space or to a real projective space, the conclusion follows by passing to the universal covering space and observing that the covering torus action must be equivalent to a linear action on $\mathbb{S}^n$.

The proof when M^n is diffeomorphic to $\mathbb{C}P^m$ is analogous to the case of the sphere. For $m \geq 2$, the equivariant disc bundle decomposition is given by

$$\mathbb{C}P^m \simeq \mathbb{S}^{2n-1} \times_{T^1} D^2 \cup_{\mathbb{S}^{2n-1}} D^{2n},$$

where T^1 is the circle subgroup of T^m fixing both $\mathbb{C}P^{m-1} \subset \mathbb{C}P^m$ and the vertex of D^{2n}, and $\mathbb{S}^{2n-1} \times_{T^1} D^2$ is the normal disc bundle of $\mathbb{C}P^{m-1}$ in $\mathbb{C}P^m$. The T^m action is equivalent to a linear T^m action on $\mathbb{C}P^m$ induced by a linear T^{m+1} action on $\mathbb{S}^{2m+1}$ via the projection map $\pi : \mathbb{S}^{2m+1} \to \mathbb{C}P^m$ of the Hopf action.   □

4   Proof of Theorem 1.4

We proceed along the lines of [12]. Let (M^n, g) be a compact, simply connected Riemannian n-manifold, $n = 4$ or 5, with quasipositive curvature and maximal symmetry rank. Then (M^n, g) has an isometric torus action whose orbit space M^* is two-dimensional. It follows from work of several authors (cf. [4, 25, 32, 34]) that the orbit space M^* of the action has the following properties: M^* is homeomorphic to a 2-disk, the boundary of M^* is the set of singular orbits and the interior of M^* consists of principal orbits. Moreover, when $n = 4$, there are at least two isolated orbits with isotropy T^2 and, when $n = 5$, there are at least three isolated orbits with isotropy T^2. In both cases, points in the arcs in the boundary of M^* joining orbits with isotropy T^2 have isotropy conjugate to a circle T^1; the angle between these arcs is $\pi/2$ and is the length of the space of directions at an orbit with isotropy T^2 in M^*.

Since (M^n, g) is a quasipositively curved Riemannian manifold, M^* is a nonnegatively curved 2-manifold with non-smooth boundary and positive curvature on an open subset. A simple comparison argument using Toponogov's theorem shows that there can be at most 4 points in M^* corresponding to orbits with isotropy T^2 and, if there are 4 such points, then M^* must be isometric to a flat rectangle. Since M^* has positive curvature on an open subset, there can be at most 3 points in M^* with isotropy T^2. Hence, for $n = 4$, the orbit space M^* has 2 or 3 points with isotropy T^2 and, for $n = 5$, M^* has exactly 3 such points. The conclusions of the theorem now follow from the Orlik-Raymond classification of compact, smooth, simply connected 4-manifolds with a smooth, effective T^2 action (cf. [34]), and from Oh's classification of compact, smooth, simply connected 5-manifolds with a smooth, effective T^3 action (cf. [32]). □

Acknowledgements This note originated from talks I gave at the *Tercer Miniencuentro de Geometría Diferencial* in December of 2010, at CIMAT, México. It is a pleasure to thank Rafael Herrera, Luis Hernández-Lamoneda and CIMAT for their hospitality during the workshop. I also wish to thank Martin Kerin, for sharing his notes with the proof of Theorem 1.3, and Xiaoyang Chen, for some conversations on quasipositively curved manifolds.

References

1. S. Aloff, N.R. Wallach, An infinite family of distinct 7-manifolds admitting positively curved Riemannian structures. Bull. Am. Math. Soc. **81**, 93–97 (1975)
2. L. Berard-Bergery, Les variétés riemanniennes homogènes simplement connexes de dimension impaire á courbure strictement positive (French). J. Math. Pures Appl. (9) **55**(1), 47–67 (1976)
3. M. Berger, Les variétés riemanniennes homogènes normales simplement connexes á courbure strictement positive (French). Ann. Scuola Norm. Sup. Pisa (3) **15**, 179–246 (1961)
4. G.E. Bredon, *Introduction to Compact Transformation Groups*. Pure and Applied Mathematics, vol. 46 (Academic, New York/London, 1972)
5. D. Burago, Y. Burago, S. Ivanov, *A Course in Metric Geometry*. Graduate Studies in Mathematics, vol. 33 (American Mathematical Society, Providence, 2001)
6. O. Dearricott, A 7-manifold with positive curvature. Duke Math. J. **158**(2), 307–346 (2011)
7. O. Dearricott, Relations between metrics of almost positive curvature on the Gromoll-Meyer sphere. Proc. Am. Math. Soc. **140**(6), 2169–2178 (2012)
8. J. DeVito, The classification of simply connected biquotients of dimension at most 7 and 3 new examples of almost positively curved manifolds. Ph.D. Thesis, University of Pennsylvania, 2011. http://repository.upenn.edu/edissertations/311.
9. J.-H. Eschenburg, M. Kerin, Almost positive curvature on the Gromoll-Meyer sphere. Proc. Am. Math. Soc. **136**(9), 3263–3270 (2008)
10. R. Fintushel, Circle actions on simply connected 4-manifolds. Trans. Am. Math. Soc. **230**, 147–171 (1977)

11. R. Fintushel, Classification of circle actions on 4-manifolds. Trans. Am. Math. Soc. **242**, 377–390 (1978)
12. F. Galaz-Garcia, C. Searle, Low-dimensional manifolds with non-negative curvature and maximal symmetry rank. Proc. Am. Math. Soc. **139**, 2559–2564 (2011)
13. F. Galaz-Garcia, C. Searle, Non-negatively curved 5-manifolds with almost maximal symmetry rank. Geom. Topol. Preprint: arXiv:1111.3183 [math.DG] (2014, in press)
14. F. Galaz-Garcia, M. Kerin, Cohomogeneity-two torus actions on non-negatively curved manifolds of low dimension. Math. Z. **276**(1–2), 133–152 (2014)
15. D. Gromoll, W. Meyer, An exotic sphere with nonÐnegative sectional curvature. Ann. Math. **100**, 401–408 (1974)
16. K. Grove, Geometry of, and via, symmetries, in *Conformal, Riemannian and Lagrangian Geometry*, Knoxville, 2000. University Lecture Series, vol. 27 (American Mathematical Society, Providence, 2002), pp. 31–53
17. K. Grove, C. Searle, Positively curved manifolds with maximal symmetry-rank. J. Pure Appl. Algebra **91**(1–3), 137–142 (1994)
18. K. Grove, C. Searle, Differential topological restrictions curvature and symmetry. J. Differ. Geom. **47**(3), 530–559 (1997)
19. K. Grove, W. Ziller, Curvature and symmetry of Milnor spheres. Ann. Math. (2) **152**(1), 331–367 (2000)
20. K. Grove, B. Wilking, W. Ziller, Positively curved cohomogeneity one manifolds and 3-Sasakian geometry. J. Differ. Geom. **78**(1), 33–111 (2008)
21. K. Grove, L. Verdani, W. Ziller, An exotic T^1S^4 with positive curvature. Geom. Funct. Anal. **21**(3), 499–524 (2011)
22. R.S. Hamilton, Four-manifolds with positive curvature operator. J. Differ. Geom. **24**(2), 153–179 (1986)
23. W.Y. Hsiang, B. Kleiner, On the topology of positively curved 4-manifolds with symmetry. J. Differ. Geom. **29**(3), 615–621 (1989)
24. M. Kerin, Some new examples with almost positive curvature. Geom. Topol. **15**(1), 217–260 (2011)
25. S. Kim, D. McGavran, J. Pak, Torus group actions on simply connected manifolds. Pac. J. Math. **53**, 435–444 (1974)
26. B. Kleiner, Riemannian four-manifolds with nonnegative curvature and continuous symmetry. Ph.D. Thesis, U.C. Berkeley, 1989
27. S. Kobayashi, *Transformation Groups in Differential Geometry*. Ergebnisse der Mathematik und ihrer Grenzgebiete, Band 70 (Springer, New York/Heidelberg, 1972)
28. J. Kollár, Circle actions on simply connected 5-manifolds. Topology **45**(3), 643–671 (2006)
29. S.B. Myers, N. Steenrod, The group of isometries of a Riemannian manifold. Ann. Math. (2) **40**(2), 400–416 (1939)
30. W.D. Neumann, 3-Dimensional G-manifolds with 2-dimensional orbits. in *1968 Proceedings Conference on Transformation Groups*, New Orleans, 1967, pp. 220–222
31. H.S. Oh, 6-Dimensional manifolds with effective T^4-actions. Topol. Appl. **13**(2), 137–154 (1982)
32. H.S. Oh, Toral actions on 5-manifolds. Trans. Am. Math. Soc. **278**(1), 233–252 (1983)
33. P. Orlik, F. Raymond, Actions of SO(2) on 3-manifolds, in *Proceeding Conference on Transformation Groups*, New Orleans, 1967 (Springer, New York, 1968) pp. 297–318
34. P. Orlik, F. Raymond, Actions of the torus on 4-manifolds. I. Trans. Am. Math. Soc. **152**, 531–559 (1970)
35. P. Orlik, F. Raymond, Actions of the torus on 4-manifolds. II. Topology **13**, 89–112 (1974)
36. J. Pak, Actions of torus T^n on $(n+1)$-manifolds M^{n+1}. Pac. J. Math. **44**, 671–674 (1973)
37. G. Perelman, Proof of the soul conjecture of Cheeger and Gromoll. J. Differ. Geom. **40**, 209–212 (1994)
38. P. Petersen, F. Wilhelm, Examples of Riemannian manifolds with positive curvature almost everywhere. Geom. Topol. **3**, 331–367 (1999)

39. P. Petersen, F. Wilhelm, An exotic sphere with positive sectional curvature (2008). arXiv:0805.0812v3 [math.DG]
40. F. Raymond, Classification of the actions of the circle on 3-manifolds. Trans. Am. Math. Soc. **131**, 51–78 (1968)
41. X. Rong, Positively curved manifolds with almost maximal symmetry rank, in *Proceedings of the Conference on Geometric and Combinatorial Group Theory, Part II*, Haifa, 2000. Geometriae Dedicata, vol. 95 (2002), pp. 157–182
42. C. Searle, D. Yang, On the topology of non-negatively curved simply-connected 4-manifolds with continuous symmetry. Duke Math. J. **74**(2), 547–556 (1994)
43. K. Tapp, Quasi-positive curvature on homogeneous bundles. J. Differ. Geom. **65**(2), 273–287 (2003)
44. L. Verdiani, Cohomogeneity one Riemannian manifolds of even dimension with strictly positive sectional curvature, I. Math. Z. **241**, 329–339 (2002)
45. L. Verdiani, Cohomogeneity one manifolds of even dimension with strictly positive sectional curvature. J. Differ. Geom. **68**(1), 31–72 (2004)
46. N.R. Wallach, Compact homogeneous Riemannian manifolds with strictly positive curvature. Ann. Math. (2) **96**, 277–295 (1972)
47. F. Wilhelm, An exotic sphere with positive curvature almost everywhere. J. Geom. Anal. **11**(3), 519–560 (2001)
48. B. Wilking, The normal homogeneous space has positive sectional curvature. Proc. Am. Math. Soc. **127**(4), 1191–1194 (1999)
49. B. Wilking, Manifolds with positive sectional curvature almost everywhere. Invent. Math. **148**(1), 117–141 (2002)
50. B. Wilking, Torus actions on manifolds of positive sectional curvature. Acta Math. **191**(2), 259–297 (2003)
51. B. Wilking, Positively curved manifolds with symmetry. Ann. Math. (2) **163**(2), 607–668 (2006)
52. B. Wilking, Nonnegatively and positively curved manifolds, in *Surveys in Differential Geometry*, vol. XI (International Press, Somerville, 2007), pp. 25–62
53. W. Ziller, Examples of Riemannian manifolds with non-negative sectional curvature, in *Surveys in Differential Geometry*, vol. XI (International Press, Somerville, 2007), pp. 63–102
54. B. Wilking, Torus actions on manifolds with quasipositive curvature. Notes by M. Kerin (2008) Unpublished

Lectures on n-Sasakian Manifolds

Owen Dearricott

1 Progress Toward the Classification of Isoparametric Hypersurfaces of Spheres

Given an oriented hypersurface, M^n, of a Riemannian manifold, $\bar{M}^{n+1}$, there is a well defined unit length normal field, N, along M^n. The *Weingarten map*, $A : TM \to TM$, is defined to be the negative of the component of the covariant derivative of the unit normal tangent to the hypersurface,

$$AX \equiv -(\nabla_X N)^{\|}.$$

A quick calculation reveals that at each point of M the Weingarten map is a symmetric endomorphism,

$$
\begin{aligned}
\langle AX, Y \rangle &= -\langle \nabla_X N, Y \rangle = \langle N, \nabla_X Y \rangle \text{ compatibility and } Y \perp N \\
&= \langle N, [X, Y] + \nabla_Y X \rangle \text{ symmetry} \\
&= \langle N, [X, Y] \rangle + \langle N, \nabla_Y X \rangle = \langle N, \nabla_Y X \rangle \text{ integrability} \\
&= \langle -\nabla_Y N, X \rangle = \langle AY, X \rangle.
\end{aligned}
$$

Since the Weingarten map is symmetric it has real eigenvalues at each point. These are the *principal curvatures* of the hypersurface.

Typically the principal curvatures and their multiplicities vary from point to point as one moves around M. In the late 1930s Elié Cartan became interested in classifying hypersurfaces of spheres whose principal curvatures remained constant over the hypersurface, [8]. He observed:

O. Dearricott (✉)
Centro de Investigación en Matemíicas, Guanajuato, Mexico
e-mail: owen.dearricott@gmail.com

O. Dearricott et al., *Geometry of Manifolds with Non-negative Sectional Curvature*,
Lecture Notes in Mathematics 2110, DOI 10.1007/978-3-319-06373-7_4,
© Springer International Publishing Switzerland 2014

Proposition 1. *A hypersurface, M, of a sphere has constant principal curvatures if and only if it is the level set of a function, $f : S^{n+1} \to \mathbb{R}$, such that the Laplacian, Δf, and square length of the gradient, $|\nabla f|^2$, are purely f dependent.*

For such f Cartan coined the term isoparametric function. For a level set of a regular value of an isoparametric function, i.e., a hypersurface with constant principal curvatures, he coined the term isoparametric hypersurface.

Cartan classified such hypersurfaces with g distinct principal curvatures where $g = 1, 2$ or 3. He found that each sphere was acted upon transitively by isometries fixing the levels of the isoparametric function, that is, fixing a whole family of hypersurfaces with constant principal curvatures. In general any family of hypersurfaces realised as the orbits of an isometric action on the sphere must necessarily preserve the principal curvatures along each hypersurface since the Weingarten map is equivariant under isometry and hence must be isoparametric. Cartan surmised that all isoparametric hypersurfaces were homogeneous in this way.

Although unresolved the problem remained essentially completely abandoned in the postwar years until Nomizu sparked interest in the study among the Japanese school in the 1970s. Much to the surprise of everyone, Ozeki and Takeuchi demonstrated Cartan to be wrong in his conjecture by finding infinite families of inhomogeneous examples of isoparametric hypersurface families with four principal curvatures [28]. A few years later Ferus, Karcher and Münzner extended this inhomogeneous family [20]. We will discuss this in greater detail in a moment.

A key advance toward a classification was Münzner's discovery that the number of possible distinct principal curvatures, g, is highly restricted [26]. He observed the sphere can be reconstructed as the union of two disk bundles over the critical levels (focal sets), $M_- = f^{-1}(-1)$ and $M_+ = f^{-1}(1)$, glued along a common boundary hypersurface, M. This allowed Münzner to argue the rank of the cohomology group of M is $2g$ and is restricted to $2, 4, 6, 8, 12$. Hence the possible numbers of distinct principal curvature are $g = 1, 2, 3, 4, 6$. Much later Stolz used similar information to classify the possible multiplicities and to classify isoparametric hypersurfaces in spheres up to multiplicity [29].

Armed with this topological data the stage was set to explicitly conjecture the nature of a classification.

Conjecture 1. Isoparametric hypersurface families in spheres fall under at least one of two schemes:

(i) the Ferus-Karcher-Münzner description, i.e., the FKM examples,
(ii) hypersurface orbits of isometric actions on spheres.

The principal orbits of the actions described in (ii) are submanifolds of the sphere of codimension one, that is the action is *cohomogeneity one*. Such actions on spheres were classified by Hsiang and Lawson [21].

By the close the first decade of the twenty first century Cartan's classification problem is finally coming close to a resolution. The range of techniques used to attack the problem have been unusually broad. In this lecture we only present an informal outline.

Cartan observed that any isoparametric function, $f : S^n \to \mathbb{R}$, was the restriction of a homogeneous polynomial, $F : \mathbb{R}^{n+1} \to \mathbb{R}$, $f = F|_{S^n}$. In particular for his classification of isoparametric hypersurfaces with three principal curvatures he observed that the Cartan polynomial must obey a pair of partial differential equations,

$$\Delta F = 0, \qquad\qquad |\nabla F|^2 = 9|x|^4.$$

Münzner's articulated this situation in full generality in the late 1970s:

Proposition 2. $f : S^{n-1} \to \mathbb{R}$ *is an isoparametric function if and only if* $f = F|_{S^{n-1}}$ *for a homogeneous polynomial of degree g,* $F : \mathbb{R}^n \to \mathbb{R}$*, satisfying*

$$2\Delta F = (m_1 - m_2)g^2|x|^{g-2}, \qquad\qquad |\nabla F|^2 = g^2|x|^{2g-2} \qquad (1)$$

for non-negative integers, m_1, m_2.

Following this Ferus, Karcher and Münzner found elegant solutions to (1) which we now describe. Let $P_0, P_1, \ldots, P_m$ be symmetric $n \times n$ matrices such that

$$P_i P_j + P_j P_i = 2\delta_{ij}I,$$

that is, P_i and P_j anti-commute if $i \neq j$ and $P_i^2 = I$ for each i. Define $F : \mathbb{R}^n \to \mathbb{R}$ by

$$F(x) = |x|^4 - 2\sum_{k=0}^m \langle P_k x, x \rangle^2. \qquad (2)$$

Exercise 1. Verify that (2) satisfies (1) with $m_1 = m$ and $m_2 = \ell - m - 1$ where $n = 2\ell$.

Exercise 2. Verify that the isoparametric function, $f = F|_{S^{n-1}}$, for (2) has that the focal submanifolds, $M_+ = f^{-1}(1)$ and $M_- = f^{-1}(-1)$, are

$$M_+ = \left\{ x \in S^{n-1} : \langle P_k x, x \rangle = 0, k = 0, 1, \ldots, m \right\}, \qquad (3)$$

$$M_- = \left\{ x \in S^{n-1} : Px = x \text{ for some } P = \sum_{k=0}^m a_k P_k \text{ s.t. } \sum_{k=0}^m a_k^2 = 1 \right\}. \qquad (4)$$

In the case of six principal curvatures Abresch classified the hypersurfaces up to multiplicity some years before Stolz [2]. He found the multiplicity of each of the principal curvatures must be either $m = 1$ or $m = 2$. In the mid 1980s Dorfmeister and Neher classified the $m = 1$ case using what they called E-families.

Definition 1. An E-family is a linear map, E, from $\mathbb{R}^2$ to the symmetric 5×5 matrices with distinct eigenvalues that satisfies

$$E(w)^5 - 120|w|^2 E(w)^3 + 1296|w|^4 E(w) = 0.$$

They argued that the hypersurface family associated with the E-family naturally associated with the $m = 1$ case must be homogeneous and thus classified [18]. Quite recently Miyaoka gave arguments to deal with $m = 1$ by different means, [23], and extended the approach to deal with $m = 2$, [24], completing the classification for $g = 6$. Recall the shape operator of a submanifold, M, of a Riemannian manifold, $\bar{M}$, is defined by taking minus the tangential component of the covariant derivative of a normal field along a submanifold,

$$S_N X = - (\nabla_X N)^{\parallel}.$$

Definition 2. An isoparametric hypersurface family is said to obey *Condition A* at $x \in M_+$ if $\ker(S_N) = \ker(S_{N'})$ for all normal vectors, N, N'.

In her arguments, using Lax pairs, Miyaoka shows that in the $g = 6$ case that Condition A holds at each point of M_+. She argues this implies the existence of a linear isometry between any two points on each hypersurface that preserves all covariant derivatives of the curvature tensor. With simple connectivity it then follows from a theorem of Singer that the hypersurfaces are homogeneous and thus fall under the cohomogeneity one classification of Hsiang and Lawson.

The $g = 4$ case remains open, but now is largely done. This is thanks to a result proven by Cecil, Chi and Jensen [9], and again shortly thereafter by Immervoll [22]. Cecil, Chi and Jensen's exhaustive work is based on Cartan's method of moving frames and complete intersections from commutative algebra. Immervoll's brief proof uses the algebraic framework of ternary products developed by Dorfmeister and Neher in conjunction with his original ideas from incidence geometry.

Proposition 3. *If an isoparametric hypersurface family with* $g = 4$ *has multiplicities,* $m_2 \geq 2m_1 - 1$, *then the family is of FKM type.*

Here m_1 and m_2 are two repeated multiplicities of the four eigenspaces of the Weingarten map on a hypersurface. Since these proofs originally appeared Quo-Shin Chi has proven the above result by yet two more different means [10, 12].

This leaves only multiplicities, (2,2), (3,4), (4,5), (6,9) and (7,8), uncovered by this theorem. Ozeki and Takeuchi dealt with (2,2) back in the 1970s. For the multiplicity pair, (3,4), Chi used the notion of a regular sequence from commutative algebra to show that Condition A holds at a point of M_+ and hence an old result of Dorfmeister and Neher classifies it as FKM type [11]. Since the meeting in 2010 Chi has built on this result to settle multiplicities (4,5) and (6,9), leaving only the (7,8) case unresolved as of June 2014 [13].

The focus of these lectures is not so much the resolution of Cartan's problem but rather explore interesting geometric structure carried by some focal sets among isoparametric hypersurface families with $g = 4$. For this it is useful for us to reiterate some elements of the algebraic framework built up by Dorfmeister and Neher [17], as a convenient language to build up an understanding of the curvature tensor of a focal submanifold through the shape operator. Since Dorfmeister and Neher's formalism is somewhat intimidating to the uninitiated we largely avoid using it in the middle lectures. We return to it in the final lecture when the reader is in a position to more fully appreciate its usefulness.

Definition 3. A ternary product, $\{\ldots\} : V \times V \times V \to V$, or triple, on an inner product space, V, is a multi-linear map sending triples of vectors to vectors. A triple is said to be symmetric if:

(i) $\{xyz\}$ is unchanged if its arguments are permuted.
(ii) $\langle\{xyz\}, w\rangle = \langle z, \{xyw\}\rangle$ for each quadruple of vectors, x, y, z, w, i.e., $T(x, y) :$
$V \to V$ defined by $T(x, y)z = \{xyz\}$ is self adjoint, for each, x, y.

Lemma 1. *The set of symmetric triples on V is in one to one correspondence with the set of homogeneous symmetric quartics, $F : V \to \mathbb{R}$, via*

$$F(x) = 3|x|^4 - \frac{2}{3}\langle\{xxx\}, x\rangle. \tag{5}$$

Proof. Let $\{e_i | i = 1, .., n\}$ be an orthonormal basis for V and $x = \sum_i x_i e_i$. Let $a_{ijkl} = \langle\{e_i e_j e_k\}, e_l\rangle$. Note conditions (i) and (ii) amount to the invariance of a_{ijkl} under permutation of the indices, i, j, k, l. Hence

$$\langle\{xxx\}, x\rangle = \sum_{ijkl} a_{ijkl} x_i x_j x_k x_l$$

is a homogeneous symmetric quartic if and only if we have (i) and (ii). Since $|x|^4$ is also a homogeneous symmetric quartic, $F(x) = 3|x|^4 - \frac{2}{3}\langle\{xxx\}, x\rangle$ is as well.

Conversely if $F(x)$ is a homogeneous symmetric quartic it follows that $\frac{9}{2}|x|^4 - \frac{3}{2}F(x)$ is as well. Let

$$\frac{9}{2}|x|^4 - \frac{3}{2}F(x) = \sum_{ijkl} a_{ijkl} x_i x_j x_k x_l.$$

where a_{ijkl} are invariant under permutation of the indices, i, j, k, l Define $\{e_i e_j e_k\} \equiv \sum_l a_{ijkl} e_l$ and extend multi-linearly. Verify that (i) and (ii) follow immediately.

As both sets are in correspondence to choices of a_{ijkl} invariant under permutation of i, j, k, l it now follows that the correspondence is established.

Example 1. Let $V = M_{2 \times r}(\mathbb{F})$ be the vector space of $2 \times r$ matrices with entries in $\mathbb{F}$, where $\mathbb{F}$ is one of the associative division algebras, $\mathbb{R}, \mathbb{C}$ or $\mathbb{H}$. Let $|x|^2 = \text{tr}(xx^*)$ give the norm. Consider the Cartan polynomial found by Ozeki and Takuechi,

$$F(x) = 3\left(\text{tr}(xx^*)\right)^2 - 4\text{tr}\left((xx^*)^2\right). \tag{6}$$

Let us find the associated symmetric triple. Comparing (5) with (6) we have

$$3|x|^4 - \frac{2}{3}\langle\{xxx\}, x\rangle = 3\left(\text{tr}(xx^*)\right)^2 - \frac{2}{3}\langle\{xxx\}, x\rangle = 3(\text{tr}(xx^*))^2 - 4\text{tr}((xx^*)^2).$$

Hence

$$\langle\{xxx\}, x\rangle = 6\text{tr}(xx^*xx^*). \tag{7}$$

Exercise 3. Show that in general, $y\langle\{xxx\}, x\rangle = 4\langle\{xxx\}, y\rangle$, where $yh \equiv dh(y) \in V$ for $h : V \to \mathbb{R}$ is understood.

$y(7)$, where by this we mean Eq. (7) differentiated in direction y, becomes

$$\begin{aligned}
4\langle\{xxx\}, y\rangle &= 6\left(\text{tr}(yx^*xx^*) + \text{tr}(xy^*xx^*) + \text{tr}(xx^*yx^*) + \text{tr}(xx^*xy^*)\right) \\
&= 6\left(\text{tr}(yx^*xx^*) + \text{tr}(xx^*xy^*) + \text{tr}(yx^*xx^*) + \text{tr}(xx^*xy^*)\right) \\
&= 12\left(\text{tr}\left((xx^*x)y^*\right) + \text{tr}\left(y(xx^*x)^*\right)\right) = 24\langle xx^*x, y\rangle.
\end{aligned}$$

Hence

$$\{xxx\} = 6xx^*x. \tag{8}$$

Likewise $y(8)$ becomes

$$3\{xxy\} = 6\left(yx^*x + xy^*x + xx^*y\right). \tag{9}$$

Moreover since y is a constant vector $z(9)$ becomes

$$3\left(2\{xzy\}\right) = 6\left(yz^*x + yx^*z + zy^*x + xy^*z + zx^*y + xz^*y\right).$$

Hence for Ozeki and Takuechi's Cartan polynomial, F, we have the symmetric triple,

$$\{xyz\} = xy^*z + yx^*z + xz^*y + yz^*x + zx^*y + zy^*x. \tag{10}$$

Given this equivalence between symmetric triples and homogeneous symmetric quartics Dorfmeister and Neher reformulated Münzner's equations in terms of conditions on the symmetric triple.

Lemma 2. $|\nabla F|^2 = 16|x|^6$ if and only if $|\{xxx\}|^2 - 9|x|^2\langle\{xxx\}, x\rangle + 18|x|^4 = 0.$

Proof. Recall

$$F(x) = 3\langle x, x\rangle^2 - \frac{2}{3}\langle\{xxx\}, x\rangle.$$

Hence $y(5)$ becomes

$$\langle \nabla F, y \rangle = 12\langle x, x \rangle \langle x, y \rangle - \frac{8}{3}\langle \{xxx\}, y \rangle.$$

Hence

$$\nabla F = 12|x|^2 x - \frac{8}{3}\{xxx\}. \tag{11}$$

$|(11)|^2$ is now

$$|\nabla F|^2 = \left|12|x|^2 x - \frac{8}{3}\{xxx\}\right|^2 = \left|4\left(3|x|^2 x - \frac{2}{3}\{xxx\}\right)\right|^2$$

$$= 16\left|3|x|^2 x - \frac{2}{3}\{xxx\}\right|^2.$$

$$\left|3|x|^2 x - \frac{2}{3}\{xxx\}\right|^2 = 9|x|^4|x|^2 - 4|x|^2\langle \{xxx\}, x \rangle + \frac{4}{9}|\{xxx\}|^2$$

$$= 9|x|^6 - 4|x|^2\langle \{xxx\}, x \rangle + \frac{4}{9}|\{xxx\}|^2.$$

Hence $|\nabla F| = 16|x|^6$ if and only if

$$9|x|^6 - 4|x|^2\langle \{xxx\}, x \rangle + \frac{4}{9}|\{xxx\}|^2 = |x|^6.$$

That is

$$\frac{4}{9}|\{xxx\}|^2 - 4|x|^2\langle \{xxx\}, x \rangle + 8|x|^6 = 0. \tag{12}$$

Equivalently $\frac{9}{4}$ (12) reads

$$|\{xxx\}|^2 - 9|x|^2\langle \{xxx\}, x \rangle + 18|x|^6 = 0. \tag{13}$$

Exercise 4. Prove that $\Delta F = 8(m_1 - m_2)|x|^2$ if and only if both

(i) $\mathrm{tr}\,(T(x, y)) = 2(3 + 2m_1 + m_2)$,
(ii) $\dim(V) = 2(1 + m_1 + m_2)$.

Lemma 3.

$$M_+ = \left\{x \in V : |x|^2 = 1 \text{ and } \{xxx\} = 3x\right\},$$

$$M_- = \left\{x \in V : |x|^2 = 1 \text{ and } \{xxx\} = 6x\right\}.$$

Proof. Recall the focal sets correspond to critical values of $f = F|_{S^{n-1}}$. Thus to find them amounts to a Lagrange multiplier problem for F subject to the constraint, $|x|^2 = 1$.

$$\nabla F = 12|x|^2 x - \frac{8}{3}\{xxx\}, \qquad\qquad \nabla|x|^2 = 2x.$$

Hence the Lagrange multiplier condition, $\nabla F = \lambda \nabla|x|^2$, can be written

$$12x - \frac{8}{3}\{xxx\} = 2\lambda x, \qquad\qquad \{xxx\} = \mu x.$$

The condition (13) subject to $|x|^2 = 1$ is

$$|\{xxx\}|^2 - 9\langle\{xxx\}, x\rangle + 18 = 0. \tag{14}$$

Hence $\mu^2 - 9\mu + 18 = 0$. That is $(\mu - 3)(\mu - 6) = 0$. So either $\mu = 3$ or $\mu = 6$. Cauchy-Schwarz gives $|\langle\{xxx\}, x\rangle| \le |\{xxx\}||x| = |\{xxx\}|$. Hence

$$|\{xxx\}|^2 - 9|\{xxx\}| + 18 \le 0. \tag{15}$$

Hence $3 \le |\{xxx\}| \le 6$.

$$-1 = 3 - \frac{2}{3} \cdot 6 \le F(x) = 3 - \frac{2}{3}\langle\{xxx\}, x\rangle \le 3 - \frac{2}{3} \cdot 3 = 1.$$

Hence x with $|x|^2 = 1$ and $\{xxx\} = 3x$ correspond to elements of M_+ and x with $|x|^2 = 1$ and $\{xxx\} = 6x$ correspond to elements of M_-.

Definition 4. For $x \in M_-$. Define

$$V_\kappa(x) \equiv \{v \in V | \langle x, v\rangle = 0 \text{ and } T(x)v \equiv T(x, x)v = \kappa v\}.$$

Exercise 5. Show for $x \in M_-$ that $T_x M_- = V_2(x)$.

Lemma 4. *The normal space, $(T_x M_-)^\perp = V_0(x)$.*

Proof. Since $T(x)$ is a symmetric endomorphism of the vector space, V, it follows from the spectral theorem that V orthogonally decomposes into a sum of spaces, $V_\kappa(x)$ and $\mathbb{R}x$. To understand what κ occur consider y (13).

$$6\langle\{xxx\}, \{xxy\}\rangle - 36|x|^2\langle\{xxx\}, y\rangle - 18\langle x, y\rangle\langle\{xxx\}, x\rangle + 128|x|^4\langle x, y\rangle = 0.$$

Hence

$$\langle\{xx\{xxx\}\}, y\rangle - 6|x|^2\langle\{xxx\}, y\rangle - 3\langle\{xxx\}, x\rangle\langle x, y\rangle + 18|x|^4\langle x, y\rangle = 0.$$

Hence

$$\{xx\{xxx\}\} - 6|x|^2\{xxx\} - 3\langle\{xxx\}, x\rangle x + 18|x|^4 x = 0. \tag{16}$$

Consider $v(16)$.

$$3\{xx\{xxv\}\} + 2\{\{xxx\}xv\} - 18|x|^2\{xxv\} - 12\langle x, v\rangle\{xxx\}$$
$$-12\langle\{xxx\}, v\rangle x - 3\langle\{xxx\}, x\rangle v + 72|x|^2\langle x, v\rangle x + 18|x|^4 v = 0.$$

Now $\{xxx\} = 6x$ and $|x|^2 = 1$.

$$3\{xx\{xxv\}\} + 12\{xxv\} - 18\{xxv\} - 72\langle x, v\rangle x$$
$$-72\langle x, v\rangle x - 18\langle x, x\rangle v + 72\langle x, v\rangle x + 18v = 0.$$

$$3\{xx\{xxv\}\} - 6\{xxv\} - 72\langle x, v\rangle x - 18v + 18v = 0.$$
$$3\{xx\{xxv\}\} - 6\{xxv\} - 72\langle x, v\rangle x = 0.$$
$$\{xx\{xxv\}\} - 2\{xxv\} - 18\langle x, v\rangle x = 0.$$
$$T(x)^2 v - 2T(x)v - 18\langle x, v\rangle x = 0.$$

Let $v \in V_\kappa(x)$. Then $\kappa^2 v - 2\kappa v = 0$. Hence $\kappa^2 - 2\kappa = 0$. Thus $\kappa = 2$ or $\kappa = 0$.

$$V = V_2(x) \oplus V_0(x) \oplus \mathbb{R}x = T_x S^{n-1} \oplus \mathbb{R}x$$
$$= T_x M_- \oplus (T_x M_-)^\perp \oplus \mathbb{R}x.$$

One can reason along similar lines to find

Lemma 5. *Let $x \in M_+$. Then the tangent space, $T_x M_+ = V_1(x)$ and the normal space, $(T_x M_+)^\perp = V_3(x)$.*

The behaviour of the curvature tensor of a submanifold, M, of a Riemannian manifold, $\bar{M}$, can be understood via the second fundamental form. Recall the Levi-Civita connection on M is the unique connection on the tangent bundle for which

(i) The metric is compatible, i.e., $X\langle Y, Z\rangle = \langle \nabla_X Y, Z\rangle + \langle Y, \nabla_X Z\rangle$, for fields X, Y, Z along M.
(ii) The connection is symmetric, i.e., $\nabla_X Y - \nabla_Y X = [X, Y]$ for fields X, Y along M.

If we set $\nabla_X Y = (\bar{\nabla}_X Y)^\|$ then compatibility is clear. Moreover since $[X, Y]$ lies along M it follows $(\bar{\nabla}_X Y)^\| - (\bar{\nabla}_Y X)^\| = [X, Y]$. Thus $\nabla_X Y = (\bar{\nabla}_X Y)^\|$. Define the second fundamental form, $B_X Y = (\bar{\nabla}_X Y)^\perp$. We have $B_X Y - B_Y X = 0$ and hence $B_X Y = B_Y X$.

Lemma 6. *The second fundamental form is a symmetric tensor.*

Proof.

$$\langle B_X(fY), N \rangle = \langle \bar{\nabla}_X(fY), N \rangle$$
$$= -\langle fY, \bar{\nabla}_X N \rangle = -f\langle Y, \bar{\nabla}_X N \rangle$$
$$= f\langle \bar{\nabla}_X Y, N \rangle = \langle fB_X Y, N \rangle.$$

Hence $B_X(fY) = fB_X Y$.

Exercise 6. The shape operator, $S_N X = -\left(\bar{\nabla}_X N\right)^{\|}$, is tensorial.

Lemma 7. *For each normal field, N, S_N defines a symmetric endomorphism.*

Proof.

$$\langle S_N X, Y \rangle = -\langle \bar{\nabla}_X N, Y \rangle = \langle N, \bar{\nabla}_X Y \rangle = \langle N, B_X Y \rangle = \langle N, B_Y X \rangle = \langle S_N Y, X \rangle.$$

Let n be a normal field to M_-. Then $\{xxn\} = 0$ for each $x \in M_-$. Take the covariant derivative along M_+ in direction z. Here D denotes the covariant derivative in V.

$$2\{zxn\} + \{xxD_z n\} = 0.$$
$$\langle D_z n, x \rangle = -\langle n, D_z x \rangle = -\langle n, z \rangle = 0.$$
$$D_z n = \bar{\nabla}_z n = \left(\bar{\nabla}_z n\right)_2 + \left(\bar{\nabla}_z n\right)_0$$

Hence

$$T(x)D_z n = 2\left(\bar{\nabla}_z n\right)_2 = -2S_n z.$$

Hence

$$2\{zxn\} - 2S_n z = 0, \qquad\qquad \text{i.e. } S_n z = \{nxz\}.$$

Definition 5. Define the binary product, $\circ : V \times V \to V$ by $y \circ z \equiv \{yxz\}$, relative to $x \in M_-$.

The second fundamental form and the shape operator of M_- are encoded by the binary product, $\circ$.

2 CR/Contact CR Submanifolds of Kähler/Sasakian Manifolds

A *Kähler manifold* is a Riemannian manifold, $\bar{M}$, equipped with an almost complex structure, J, that is compatible with the metric and is parallel relative to the Levi-Civita connection. Namely

$$J : T\bar{M} \to T\bar{M} \text{ such that } \qquad J^2(X) = -X \text{ for each vector, } X,$$

$$\langle JX, Y \rangle = -\langle X, JY \rangle, \qquad \left(\bar{\nabla}_X J\right)(Y) = 0 \text{ for each pair of vectors, } X, Y.$$

An almost complex structure can only exist on an even dimensional vector space. Hence Kähler geometry can only take place on even dimensional manifolds. The natural analogue for Kähler geometry in odd dimensions is Sasakian geometry. A *Sasakian manifold* is a Riemannian manifold, S, equipped with a unit Killing field, ξ, (i.e., $|\xi|^2 = 1$, and $\langle \bar{\nabla}_X \xi, Y \rangle = -\langle X, \bar{\nabla}_Y \xi \rangle$ for vector fields X, Y) such that

$$R(X, \xi)Y = \langle Y, \xi \rangle X - \langle X, Y \rangle \xi \tag{17}$$

for all vector fields, X, Y.

Note that the restriction of the covariant derivative of ξ, $\varphi(X) = \bar{\nabla}_X \xi$, defines an almost complex structure when restricted to the distribution, $\xi^\perp$.

Example 2. Consider $S = S^{2n-1} \subset \mathbb{C}^n$. $S^1 = \{z \in \mathbb{C} : |z| = 1\}$ acts isometrically on S via scalar multiplication on the left,

$$z(z_1, z_2, z_3, \ldots, z_n) = (zz_1, zz_2, zz_3, \ldots, zz_n).$$

The fundamental field of i under this action, $\xi = (iz_1, iz_2, \ldots, iz_n)$, is a unit length Killing field. Equation (17) follows vacuously since this is the form of the curvature tensor for a sphere. Hence standard odd dimensional spheres are Sasakian.

Exercise 7. Show that the quotient space, $\mathbb{C}P^{n-1} = S^{2n-1}/S^1$, is equipped with the Kähler structure,

$$J(v_*(x_1, x_2, x_3, \ldots, x_n)) = v_*(ix_1, ix_2, ix_3, \ldots, ix_n),$$

for tangent vectors of the sphere, $(x_1, x_2, x_3, \ldots, x_n) \perp \xi$, where v is the natural projection.

The above are the only examples we consider in these lectures.

We are interested in structure on submanifolds of Kähler and Sasakian manifolds. A fine reference for much of this material is [30].

Definition 6. A contact CR submanifold, M, of a Sasakian manifold, S, is a manifold equipped with a smooth distribution, $\mathscr{D}$, of TM such that for each $x \in M$ one has

$$\varphi(\mathscr{D}_x) \subset \mathscr{D}_x \text{ (i.e., } \mathscr{D} \text{ holomorphic)},$$

$$\varphi(\mathscr{D}_x^{\perp}) \subset (T_x M)^{\perp} \text{ (i.e., } \mathscr{D}^{\perp} \text{ anti-holomorphic)}$$

where $\varphi(X) \equiv \bar{\nabla}_X \xi$. When $\varphi(\mathscr{D}_x^{\perp}) = (T_x M)^{\perp}$ for each x the contact CR structure is called generic.

In this treatment we also require the innocuous assumption that ξ is anti-homomorphic, i.e., $\xi_x \in \mathscr{D}^{\perp}$ for each $x \in M$.

Proposition 4. *The anti-holomorphic distribution of a contact CR-submanifold of a Sasakian manifold is tangent to a foliation of the submanifold.*

Proof. We show $\mathscr{D}^{\perp}$ is involutive, i.e., show that for fields, U, V, that lie along $\mathscr{D}^{\perp}$ that $[U, V]$ lies along $\mathscr{D}^{\perp}$. Recall that $(\bar{\nabla}_X \varphi) Y = R(X, \xi) Y$ since ξ is a Killing field.

Let Z lie along $\mathscr{D}$.

$$\left(\bar{\nabla}_Z \varphi\right) \xi = \langle \xi, \xi \rangle Z - \langle Z, \xi \rangle \xi = Z, \quad \left(\bar{\nabla}_Z \varphi\right) \xi = \bar{\nabla}\left(\varphi(\xi)\right) - \varphi(\bar{\nabla}_Z \xi) = -\varphi^2(Z).$$

Hence $\varphi^2(Z) = -Z$ for each $Z \in \mathscr{D}$. φ defines an almost complex structure on $\mathscr{D}$.

$$\langle [U, V], \varphi(Z) \rangle = \langle \bar{\nabla}_U V, \varphi(Z) \rangle - \langle \bar{\nabla}_V U, \varphi(Z) \rangle$$
$$= -\langle \varphi\left(\bar{\nabla}_U V\right), Z \rangle + \langle \varphi\left(\bar{\nabla}_V U\right), Z \rangle.$$

The Sasakian structure gives

$$\langle V, \xi \rangle U - \langle U, V \rangle \xi = \bar{R}(U, \xi) V = \bar{\nabla}_U \left(\varphi(V)\right) - \varphi\left(\bar{\nabla}_U V\right).$$

Hence $\langle \varphi\left(\bar{\nabla}_U V\right), Z \rangle = \langle \bar{\nabla}_U \left(\varphi(V)\right), Z \rangle$.

Since $\varphi(V)$ is a normal it follows

$$\langle \bar{\nabla}_U \left(\varphi(V)\right), Z \rangle = -\langle \varphi(V), \bar{\nabla}_U Z \rangle = -\langle \varphi(V), \bar{\nabla}_Z U \rangle.$$

$$\langle [U, V], \varphi(Z) \rangle = \langle \varphi(V), \bar{\nabla}_Z U \rangle - \langle \varphi(U), \bar{\nabla}_Z V \rangle$$
$$= -\langle V, \varphi\left(\bar{\nabla}_Z U\right) \rangle + \langle V, \bar{\nabla}_Z \left(\varphi(U)\right) \rangle = -\langle V, \left(\bar{\nabla}_Z \varphi\right)(U) \rangle.$$

Since $\left(\bar{\nabla}_Z \varphi\right)(U) = \langle Z, \xi \rangle U - \langle Z, U \rangle \xi = 0$ we conclude that $[U, V] \perp \mathscr{D}$. As $[U, V]$ is tangent to M if follows $[U, V]$ lies along $\mathscr{D}^{\perp}$ and $\mathscr{D}^{\perp}$ is involutive.

Example 3. Let $S = S^{4n-1} \subset \mathbb{H}^n$. Let

$$M = \{(q_1, q_2, q_3, \dots, q_n) \in S^{4n-1} : \bar{q}_1 i q_1 + \bar{q}_2 i q_2 + \bar{q}_3 i q_3 + \dots + \bar{q}_n i q_n = 0\}. \tag{18}$$

It follows

$$T_q M = \left\{ (v_1, v_2, v_3, \ldots, v_n) \in \mathbb{H}^n : \begin{array}{l} \bar{v}_1 i q_1 + \bar{v}_2 i q_2 + \bar{v}_3 i q_3 + \ldots \bar{v}_n i q_n \text{ is real} \\ \bar{v}_1 q_1 + \ldots \bar{v}_n q_n \text{ is purely imaginary} \end{array} \right\}. \tag{19}$$

Let

$$\mathscr{D}_q^\perp = \{ (q_1 v, q_2 v, \ldots, q_n v) : \operatorname{Re}(v) = 0 \} \oplus \mathbb{R}(i q_1, i q_2, \ldots, i q_n).$$

We argue the above sum is orthogonal. If $(v_1, v_2, \ldots, v_n)$ is in both summands, then $v_k = q_k v = \lambda i q_k$. Thus $\lambda \bar{q}_k i q_k = v$ and thus $nv = \lambda \left(\bar{q}_1 i q_1 + \bar{q}_2 i q_2 + \bar{q}_3 i q_3 + \ldots + \bar{q}_n i q_n \right) = 0$. That is $v = 0$. $v_k = i q_k$ gives $\sum_k \bar{v}_k i q_k = -1$ is real and $\sum_k \bar{v}_k q_k = 0$ is purely imaginary. $v_k = q_k v$ gives $\sum_k \bar{v}_k i q_k = -v \left(\sum_k \bar{q}_k i q_k \right) = 0$ is real and $\sum_k \bar{v}_k q_k = -v$ is purely imaginary. Hence $\mathscr{D}^\perp$ is a smooth distribution on TM.

Since $\varphi(\xi) = 0$ it suffices to show for $V = (q_1 v, q_2 v, \ldots, q_n v)$ that $\varphi(V) \in \left(T_q M \right)^\perp$.

Let $W \in T_q M$ be arbitrary. Then $W = (w_1, w_2, w_3, \ldots, w_n)$ and from (19), $\bar{w}_1 i q_1 + \bar{w}_2 i q_2 + \bar{w}_3 i q_3 + \ldots + \bar{w}_n i q_n$ is real. Multiply on the right by v and observe $\bar{w}_1 i q_1 v + \bar{w}_2 i q_2 v + \bar{w}_3 i q_3 v + \ldots + \bar{w}_n i q_n v$ is purely imaginary, hence $\operatorname{Re}(\bar{w}_1 i q_1 v + \bar{w}_2 i q_2 v + \bar{w}_3 i q_3 v + \ldots + \bar{w}_n i q_n v) = 0$. Namely $\varphi(V) \in \left(T_q M \right)^\perp$. Since $\dim \left((T_q M)^\perp \right) = 3$ and $\dim \left(\mathscr{D}^\perp \right) = 4$ it follows $\varphi(\mathscr{D}) \subset \mathscr{D}$ and M is a generic contact CR submanifold.

As it happens this contact CR submanifold also occurs as the focal set of the isoparametric hypersurface family discussed in Example 1. To see this let $q_k = z_k + w_k j$ where $z_k, w_k \in \mathbb{C}$ for $k = 1, 2, 3, \ldots, n$. Consider the map,

$$(q_1, q_2, q_3, \ldots, q_n) \in M \mapsto x = \begin{pmatrix} z_1 & z_2 & z_3 & \cdots & z_n \\ w_1 & w_2 & w_3 & \cdots & w_n \end{pmatrix} \in M_{2 \times n}(\mathbb{C}).$$

$$\bar{q}_k i q_k = (\bar{z}_k - j \bar{w}_k) i (z_k + w_k j) = (\bar{z}_k - j \bar{w}_k) (i z_k + (i w_k) j)$$

$$= \bar{z}_k i z_k - j \bar{w}_k i z_k + \bar{z}_k i w_k j - j \bar{w}_k i w_k j = \left(|z_k|^2 - |w_k|^2 \right) i + (2 \bar{z}_k w_k) i j.$$

Hence (18) implies

$$|z_1|^2 + |z_2|^2 + |z_3|^2 + \ldots + |z_n|^2 = |w_1|^2 + |w_2|^2 + |w_3|^2 + \ldots + |w_n|^2 = \frac{1}{2}$$

and

$$\bar{z}_1 w_1 + \bar{z}_2 w_2 + \bar{z}_3 w_3 + \ldots + \bar{z}_n w_n = 0.$$

Hence

$$
xx^* = \begin{pmatrix} z_1 & z_2 & z_3 & \cdots & z_n \\ w_1 & w_2 & w_3 & \cdots & w_n \end{pmatrix} \begin{pmatrix} \bar{z}_1 & \bar{w}_1 \\ \bar{z}_2 & \bar{w}_2 \\ \bar{z}_3 & \bar{w}_3 \\ \vdots & \vdots \\ \bar{z}_n & \bar{w}_n \end{pmatrix} = \begin{pmatrix} \frac{1}{2} & 0 \\ 0 & \frac{1}{2} \end{pmatrix}.
$$

Thus $|x|^2 = \mathrm{tr}\,(xx^*) = \frac{1}{2} + \frac{1}{2} = 1$ and $\{xxx\} = 6xx^*x = 3x$. We conclude $M = M_+$ from Example 1.

Example 4. Consider $S = S^{15} \subset \mathbb{O}^2$, where $\mathbb{O}$ denotes the set of Cayley numbers.

Note $\mathbb{O}^2$ has an almost complex structure defined by $J(a,b) \equiv (-b,a)$. Hence we may take $\xi = (-b,a)$ at $(a,b) \in S^{15}$.

Let

$$
M \equiv \{(a,b) \in S^{15} : \bar{a}b \text{ is real}\}.
$$

It follows

$$
T_{(a,b)}M = \left\{ (r,s) \in \mathbb{O}^2 : \begin{array}{l} \bar{r}a + \bar{s}b \text{ is purely imaginary,} \\ \bar{r}b + \bar{a}s \text{ is real} \end{array} \right\}.
$$

Let

$$
\mathscr{D}^{\perp}_{(a,b)} = \{(av, bv) : \mathrm{Re}(v) = 0\} \oplus \mathbb{R}(b, -a) = T_{(a,b)}M.
$$

We show the sum is orthogonal. If (r,s) is in both summands, then $r = av = \lambda b$ so $v = \lambda \bar{a}b$ is both real and purely imaginary. Hence $v = 0$. Now $(r,s) = (av,bv)$ has $\bar{r}a + \bar{s}b = -(v\bar{a})a - (v\bar{b})b = -v(|a|^2 + |b|^2) = -v$ by a Moufang identity, hence $\bar{r}a + \bar{s}b$ is purely imaginary. $\bar{r}b + \bar{a}s = -(v\bar{a})b + \bar{a}(bv)$. Since $\bar{a}b$ is real $b = \lambda a$ for some $\lambda \in \mathbb{R}$. $\bar{r}b + \bar{a}s = \lambda(-(v\bar{a})a + \bar{a}(av)) = \lambda(-v|a|^2 + |a|^2v) = 0$ is real. If $(r,s) = (b,-a)$ then $\bar{r}a + \bar{s}b = \bar{b}a - \bar{a}b = 0$, since $\bar{a}b$ is real, and $\bar{r}b + \bar{a}s = -|b|^2 + |a|^2$ is real. Hence $\mathscr{D}^{\perp}_{(a,b)} \subset T_{(a,b)}M$.

Note that for the anti-holomorphic field, $V = (av, bv)$, with $\bar{a}b = \lambda \in \mathbb{R}$ that

$$
\varphi(V) = (bv, -av) = (\lambda av, -av).
$$

Given another anti-holomorphic field, $W = (aw, bw) = (aw, \lambda aw)$, one has

$$
\langle (aw, \lambda aw), (\lambda av, -av) \rangle = \lambda \mathrm{Re}\,(-(aw)(v\bar{a}) + (aw)(v\bar{a})) = 0.
$$

Hence we have a generic contact CR structure. Note that, since $\mathscr{D}^{\perp}_{(a,b)} = T_{(a,b)}M$ for each (a,b), that M is a totally real submanifold of S^{15}.

This example is the focal submanifold of an isoparametric hypersurface family of FKM type. To see this consider maps, $P_k(a, b) \equiv (-be_k, ae_k)$.

$$P_k^2(a, b) = P_k(-be_k, ae_k) = (-(ae_k)e_k, (-be_k)e_k) = (-ae_k^2, -be_k^2) = (a, b).$$

Hence $P_k^2 = \mathrm{Id}_{\mathbb{O}^2}$. Consider $v = e_i + e_j$ where $e_0, e_1, \ldots, e_6$ is an orthonormal basis for $\mathrm{Im}(\mathbb{H})$ and $i \neq j$. $v^2 = -2$. Hence $(-av^2, -bv^2) = 2(a, b)$.

$(-av^2, -bv^2)$

$\quad = (-(av)v, (-bv)v))$

$\quad = (-(a(e_i + e_j))(e_i + e_j), (-b(e_i + e_j))(e_i + e_j))$

$\quad = (-(ae_i)e_i, (-be_i)e_i) + \big((-(ae_i)e_j, (-be_i)e_j) + (-(ae_j)e_i, (-be_j)e_i)\big)$

$\qquad + (-(ae_j)e_j, (-be_j)e_j)$

$\quad = (a, b) + \big((-(ae_i)e_j, (-be_i)e_j) + (-(ae_j)e_i, (-be_j)e_i)\big) + (a, b)$

$\quad = 2(a, b) + \big((-(ae_i)e_j, (-be_i)e_j) + (-(ae_j)e_i, (-be_j)e_i)\big),$

$$(-(ae_i)e_j, (-be_i)e_j) + (-(ae_j)e_i, (-be_j)e_i) = (0, 0),$$
$$P_i P_j(a, b) + P_j P_i(a, b) = (0, 0).$$

Hence $P_i P_j = -P_j P_i$. Thus $P_0, P_1, \ldots, P_6$ give an FKM system. Hence for $(a, b) \in M_+$ we have

$$\langle P_k(a, b), (a, b) \rangle = \langle (-be_k, ae_k), (a, b) \rangle$$
$$= \mathrm{Re}\left((e_k \bar{b})a\right) - \mathrm{Re}\left((e_k \bar{a})b\right)$$
$$= \mathrm{Re}\left(e_k(\bar{b}a)\right) - \mathrm{Re}\left(e_k(\bar{a}b)\right)$$
$$= \mathrm{Re}\left(e_k(\bar{b}a - \bar{a}b)\right) = 0$$

for $k = 0, 1, 2, \ldots, 6$. Hence $\mathrm{Im}(\bar{b}a) = 0$. Namely $\bar{a}b$ is real. Hence $M = M_+$.

Example 5. Consider $S^{19} \subset V = \{x \in M_{5 \times 5}(\mathbb{C}) : x^T = -x\}$. $U(5)$ acts isometrically on V via $A \cdot x \equiv AxA^T$.

Exercise 8. Show that (10) defines an isoparametric triple on V and the action of $U(5)$ is transitive on the levels of f.

Accordingly $M_+ = \{x \in V : \frac{1}{2}\mathrm{tr}(xx^*) = 1$ and $\{xxx\} = 3x\}$.

$$\text{Consider } x = \frac{1}{\sqrt{2}} \begin{pmatrix} 0 & 0 & 1 & 0 & 0 \\ 0 & 0 & 0 & 1 & 0 \\ -1 & 0 & 0 & 0 & 0 \\ 0 & -1 & 0 & 0 & 0 \\ 0 & 0 & 0 & 0 & 0 \end{pmatrix}. \text{ Then } xx^* = \frac{1}{2} \begin{pmatrix} 1 & 0 & 0 & 0 & 0 \\ 0 & 1 & 0 & 0 & 0 \\ 0 & 0 & 1 & 0 & 0 \\ 0 & 0 & 0 & 1 & 0 \\ 0 & 0 & 0 & 0 & 0 \end{pmatrix}.$$

Hence $\{xxx\} = 6xx^*x = 3x$ and $\frac{1}{2}\text{tr}(xx^*) = 1$. Hence $x \in M_+$. To understand the tangent and normal spaces to M_+ at x, recall from last lecture, amounts to understanding the spectrum of the operator, $T(x) : V \to V$, defined by

$$T(x)y = 2xx^*y + 2xy^*x + 2yx^*x.$$

Let $y = \begin{pmatrix} A & b \\ -b^T & 0 \end{pmatrix}$ where A is an antisymmetric 4×4 matrix and b is a column vector in $\mathbb{C}^4$.

$$T(x)y = \begin{pmatrix} I & 0 \\ 0 & 0 \end{pmatrix} y + y \begin{pmatrix} I & 0 \\ 0 & 0 \end{pmatrix} + \begin{pmatrix} J & 0 \\ 0 & 0 \end{pmatrix} y \begin{pmatrix} J & 0 \\ 0 & 0 \end{pmatrix}$$

$$= \begin{pmatrix} A & b \\ 0 & 0 \end{pmatrix} + \begin{pmatrix} A & 0 \\ -b^T & 0 \end{pmatrix} + \begin{pmatrix} JA^*J & 0 \\ 0 & 0 \end{pmatrix} = \begin{pmatrix} 2A + JA^*J & b \\ -b^T & 0 \end{pmatrix}$$

where $J = \begin{pmatrix} 0 & 0 & 1 & 0 \\ 0 & 0 & 0 & 1 \\ -1 & 0 & 0 & 0 \\ 0 & -1 & 0 & 0 \end{pmatrix}.$

Hence the subspace, $\mathscr{D}_x = \left\{ \begin{pmatrix} 0 & b \\ -b^T & 0 \end{pmatrix} \right\} \subset V_1(x) = T_x M_+$. The map, $A \mapsto JA^*J$, is self adjoint,

$$\langle JA^*J, B \rangle = \text{tr}((JAJ)B) = \text{tr}((JA)(JB)) = \text{tr}((JB)(JA))$$
$$= \text{tr}((JBJ)A) = \langle JB^*J, A \rangle.$$

Thus the eigenvalues of the map are real. Note that $J(JA^*J)^*J = JJAJJ = A$. Hence the eigenvalues of this map are ± 1. The subspace,

$$\mathscr{D}_x^\perp = \left\{ \begin{pmatrix} A & 0 \\ 0 & 0 \end{pmatrix} : A \text{ is in the } -1 \text{ eigenspace} \right\},$$

is in $V_1(x) = T_x M_+$ since on this subspace,

$$T(x)y = \begin{pmatrix} 2A + JA^*J & 0 \\ 0 & 0 \end{pmatrix} = \begin{pmatrix} 2A + (-A) & 0 \\ 0 & 0 \end{pmatrix} = \begin{pmatrix} A & 0 \\ 0 & 0 \end{pmatrix}.$$

For the $+1$-eigenspace,

$$T(x)y = \begin{pmatrix} 2A + JA^*J & 0 \\ 0 & 0 \end{pmatrix} = \begin{pmatrix} 2A + A & 0 \\ 0 & 0 \end{pmatrix} = \begin{pmatrix} 3A & 0 \\ 0 & 0 \end{pmatrix}.$$

Now scalar multiplication by i in V induces the standard Sasakian structure on S^{19}.
Note if $y \in \mathscr{D}_x$ that $iy = \begin{pmatrix} 0 & ib \\ -ib^T & 0 \end{pmatrix} \in \mathscr{D}_x$. Note if $y \in \mathscr{D}_x^{\perp}$ such that $\langle y, ix \rangle = 0$, then $\langle iy, x \rangle = 0$ and $J(iA)^*J = -JA^*J = -(-A) = A$. Hence $iy \in V_3(x) = (T_x M_+)^{\perp}$. Note $\dim \left(\mathscr{D}_x^{\perp} \right) = \dim \left((T_x M_+)^{\perp} \right) + 1$. Hence M_+ is a contact CR submanifold of S^{19}.

Definition 7. Let $\pi : S \to B$ be a submersion between Riemannian manifolds. Let $\mathscr{V}$ be the distribution defined by $\ker(\pi_*)$. Since π_* is onto this distribution is tangent to a foliation, $\mathscr{F}$, of S by submanifolds defined by pre-images of points of B. Let $\mathscr{H} = \mathscr{V}^{\perp}$. π is said to be a Riemannian submersion if $\pi_*|\mathscr{H}$ is a linear isometry at each point,

$$\text{i.e. } \langle \pi_*(X), \pi_*(Y) \rangle = \langle X, Y \rangle$$

for fields, X, Y, orthogonal to the leaves of $\mathscr{F}$ (such fields are called horizontal).

It will be useful for us to consider the analogous structure equations for a Riemannian submersion to those we considered for submanifolds earlier [27].

From here on out we will often have occasion to think about the projection of vector fields onto distributions. Given a vector field, E, on M we denote its projection onto the distribution, $\mathscr{D}$, by $\mathscr{D}E$.

Let $\tilde{X}, \tilde{Y}$ be horizontal fields such that $\pi_*(\tilde{X}) = X$ and $\pi_*(\tilde{Y}) = Y$ are vector fields on B, such fields are called basic. For any vector field, X, on B there is a unique basic field, $\tilde{X}$, called its horizontal lift.

Proposition 5. *For vector fields, X, Y, on B the connection defined by*

$$\nabla_X^* Y = \pi_* \left(\nabla_{\tilde{X}} \tilde{Y} \right)$$

is the Levi-Civita connection.

Proof. We take $\nabla_X^* Y = \pi_* \left(\nabla_{\tilde{X}} \tilde{Y} \right)$. Check symmetry and compatibility.
Recall $\pi_*([X, Y]) = [\pi_*(X), \pi_*(Y)]$.

$$\nabla_X^* Y - \nabla_Y^* Y = \pi_* \left(\nabla_{\tilde{X}} \tilde{Y} \right) - \pi_* \left(\nabla_{\tilde{Y}} \tilde{X} \right)$$
$$= \pi_* \left(\nabla_{\tilde{X}} \tilde{Y} - \nabla_{\tilde{Y}} \tilde{X} \right) = \pi_*([\tilde{X}, \tilde{Y}]) = [X, Y].$$

$$X \langle Y, Z \rangle = \tilde{X} \langle \tilde{Y}, \tilde{Z} \rangle = \langle \nabla_{\tilde{X}} \tilde{Y}, \tilde{Z} \rangle + \langle \tilde{Y}, \nabla_{\tilde{X}} \tilde{Y} \rangle = \langle \mathscr{H} \nabla_{\tilde{X}} \tilde{Y}, \tilde{Z} \rangle + \langle \tilde{Y}, \mathscr{H} \nabla_{\tilde{X}} \tilde{Y} \rangle$$
$$= \langle \pi_*(\nabla_{\tilde{X}} \tilde{Y}), Z \rangle + \langle Y, \pi_*(\nabla_{\tilde{X}} \tilde{Z}) \rangle = \langle \nabla_X^* Y, Z \rangle + \langle Y, \nabla_X^* Z \rangle.$$

Lemma 8. *The operation on horizontal vector fields,*

$$A_X Y = \mathscr{V} \nabla_X Y$$

is tensorial.

Proof. Let $f : M \to \mathbb{R}$.

$$\langle A_X(fY), V \rangle = \langle \nabla_X(fY), V \rangle = -\langle fY, \nabla_X V \rangle = -f \langle Y, \nabla_X V \rangle$$
$$= f \langle \nabla_X Y, V \rangle = \langle f A_X Y, V \rangle.$$

Lemma 9. *Let X be a basic vector field and V be vertical. Then $[X, V]$ is vertical*

Proof. $\pi_*([X, V]) = [\pi_*(X), \pi_*(V)] = [\pi_*(X), 0] = 0.$

Lemma 10. *For basic vector fields, X, Y, one has $A_X Y = \frac{1}{2} \mathscr{V}[X, Y]$.*

Proof. $\langle A_X Y, V \rangle = \langle \nabla_X Y, V \rangle = -\langle Y, \nabla_X V \rangle = -\langle Y, [X, V] + \nabla_V X \rangle = -\langle Y, \nabla_V X \rangle = -V \langle X, Y \rangle + \langle \nabla_V Y, X \rangle = \langle \nabla_V Y, X \rangle = \langle \nabla_Y V, X \rangle = -\langle V, \nabla_Y X \rangle = \langle -A_Y X, V \rangle.$ Hence $A_X Y = -A_Y X.$

Since $\nabla_X Y - \nabla_Y X = [X, Y]$. The vertical part is

$$A_X Y - A_Y X = \mathscr{V}[X, Y].$$

Thus $2 A_X Y = \mathscr{V}[X, Y].$

Proposition 6. *Suppose a Sasakian manifold, S, has that ξ generates a free S^1-action. Then the orbit space, S/ξ, is a Kähler manifold and $\pi : S \to S/\xi$ is a Riemannian submersion with totally geodesic circular fibres.*

Proof. Define $JX = \pi_*(\varphi(\tilde{X}))$. The proof is left as an exercise.

Definition 8. A CR submanifold, N, of a Kähler manifold, M, is a manifold equipped with a smooth distribution, $\mathscr{D}$, of TN such that for each $x \in N$ one has

$$J(\mathscr{D}_x) \subset \mathscr{D}_x \text{ (i.e., } \mathscr{D} \text{ holomorphic)},$$

$$J(\mathscr{D}_x^\perp) \subset (T_x N)^\perp \text{ (i.e., } \mathscr{D}^\perp \text{ anti-holomorphic)}.$$

When $\varphi(\mathscr{D}_x^\perp) = (T_x N)^\perp$ for each x the CR structure is called generic.

Exercise 9. Let M be a contact CR submanifold of a Sasakian manifold, S. Then M/ξ is a CR submanifold of the Kähler manifold, S/ξ.

Lemma 11. *The anti-holomorphic distribution, $\mathscr{D}^\perp$, of a CR submanifold is tangent to a foliation.*

The proof is analogous to that of Proposition 4.

Lemma 12. *The leaves of the integral foliation of $\mathscr{D}^\perp$ are totally geodesic in the CR submanifold if and only if*

$$S_{J\mathscr{D}^\perp}\mathscr{D}^\perp \subset \mathscr{D}^\perp \text{ or equivalently } S_{J\mathscr{D}^\perp}\mathscr{D} \subset \mathscr{D}.$$

Proof. Let U, V be anti-holomorphic fields. i.e., lie along $\mathscr{D}^\perp$. One has $\bar{\nabla}_U JV = J\left(\bar{\nabla}_U V\right)$. Let X be a normal field, i.e., lie along $\mathscr{D}$.

$$\langle S_{JV}U, X \rangle = -\langle \bar{\nabla}_U JV, X \rangle = -\langle J\left(\bar{\nabla}_U V\right), X \rangle = \langle \bar{\nabla}_U V, JX \rangle$$
$$= \langle \nabla_U V, JX \rangle = \langle T_U V, JX \rangle$$

where T denotes the second fundamental form of the leaves of $\mathscr{F}$ in the CR submanifold.

Hence $T = 0$ if and only if $S_{J\mathscr{D}^\perp}\mathscr{D}^\perp \subset \mathscr{D}^\perp$. The equivalence of the remaining condition is left as an exercise.

Lemma 13. *A generic CR submanifold has that the leaves of the integral foliation of $\mathscr{D}^\perp$ are totally geodesic if and only if $B_{\mathscr{D}}\mathscr{D}^\perp = 0$.*

Proof. Let X be a holomorphic vector field, U be anti-holomorphic and N be a normal field. Since the CR-structure is generic there is an anti-holomorphic field, V, with $JV = N$.

$\langle B_X U, N \rangle = \langle B_X U, JV \rangle = \langle B_U X, JV \rangle = \langle S_{JV}U, X \rangle$. Hence $B_X U = 0$ for each holomorphic X and anti-holomorphic U precisely when the leaves are totally geodesic by Lemma 12.

Lemma 14. *For a CR submanifold the leaves of the integral foliation of $\mathscr{D}^\perp$ are totally geodesic in the ambient Kähler manifold if and only if the leaves are totally geodesic in the submanifold and $B_{\mathscr{D}^\perp}\mathscr{D}^\perp = 0$.*

Proof.

$$\langle \bar{\nabla}_U V, X + N \rangle = \langle \bar{\nabla}_U V, X \rangle + \langle \bar{\nabla}_U V, N \rangle$$
$$= \langle \nabla_U V, X \rangle + \langle B_U V, N \rangle = \langle T_U V, X \rangle + \langle B_U V, N \rangle.$$

Thus the leaves are totally geodesic if and only if $T = 0$ and $B_{\mathscr{D}^\perp}\mathscr{D}^\perp = 0$.

Definition 9. The leaves of the integral foliation of $\mathscr{D}^\perp$ for a generic contact CR submanifold are totally contact geodesic precisely if

$$\sigma_U V = \langle U, \xi \rangle \varphi(V) + \langle V, \xi \rangle \varphi(U)$$

for each pair of anti-holomorphic fields U, V, where σ denotes the second fundamental form of each leaf sitting in the ambient Sasakian manifold.

Corollary 1. *A contact CR submanifold has that the leaves of the integral foliation of $\mathcal{D}^\perp$ are totally contact geodesic if and only if the corresponding Kähler manifold has that the leaves of its integral foliation of $\mathcal{D}^\perp \subset TM$ are totally geodesic.*

In discussions that follow it will be useful for us to consider manifolds that are foliated in such a way that analogues of O'Neill's structure equations for a submersion hold in the absence of the existence a bonafide Riemannian submersion. Much of the following material can be found in [25].

Definition 10. A foliation, $\mathcal{F}$, of a Riemannian manifold, M, is said to be equidistant if for any two leaves of $\mathcal{F}$, L and L', one has for any two points, $x, y \in L$, that the distance from x to L' is the same as the distance from y to L'.

The following is a useful local characterisation of an equidistant foliation.

Proposition 7. *A Riemannian foliation is equidistant if and only if one has that for any field, V, tangent to $\mathcal{F}$ and any pair of fields, X, Y, normal to $\mathcal{F}$ that*

$$\langle \nabla_X V, Y \rangle = -\langle \nabla_Y V, X \rangle. \tag{20}$$

We will break up the proof into a string of Lemmata.

Lemma 15. *The condition that for any field, V, tangent to $\mathcal{F}$ and any pair of fields, X, Y, normal to $\mathcal{F}$ that*

$$\langle \nabla_X V, Y \rangle = -\langle \nabla_Y V, X \rangle$$

is equivalent to the condition that $L_V g_T = 0$ where $g_T(E, F) \equiv \langle (T\mathcal{F})^\perp E, (T\mathcal{F})^\perp F \rangle$ for each V tangent to $\mathcal{F}$.

Proof. Assume $L_V g_T = 0$ where $g_T(E, F) \equiv \langle (T\mathcal{F})^\perp E, (T\mathcal{F})^\perp F \rangle$ for each V. Let X, Y be fields normal to $\mathcal{F}$. By assumption,

$$(L_V g_T)(X, Y) = V\langle X, Y \rangle - \langle [V, X], Y \rangle - \langle X, [V, Y] \rangle = 0.$$

Thus by symmetry

$$V\langle X, Y \rangle - \langle \nabla_V X - \nabla_X V, Y \rangle - \langle X, \nabla_V Y - \nabla_Y V \rangle = 0.$$

Hence

$$\langle \nabla_X V, Y \rangle + \langle X, \nabla_Y V \rangle = V\langle X, Y \rangle - \langle \nabla_V X, Y \rangle - \langle X, \nabla_V Y \rangle = 0$$

by metric compatibility. Thus (20) holds. Conversely

$$(L_V g_T)(E, F) = V\langle X, Y \rangle - \langle (T\mathcal{F})^\perp [V, E], Y \rangle - \langle X, (T\mathcal{F})^\perp [V, F] \rangle$$

where X, Y denote the components of E, F normal to $\mathscr{F}$.

Let U denote the component of E tangent to $\mathscr{F}$. Then $[V, U]$ is tangent to $\mathscr{F}$. Hence

$$(L_V g_T)(E, F) = V\langle X, Y\rangle - \langle [V, X], Y\rangle - \langle X, [V, Y]\rangle$$
$$= (V\langle X, Y\rangle - \langle \nabla_V X, Y\rangle - \langle X, \nabla_V Y\rangle) + (\langle \nabla_X V, Y\rangle + \langle X, \nabla_Y V\rangle)$$
$$= 0 + 0 = 0$$

by metric compatibility and (20) respectively.

Definition 11. A vector field, X, is said to be basic with respect to the foliation, $\mathscr{F}$, if it is normal to $\mathscr{F}$ and $[X, V]$ is tangent to $\mathscr{F}$ for any vector field V tangent to $\mathscr{F}$.

Lemma 16. $L_V g_T = 0$ for each field, V, tangent to $\mathscr{F}$ if and only if for each pair of basic vector fields, X, Y, the function, $\langle X, Y\rangle$, restricted to each leaf is constant.

Proof.

$$(L_V g_T)(E, F) = V\langle X, Y\rangle - \langle [V, X], Y\rangle - \langle X, [V, Y]\rangle.$$

For vectors, E, F, at a point, x, we may find basic vector fields, X, Y, that are perpendicular components of E, F at the point, x. Hence $(L_V g_T)(E, F) = V\langle X, Y\rangle = 0$ by assumption. Conversely assume $L_V g_T = 0$ for each V tangent to $\mathscr{F}$. Let X, Y be basic fields. Then $[V, X]$ and $[V, Y]$ are tangent to $\mathscr{F}$.

$$0 = (L_V g_T)(E, F) = V\langle X, Y\rangle - \langle [V, X], Y\rangle - \langle X, [V, Y]\rangle = V\langle X, Y\rangle.$$

Hence $\langle X, Y\rangle$ is constant along each leaf of $\mathscr{F}$.

Lemma 17. If for each pair of basic vector fields, X, Y, the function, $\langle X, Y\rangle$, restricted to each leaf is constant, then each geodesic that meets a leaf orthogonally meets all leaves in its path orthogonally.

Proof. There is a neighbourhood, U of $x \in L$ in for which there is a chart $\varphi : \mathbb{R}^p \times \mathbb{R}^q \to U$ so that $\varphi(0) = x$ and each $\varphi(\mathbb{R}^p \times \{z\})$ is the intersection of U with an open subset in a leaf. Let $(x_1, \ldots, x_p, y_1, \ldots, y_q)$ denote the coordinates. Suppose X is basic. Then $\left[\frac{\partial}{\partial x_i}, X\right]$ is tangent to $T\mathscr{F}$. Let $X = \sum_i X_i \frac{\partial}{\partial x_i} + \sum_j \bar{X}_j \frac{\partial}{\partial y_j}$. Then this occurs if and only if $\bar{X} = \sum_j \bar{X}_j \frac{\partial}{\partial y_j}$ has $\left[\frac{\partial}{\partial x_i}, \bar{X}\right]$ is tangent to $\mathscr{F}$. Now the expression

$$\left[\frac{\partial}{\partial x_i}, \bar{X}\right] = \sum_j \frac{\partial \bar{X}_j}{\partial x_i} \frac{\partial}{\partial y_j}$$

is transverse to $\mathscr{F}$, so can only be basic if $\frac{\partial \bar{X}_j}{\partial x_i} = 0$. That is if the $\{\bar{X}_j\}$ depend purely upon $(y_1, \ldots, y_q)$. Note that the perpendicular component, $\overline{\frac{\partial}{\partial y_j}}$, of $\frac{\partial}{\partial y_j}$ is basic. Hence for X, Y basic we have

$$X = \sum_j \bar{X}_j \overline{\frac{\partial}{\partial y_j}}, \qquad\qquad Y = \sum_j \bar{X}_j \overline{\frac{\partial}{\partial y_j}}.$$

Assume that for each pair of basic vector fields, X, Y, the function, $\langle X, Y \rangle$, restricted to each leaf is constant. Define an associated Riemannian metric on $\mathbb{R}^q$ by extending

$$\left\langle \frac{\partial}{\partial y_i}, \frac{\partial}{\partial y_j} \right\rangle_* \equiv \left\langle \overline{\frac{\partial}{\partial y_i}}, \overline{\frac{\partial}{\partial y_j}} \right\rangle$$

Note that each field $X_* = \sum_j \bar{X}_j \frac{\partial}{\partial y_j}$ of $\mathbb{R}^q$ uniquely lifts to the basic vector field, $X = \sum_j \bar{X}_j \overline{\frac{\partial}{\partial y_j}}$.

Let $x \in L$ and $v \in (T_x L)^\perp$. Let c be a geodesic for $\langle , \rangle_*$ starting at $0 \in \mathbb{R}^q$ in direction, $\bar{v} = (\pi_2 \circ \varphi)_*(v)$. Let $\tilde{c}$ be the unique lift of c starting at x with direction, v, that is everywhere perpendicular to $\mathscr{F}$. Let y be the other endpoint of $\tilde{c}$. Let $\hat{c}$ be a geodesic starting with x and ending at y. $L(\hat{c}) \leq L(\tilde{c}) = L(c)$ since both $\hat{c}$ and $\tilde{c}$ link x and y and the integral expressions for $L(c)$ and $L(\tilde{c})$ are equal by change of variables. However $L(c) \leq L(\pi_2 \circ \varphi \circ \hat{c}) \leq L(\hat{c})$ as well since both c and $\pi_2 \circ \varphi \circ \hat{c}$ share the same endpoints and $\hat{c}$ is a lift of $\pi_2 \circ \varphi \circ \hat{c}$. Hence $L(\hat{c}) = L(\tilde{c})$. By uniqueness of the locally minimising curve, $\hat{c} = \tilde{c}$, is the geodesic that starts with direction, v, which we observe remains perpendicular to each leaf of $\mathscr{F}$ in its trajectory. This gives the desired property.

Lemma 18. *The condition that any geodesic that meets a leaf orthogonally meets all leaves in its path orthogonally is equivalent to the condition that the leaves of the foliation are equidistant.*

Proof. Let x and y be two points of a leaf, L, and L' be a nearby leaf. Connect x and y by a path $\gamma : [0, 1] \to L$. Consider an appropriately parametrised constant speed geodesic, $c_s : [0, 1] \to M$, that connects $\gamma(s)$ to the nearest point of L'. $F : [0, 1] \times [0, 1] \to M$ defined by $F(s, t) = c_s(t)$ defines a smooth variation of geodesics. The arc-length, $\ell(s) = \int_0^1 |\dot{c}_s| dt$, defines the distance from $\gamma(s)$ to L'. The derivative is given via the first variation of arc-length.

$$\frac{d\ell}{ds} = \frac{1}{\ell} \left(\left\langle \frac{\partial F}{\partial s}(s, 1), \dot{c}_s(1) \right\rangle - \left\langle \frac{\partial F}{\partial s}(s, 0), \dot{c}_s(0) \right\rangle - \int_0^1 \left\langle \frac{\partial F}{\partial s}, \nabla_{\dot{c}_s} \dot{c}_s \right\rangle dt \right).$$

$$\frac{d}{ds} \left(\frac{1}{2} \ell^2 \right) = \left\langle \frac{\partial F}{\partial s}(s, 1), \dot{c}_s(1) \right\rangle - \left\langle \frac{\partial F}{\partial s}(s, 0), \dot{c}_s(0) \right\rangle - \int_0^1 \left\langle \frac{\partial F}{\partial s}, \nabla_{\dot{c}_s} \dot{c}_s \right\rangle dt.$$

Since c_s are geodesics and c_s meets L' orthogonally we have

$$\frac{d}{ds}\left(\frac{1}{2}\ell^2\right) = -\langle\gamma'(s), \dot{c}_s(0)\rangle. \tag{21}$$

Assuming the first condition in the statement, since c_s meets L' orthogonally it must also meet L orthogonally. However γ' must then be perpendicular to $\dot{c}_s(0)$. Hence ℓ must be constant. If L' is not nearby then we use the fact that we may approximate the distance between x in L and L' by broken geodesics that meet each leaf orthogonally to make this local result global. We leave this as an exercise.

Conversely for nearby L' we have (21). Assume ℓ is constant then $\frac{d}{ds}\left(\frac{1}{2}\ell^2\right) = 0$ and $\gamma'(0)$ may be chosen to be an arbitrary vector of T_xL to see that $\dot{c}(0)$ is perpendicular to L.

Lemma 19. *If the leaves of a foliation are equidistant, then for any two basic fields, X, Y, one has that $\langle X, Y\rangle$ is constant along each leaf.*

Proof. Assume the leaves are equidistant. It suffices to show that for any basic field, X, that $|X|$ is constant along a leaf.

Suppose X is basic. Then for each vector field, V, along a leaf, L, we may locally extend to a vector field, V, tangent to the foliation, $\mathscr{F}$, with $[X, V] = 0$. Let ϕ_t be the flow of X and ψ_s be the flow of V. Then $\phi_t \circ \psi_s = \psi_s \circ \phi_t$. Let $x \in L$. Then $\psi_s(x) \in L$. Suppose $\phi_t(x) \in L_t$. Then $\phi_t(\psi_s(x)) = \psi_s(\phi_t(x)) \in L_t$. Thus $\phi_t(L) \subset L_t$. Note dist$(x, L_t) = $ dist$(\psi_s(x), L_t)$ by hypothesis.

Let $f(t) = $ dist(x, L_t). We compute $f'(0)$. To do this it is simplest to think in coordinates. Think of $(T_xL)^\perp$ as a Euclidean subspace of T_xM. For small t the geodesic linking x to L_t for small t is entirely in $\exp(B_r(0))$ where $B_r(0)$ is a Euclidean ball of small radius r about 0 in $(T_xL)^\perp$. Pulling the geodesics back gives straight line segments secant to the curve, c, at 0 and $c(t)$ defined by $\{\exp(c(t))\} = L_t \cap \exp(B_r(0))$. Hence $f(t) = |c(t)|$. Now we have a standard calculation in Euclidean space,

$$\lim_{t\to 0+}\frac{|c(t)|}{t} = \left|\lim_{t\to 0+}\frac{1}{t}c(t)\right| = |c'(0)|$$

Hence $f'(0) = |c'(0)|$. Consider a trivialisation of a neighbourhood of N of L sending y to (a, b) where $\{a\} = \exp(B_r(0)) \cap L'$ and b is the point of L closest to y. Then $\phi_t(x)$ is sent to $(c(t), \gamma(t))$. At time zero the tangent vector, X_x, is sent to $(c'(0), 0)$. Hence $|X_x| = |c'(0)| = f'(0)$. Likewise $|X_{\psi_s(x)}| = f'(0)$.

We conclude $|X_x| = |X_{\psi_s(x)}|$. Hence the result.

Lemma 20. *For a contact CR submanifold, the condition that the leaves of the integral foliation of $\mathscr{D}^\perp$ are equidistant is equivalent to the condition that*

$$\mathscr{D}S_{\varphi(V)}\varphi(X) = \mathscr{D}\varphi(S_{\varphi(V)}X)$$

for each anti-holomorphic field, V, perpendicular to ξ and each holomorphic field, X.

Proof. Since $\langle \nabla_X \xi, Y \rangle = \langle \varphi(X), Y \rangle = -\langle X, \varphi(Y) \rangle = -\langle X, \nabla_Y \xi \rangle$ it suffices to consider $V \perp \xi$.

$$\langle \nabla_X V, Y \rangle = \langle \bar{\nabla}_X V, Y \rangle = \langle \bar{\nabla}_X \left(-\varphi^2(V) \right), Y \rangle = \langle \varphi \left(-\bar{\nabla}_X \varphi(V) \right), Y \rangle$$
$$= \langle \varphi \left(S_{\varphi(V)} X \right), Y \rangle.$$
$$\langle X, \nabla_Y V \rangle = \langle X, \varphi(S_{\varphi(V)} Y) \rangle = -\langle \varphi(X), S_{\varphi(V)} Y \rangle = -\langle S_{\varphi(V)} \varphi(X), Y \rangle.$$

Hence $\langle \nabla_X V, Y \rangle = -\langle X, \nabla_Y V \rangle$ for holomorphic X, Y and anti-holomorphic V if and only if the $\mathcal{D}$ components of $\varphi \left(S_{\varphi(V)} X \right)$ and $S_{\varphi(V)} \varphi(X)$ are equal.

Example 6. Consider $M_+ \in S^{19}$. Recall

$$S_n z = n \circ z = \{nxz\} = nx^* z + xz^* n + zn^* x + zx^* n + xn^* z + nz^* x.$$

Let $x \in M_+$ be the specially chosen point as before. Note that $nz^* x = xz^* n = 0$.

$$S_n z = (nx^* + xn^*)z + z(x^* n + n^* x)$$

$$= \frac{1}{\sqrt{2}} \begin{pmatrix} -AJ + JA^* & 0 \\ 0 & 0 \end{pmatrix} \begin{pmatrix} 0 & b \\ -b^T & 0 \end{pmatrix} + \begin{pmatrix} 0 & b \\ -b^T & 0 \end{pmatrix} \frac{1}{\sqrt{2}} \begin{pmatrix} A^*J - JA & 0 \\ 0 & 0 \end{pmatrix}$$

$$= \frac{1}{\sqrt{2}} \begin{pmatrix} 0 & (-AJ + JA^*)b \\ -b^T(A^*J - JA) & 0 \end{pmatrix} \in \mathcal{D}_x$$

since $(-AJ + JA^*)^T = -J^T A^T + (A^*)^T J^T = -JA - (A^T)^* J = -JA + A^* J.$
That is

$$-b^T (A^*J - JA) = -\left((-AJ + JA^*)b \right)^T.$$

Hence the leaves of the integral foliation of $\mathcal{D}^\perp$ are totally geodesic. Note

$$S_n(iz) = (nx^* + xn^*)(iz) + (iz)(x^* n + n^* x)$$
$$= i \left((nx^* + xn^*)z + z(x^* n + n^* x) \right) = i S_n z.$$

Hence the leaves of the foliation are equidistant.

Exercise 10. Show that the leaves integral foliation of $\mathcal{D}^\perp$ are totally contact geodesic in S^{19}.

Example 7. Consider the contact CR submanifold, $M \subset S^{4n-1}$, defined in (18). Recall

$$\mathcal{D}_q^\perp = \{(q_1 v, q_2 v, \ldots, q_n v) : \text{Re}(v) = 0\} \oplus \mathbb{R}(iq_1, iq_2, \ldots, iq_n).$$

Hence

$$\varphi\left(\mathscr{D}_q^\perp\right) = \{(iq_1v, iq_2v, \dots, iq_nv) : \text{Re}(v) = 0\}.$$

In the sphere the covariant derivative of $\varphi(W) = (iq_1w, iq_2w, \dots, iq_nw)$ in direction $V = (q_1v, q_2v, \dots, q_nv)$ is

$$(iq_1vw, iq_2vw, \dots, iq_nvw)$$
$$= \text{Re}(vw)(iq_1, iq_2, \dots, iq_n) + (iq_1\text{Im}(vw), iq_2\text{Im}(vw), \dots, iq_n\text{Im}(vw)).$$

Hence $S_{\varphi(W)}V = -\text{Re}(vw)(iq_1, iq_2, \dots, iq_n) = -\text{Re}(vw)\xi$. Note

$$\langle\varphi(W), \varphi(V)\rangle = \langle(iq_1w, iq_2w, \dots, iq_nw), (iq_1v, iq_2v, \dots, iq_nv)\rangle = -\text{Re}(vw).$$

Hence $S_{\varphi(W)}V = \langle\varphi(W), \varphi(V)\rangle\xi$. Hence the leaves of the foliation are totally contact geodesic in the sphere.

Exercise 11. Show the leaves of the foliation are equidistant for this example.

3 n-Sasakian CR Submanifolds of Complex Projective Space

Recall the full curvature tensor of a manifold as an expression in the metric and Levi-Civita connection is given by

$$\langle R(X, Y)Z, W\rangle = \langle\nabla_X\nabla_Y Z - \nabla_Y\nabla_X Z - \nabla_{[X,Y]}Z, W\rangle.$$

Using metric compatibility this may be written

$$\langle R(X, Y)Z, W\rangle = X\langle\nabla_Y Z, W\rangle - \langle\nabla_Y Z, \nabla_X W\rangle - Y\langle\nabla_X Z, W\rangle$$
$$+ \langle\nabla_X Z, \nabla_Y W\rangle - \langle\nabla_{[X,Y]}Z, W\rangle.$$

We need to collect some information about the curvature tensor of submanifolds and submersed manifolds.

Proposition 8. *The full curvature tensor of a submanifold, $M \subset \bar{M}$, is related to the curvature tensors of $\bar{M}$ by*

$$\langle\bar{R}(X, Y)Z, W\rangle = \langle R(X, Y)Z, W\rangle - \langle B_Y Z, B_X W\rangle + \langle B_X Z, B_Y W\rangle$$

where B is the second fundamental form and X, Y, Z, W are fields tangent to the submanifold.

Proof. Let X, Y, Z, W be fields tangent to the submanifold and B denote the second fundamental form of $M \subset \bar{M}$.

$$\langle \bar{R}(X, Y)Z, W \rangle$$

$$= X \langle \bar{\nabla}_Y Z, W \rangle - Y \langle \bar{\nabla}_X Z, W \rangle - \langle \bar{\nabla}_{[X,Y]} Z, W \rangle$$

$$\quad - \langle \bar{\nabla}_Y Z, \bar{\nabla}_X W \rangle + \langle \bar{\nabla}_X Z, \bar{\nabla}_Y W \rangle$$

$$= X \langle \nabla_Y Z, W \rangle - Y \langle \nabla_X Z, W \rangle - \langle \nabla_{[X,Y]} Z, W \rangle$$

$$\quad - \langle \nabla_Y Z + B_Y Z, \nabla_X W + B_X W \rangle + \langle \nabla_X Z + B_X Z, \nabla_Y W + B_Y W \rangle.$$

$$= X \langle \nabla_Y Z, W \rangle - Y \langle \nabla_X Z, W \rangle - \langle \nabla_{[X,Y]} Z, W \rangle$$

$$\quad - \langle \nabla_Y Z, \nabla_X W \rangle - \langle B_Y Z, B_X W \rangle + \langle \nabla_X Z, \nabla_Y W \rangle + \langle B_X Z, B_Y W \rangle$$

$$= X \langle \nabla_Y Z, W \rangle - \langle \nabla_Y Z, \nabla_X W \rangle - Y \langle \nabla_X Z, W \rangle + \langle \nabla_X Z, \nabla_Y W \rangle$$

$$\quad - \langle \nabla_{[X,Y]} Z, W \rangle - \langle B_Y Z, B_X W \rangle + \langle B_X Z, B_Y W \rangle$$

$$= \langle R(X, Y)Z, W \rangle - \langle B_Y Z, B_X W \rangle + \langle B_X Z, B_Y W \rangle.$$

Proposition 9. *The full curvature tensor of the total space of a Riemannian submersion, $\pi : M \to B$, is related to the full curvature tensor of B via*

$$\langle R(X, Y)Z, W \rangle = \langle R^*(X, Y)Z, W \rangle - \langle A_Y Z, A_X W \rangle + \langle A_X Z, A_Y W \rangle$$

$$\quad + 2 \langle A_X Y, A_Z W \rangle$$

where X, Y, Z, W are horizontal fields.

Proof. With no loss of generality assume X, Y, Z are basic fields.

$$\langle \nabla_Y Z, \nabla_X W \rangle = \langle \nabla_Y^* Z, \nabla_X^* W \rangle + \langle A_Y Z, A_X W \rangle,$$

$$\langle \nabla_X Z, \nabla_Y W \rangle = \langle \nabla_X^* Z, \nabla_Y^* W \rangle + \langle A_X Z, A_Y W \rangle,$$

$$[X, Y] = \mathcal{H}[X, Y] + \mathcal{V}[X, Y] = \mathcal{H}[X, Y] + 2 A_X Y.$$

$$\langle \nabla_{[X,Y]} Z, W \rangle = \langle \nabla_{\mathcal{H}[X,Y]} Z, W \rangle + 2 \langle \nabla_{A_X Y} Z, W \rangle$$

$$= \langle \nabla_{\mathcal{H}[X,Y]} Z, W \rangle + 2 \langle \nabla_Z A_X Y + [A_X Y, Z], W \rangle$$

$$= \langle \nabla_{\mathcal{H}[X,Y]} Z, W \rangle + 2 \langle \nabla_Z A_X Y, W \rangle$$

$$= \langle \nabla_{\mathcal{H}[X,Y]} Z, W \rangle - 2 \langle A_X Y, \nabla_Z W \rangle$$

$$= \langle \nabla_{\mathcal{H}[X,Y]} Z, W \rangle - 2 \langle A_X Y, A_Z W \rangle$$

$$= \langle \nabla_{[X,Y]}^* Z, W \rangle - 2 \langle A_X Y, A_Z W \rangle.$$

$$\langle R(X,Y)Z, W \rangle$$
$$= X\langle \nabla_Y Z, W \rangle - Y\langle \nabla_X Z, W \rangle - \langle \nabla_{[X,Y]} Z, W \rangle$$
$$- \langle \nabla_Y Z, \nabla_X W \rangle + \langle \nabla_X Z, \nabla_Y W \rangle$$
$$= X\langle \nabla_Y^* Z, W \rangle - Y\langle \nabla_X^* Z, W \rangle - \langle \nabla_{[X,Y]}^* Z, W \rangle + 2\langle A_X Y, A_Z W \rangle$$
$$- \langle \nabla_Y^* Z, \nabla_X^* W \rangle - \langle A_Y Z, A_X W \rangle + \langle \nabla_X^* Z, \nabla_Y^* W \rangle + \langle A_X Z, A_Y W \rangle$$
$$= \langle R^*(X,Y)Z, W \rangle - \langle A_Y Z, A_X W \rangle + \langle A_X Z, A_Y W \rangle + 2\langle A_X Y, A_Z W \rangle.$$

Definition 12. The unnormalised sectional curvature of a Riemannian manifold is given by $K(X,Y) = \langle R(X,Y)Y, X \rangle$.

From here on out, to avoid cluttered notation, we will often abuse notation and denote a horizontal vector field and its projection under π_* by the same symbol.

Corollary 2. *The unnormalised sectional curvatures of the total and base spaces of a Riemannian submersion, $\pi : M \to B$, are related by*

$$K_*(X,Y) = K(X,Y) + 3|A_X Y|^2.$$

Proof. Let X, Y be horizontal fields.

$$\langle R(X,Y)Y, X \rangle$$
$$= \langle R^*(X,Y)Y, X \rangle - \langle A_Y Y, A_X X \rangle + \langle A_X Y, A_Y X \rangle + 2\langle A_X Y, A_Y X \rangle$$
$$= K_*(X,Y) + 3\langle A_X Y, A_Y X \rangle = K_*(X,Y) - 3\langle A_X Y, A_X Y \rangle$$
$$= K_*(X,Y) - 3|A_X Y|^2.$$

Example 8. Consider the natural Riemannian submersion, $v : S^{2n-1} \to \mathbb{C}P^{n-1}$.
Recall that $A_X Y = \langle X, JY \rangle \xi$ and that $K(X,Y) = |X|^2|Y|^2 - \langle X, Y \rangle^2$ for S^{2n-1}. Thus the unnormalised sectional curvature of $\mathbb{C}P^{n-1}$ is

$$K_*(X,Y) = |X|^2|Y|^2 - \langle X, Y \rangle^2 + 3\langle X, JY \rangle^2.$$

Note that for $|X| = |Y| = 1$ and $\langle X, Y \rangle = 0$ we have sectional curvature,

$$\sigma_*(X,Y) = K_*(X,Y) = 1 + 3\langle X, JY \rangle^2.$$

When $X = \pm JX$ then $\sigma_*(X,Y) = 4$, when $Y \perp X$ and $Y \perp JX$ then $\sigma_*(X,Y) = 1$. Namely $\mathbb{C}P^{n-1}$ has pinching $\frac{1}{4}$.

We now introduce the concept of an n-Sasakian manifold. This notion of the author was first introduced in [15].

Definition 13. A Riemannian manifold, M, is said to be n-Sasakian if it is foliated by totally geodesic equidistant n-dimensional submanifolds such that

$$R(X, V)Y = \langle Y, V \rangle X - \langle X, Y \rangle V$$

for all fields, X, Y, and each field, V, tangent to the foliation.

Proposition 10. *Suppose that M is a generic contact CR submanifold of the sphere such that the integral leaves of $\mathscr{D}^{\perp}$ are equidistant and totally contact geodesic in the sphere. Then M/ξ is an n-Sasakian manifold where* $\dim(\mathscr{D}^{\perp}) = n + 1$.

Proof. Let X, Y, Z and V be fields along M perpendicular to ξ and V be antiholomorphic. Assume the leaves of the foliation are totally contact geodesic. Then it follows $\langle B_V X, N \rangle = 0$.

From the Gauss equation

$$\langle \bar{R}(X, V)Y, Z \rangle = \langle R(X, V)Y, Z \rangle + \langle B_V Y, B_X Z \rangle - \langle B_X Y, B_V Z \rangle$$
$$= \langle R(X, V)Y, Z \rangle$$

as both $B_V Y = 0$ and $B_V Z = 0$ since the leaves of the foliation are totally contact geodesic.

Now from the O'Neill structure equation it follows,

$$\langle R(X, V)Y, Z \rangle = \langle R^*(X, V)Y, Z \rangle - \langle A_V Y, A_X Z \rangle + \langle A_X Y, A_V Z \rangle$$
$$+ 2 \langle A_X V, A_Y Z \rangle.$$

Note that

$$A_V X = \langle \nabla_V X, \xi \rangle \xi = \langle \bar{\nabla}_V X, \xi \rangle \xi = -\langle X, \bar{\nabla}_V \xi \rangle \xi = -\langle X, \varphi(V) \rangle \xi = 0.$$

Hence $\langle R^*(X, V)Y, Z \rangle = \langle R(X, V)Y, Z \rangle = \langle \bar{R}(X, V)Y, Z \rangle = \langle Y, V \rangle \langle X, Z \rangle - \langle X, Y \rangle \langle V, Z \rangle$. Thus $R^*(X, V)Y = \langle Y, V \rangle X - \langle X, Y \rangle V$.

To discuss other structure equations of a Riemannian submersion it is necessary to extend the definition of the O'Neill tensor and second fundamental form of the fibres.

Definition 14. Relative to the vector fields, E, F, the O'Neill and second fundamental form are defined by

$$A_E F = {}^{\mathscr{V}} \nabla_{\mathscr{H} E} \mathscr{H} F + \mathscr{H} \nabla_{\mathscr{H} E} {}^{\mathscr{V}} F,$$
$$T_E F = \mathscr{H} \nabla_{\mathscr{V} E} {}^{\mathscr{V}} F + {}^{\mathscr{V}} \nabla_{\mathscr{V} E} \mathscr{H} F$$

respectively.

Proposition 11. *Suppose V is a vertical field and X is a horizontal field. Then the unnormalised sectional curvature is related to the structure tensors by*

$$K(X, V) = \langle (\nabla_X T)_V V, X \rangle - |T_V X|^2 + |A_X V|^2.$$

Proof. Assume X is basic.

$K(X, V)$

$$= X \langle \nabla_V V, X \rangle - \langle \nabla_V V, \nabla_X X \rangle - V \langle \nabla_X V, X \rangle + \langle \nabla_X V, \nabla_V X \rangle - \langle \nabla_{[X,V]} V, X \rangle$$

$$= X \langle T_V V, X \rangle - \langle T_V V, \nabla_X X \rangle - \langle \nabla_V V, A_X X \rangle - V \langle A_X V, X \rangle$$

$$\quad + \langle A_X V, \nabla_X V + [X, V] \rangle + \langle \nabla_X V, T_V X \rangle - \langle T_{[X,V]} V, X \rangle$$

$$= X \langle T_V V, X \rangle - \langle T_V V, \nabla_X X \rangle + \langle A_X V, A_X V \rangle - \langle T_V \nabla_X V, X \rangle - \langle T_{[X,V]} V, X \rangle$$

$$= \langle \nabla_X (T_V V), X \rangle - \langle T_V \nabla_X V, X \rangle - \langle T_{\nabla_X V - \nabla_V X} V, X \rangle + \langle A_X V, A_X V \rangle$$

$$= \langle \nabla_X (T_V V), X \rangle - \langle T_V \nabla_X V, X \rangle - \langle T_{\nabla_X V} V, X \rangle + \langle T_{\nabla_V X} V, X \rangle + |A_X V|^2$$

$$= \langle (\nabla_X T)_V V), X \rangle + \langle T_{T_V X} V, X \rangle + |A_X V|^2 = \langle (\nabla_X T)_V V), X \rangle$$

$$\quad + \langle T_V T_V X, X \rangle + |A_X V|^2$$

$$= \langle (\nabla_X T)_V V), X \rangle - \langle T_V X, T_V X \rangle + |A_X V|^2 = \langle (\nabla_X T)_V V), X \rangle$$

$$\quad - |T_V X|^2 + |A_X V|^2.$$

Corollary 3. *Suppose $\pi : M \to B$ is a Riemannian submersion with totally geodesic fibres. Then the unnormalised sectional curvature is related to the O'Neill tensor by*

$$K(X, V) = |A_X V|^2.$$

Proposition 12. *The O'Neill tensor of an n-Sasakian manifold has that the O'Neill tensor induces an antisymmetric Clifford representation of the vertical space on the horizontal space, i.e.,*

$$A_{A_X V} W + A_{A_X W} V = -2 \langle V, W \rangle X$$

for any vertical fields, V, W, and any horizontal field, X.

Proof. Note that for any horizontal field, X, and vertical field, V, we have

$$|A_X V|^2 = \langle R(X, V)V, X \rangle = \langle \langle V, V \rangle X - \langle X, V \rangle V, X \rangle = |X|^2 |V|^2.$$

$$|A_X (V + W)|^2 = |A_X V + A_X W|^2 = |A_X V|^2 + 2 \langle A_X V, A_X W \rangle + |A_X W|^2$$

$$= |X|^2 |V|^2 + 2 \langle A_X V, A_X W \rangle + |X|^2 |W|^2.$$

$$|X|^2|V + W|^2 = |X|^2(|V|^2 + 2\langle V, W \rangle + |W|^2)$$
$$= |X|^2|V|^2 + 2|X|^2\langle V, W \rangle + |X|^2|W|^2.$$

Hence

$$\langle A_X V, A_X W \rangle = |X|^2 \langle V, W \rangle.$$

$$\langle A_{X+Y} V, A_{X+Y} W \rangle$$
$$= \langle A_X V, A_X W \rangle + \langle A_X V, A_Y W \rangle + \langle A_Y V, A_X W \rangle + \langle A_Y V, A_Y W \rangle$$
$$= |X|^2 \langle V, W \rangle + \langle A_X V, A_Y W \rangle + \langle A_Y V, A_X W \rangle + |Y|^2 \langle V, W \rangle,$$

$$|X + Y|^2 \langle V, W \rangle = (|X|^2 + 2\langle X, Y \rangle + |Y|^2)\langle V, W \rangle$$
$$= |X|^2 \langle V, W \rangle + 2\langle V, W \rangle \langle X, Y \rangle + |Y|^2 \langle V, W \rangle.$$

Hence

$$\langle A_X V, A_Y W \rangle + \langle A_Y V, A_X W \rangle = 2\langle V, W \rangle \langle X, Y \rangle.$$

Now

$$\langle A_X V, A_Y W \rangle = -\langle A_Y A_X V, W \rangle = \langle A_{A_X V} Y, W \rangle = -\langle A_{A_X V} W, Y \rangle.$$

Similarly $\langle A_Y V, A_X W \rangle = -\langle A_{A_X W} V, Y \rangle$.

$$-\langle A_{A_X V} W, Y \rangle - \langle A_{A_X W} V, Y \rangle = 2\langle V, W \rangle \langle X, Y \rangle.$$

Hence

$$A_{A_X V} W + A_{A_X W} V = -2\langle V, W \rangle X.$$

Proposition 13. *Let M be a contact CR structure such that the integral foliation of $\mathscr{D}^\perp$ is equidistant. Then for each anti-holomorphic field, V, perpendicular to ξ and holomorphic field, X, one has*

$$A_X V = \mathscr{D} S_{\varphi(V)} \varphi(X) = \mathscr{D} \varphi(S_{\varphi(V)} X).$$

Moreover if the leaves of the foliation are totally geodesic then

$$A_X V = S_{\varphi(V)} \varphi(X) = \varphi(S_{\varphi(V)} X).$$

Proof.

$$\langle A_X V, Y \rangle = \langle \nabla_X V, Y \rangle = \langle \bar{\nabla}_X V, Y \rangle = -\langle \bar{\nabla}_X \varphi^2(V), Y \rangle = -\langle \varphi(\bar{\nabla}_X \varphi(V)), Y \rangle$$
$$= \langle \bar{\nabla}_X \varphi(V), \varphi(Y) \rangle = -\langle S_{\varphi(V)} X, \varphi(Y) \rangle = \langle \varphi(S_{\varphi(V)} X), Y \rangle.$$

The remainder is left as an exercise.

Corollary 4. *Suppose that M is a contact CR submanifold of the sphere as described in Proposition 10. Then the shape operator induces a symmetric Clifford representation of $\varphi(\mathscr{D}^{\perp})$ on $\mathscr{D}$, i.e.,*

$$S_{\varphi(V)} S_{\varphi(W)} X + S_{\varphi(W)} S_{\varphi(V)} X = 2 \langle \varphi(V), \varphi(W) \rangle X$$

for any anti-holomorphic fields, V, W, perpendicular to ξ and any holomorphic field, X.

Proof.

$$A_{A_X V} W = S_{\varphi(W)} \varphi(A_X V) = S_{\varphi(W)} \varphi(\varphi(S_{\varphi(V)} X)) = -S_{\varphi(W)} S_{\varphi(V)} X.$$

Since $\langle \varphi(V), \varphi(W) \rangle = \langle V, W \rangle$ it follows from Proposition 12 that

$$S_{\varphi(V)} S_{\varphi(W)} X + S_{\varphi(W)} S_{\varphi(V)} X = 2 \langle \varphi(V), \varphi(W) \rangle X.$$

Definition 15. Let $M \subset \bar{M}$ be a submanifold. Then the mean curvature vector of M is

$$H = \sum_i B_{E_i} E_i$$

where $\{E_i\}$ is an orthonormal basis.

Lemma 21. *The above definition is independent of the choice of basis, $\{E_i\}$.*

Proof. Suppose $\{F_i\}$ is another orthonormal basis. Then

$$F_i = \sum_j a_{ij} E_j$$

where (a_{ij}) is an orthogonal matrix.

$$\sum_i B_{F_i} F_i = \sum_i \sum_j \sum_k a_{ij} a_{ik} B_{E_j} E_k = \sum_j \sum_k \left(\sum_i a_{ij} a_{ik} \right) B_{E_j} E_k$$
$$= \sum_j \sum_k \delta_{jk} B_{E_j} E_k = \sum_j B_{E_j} E_j.$$

Definition 16. A submanifold, $M \subset \bar{M}$, is called minimal if $H = 0$.

Proposition 14. *Let M be a generic CR submanifold of complex projective space with the integral foliation of $\mathcal{D}^{\perp}$ with equidistant leaves that are totally geodesic in complex projective space. Then M is a minimal submanifold of complex projective space provided $\dim(\mathcal{D}^{\perp}) > 1$.*

Proof. Since the leaves are totally geodesic on complex projective space $B_V V = 0$ for each anti-holomorphic V. With this in mind we take a basis, $\{E_i\}$, that consists of the union of a basis for $\mathcal{D}$, $\{X_i\}$, and basis for $\mathcal{D}^{\perp}$, $\{V_i\}$. Hence $H = \sum_i B_{X_i} X_i$. Let N be a fixed unit normal vector. Since $S_N^2 X = X$ on $\mathcal{D}$ by Corollary 4 it follows $\mathcal{D}$ is split into ± 1 eigenspaces of S_N.

Since $\dim(\mathcal{D}^{\perp}) > 1$ there is a orthogonal unit normal vector, M. Hence by Corollary 4 again,

$$S_N S_M X + S_M S_N X = 0$$

for $X \in \mathcal{D}$. That is $S_N S_M X = -S_M S_N X$. If X is in the $+1$-eigenspace of S_N, then $S_N S_M X = -S_M X$, i.e., $S_M X$ is in the -1-eigenspace of S_N. However since $S_M^2 X = X$ we have that S_M is an isomorphism that exchanges the ± 1-eigenspaces of S_N. Thus take an orthogonal basis for $\mathcal{D}$, $\{X_i\}$, that is the union of bases of the two eigenspaces of S_N, $\mathcal{D}_{\pm}$. We have

$$\left\langle \sum_i B_{X_i} X_i, N \right\rangle = \sum_i \langle S_N X_i, X_i \rangle = \sum_{X_i \in \mathcal{D}_-} \langle S_N X_i, X_i \rangle + \sum_{X_i \in \mathcal{D}_+} \langle S_N X_i, X_i \rangle$$

$$= \sum_{X_i \in \mathcal{D}_-} \langle -X_i, X_i \rangle + \sum_{X_i \in \mathcal{D}_+} \langle X_i, X_i \rangle$$

$$= -\dim(\mathcal{D}_-) + \dim(\mathcal{D}_+) = 0.$$

since $\mathcal{D}_+$ and $\mathcal{D}_-$ have equal dimension.

This is true of any normal N we chose. Hence $H = 0$.

Definition 17. The Ricci curvature of a Riemannian manifold is defined by

$$\text{Ric}(X, Y) = \sum_i \langle R(X, E_i) E_i, Y \rangle$$

where $\{E_i\}$ is an orthonormal basis for the tangent space and X, Y are any vector fields.

Exercise 12. Show the Ricci curvature does not depend on the basis chosen.

Definition 18. A Riemannian manifold is said to be Einstein if there is a constant, λ, such that

$$\text{Ric}(X, Y) = \lambda \langle X, Y \rangle$$

for any pair of vector fields, X, Y.

Lemma 22. *A Riemannian manifold is Einstein if and only if*

$$\text{Ric}(X) \equiv \text{Ric}(X, X) = \sum_i K(X, E_i) = \lambda |X|^2$$

for each field, X, for some constant, λ.

Proof. Suppose $\text{Ric}(X, X) = \lambda |X|^2$ for each field, X.

$\text{Ric}(X + Y, X + Y)$

$\quad = \sum_i \langle R(X + Y, E_i) E_i, X + Y \rangle = \sum_i \langle R(X, E_i) E_i, X \rangle$

$\qquad + \sum_i \langle R(X, E_i) E_i, Y \rangle + \sum_i \langle R(Y, E_i) E_i, X \rangle + \sum_i \langle R(Y, E_i) E_i, Y \rangle$

$\quad = \text{Ric}(X, X) + \sum_i \langle R(X, E_i) E_i, Y \rangle + \sum_i \langle R(X, E_i) E_i, Y \rangle + \text{Ric}(Y, Y)$

$\quad = \text{Ric}(X, X) + 2\text{Ric}(X, Y) + \text{Ric}(Y, Y) = \lambda |X|^2 + 2\text{Ric}(X, Y) + \lambda |Y|^2.$

Whilst

$$\lambda |X + Y|^2 = \lambda |X|^2 + 2\lambda \langle X, Y \rangle + \lambda |Y|^2.$$

Hence $\text{Ric}(X + Y, X + Y) = \lambda |X + Y|^2$ reduces to $\text{Ric}(X, Y) = \lambda \langle X, Y \rangle$. The converse direction is vacuous.

Proposition 15. *Let M be a generic CR submanifold of complex projective space with the integral foliation of $\mathscr{D}^\perp$ with equidistant leaves that are totally geodesic in complex projective space and $\dim(\mathscr{D}^\perp) > 1$. Then the space of leaves of the foliation is Einstein.*

Proof. Let $\{X_i\}$ be a orthonormal basis for $\mathscr{D}$. Let X be a holomorphic field, i.e., a horizontal vector field. Then the unnormalised sectional curvature of M in complex projective space is given by the Gauss equation.

$$K(X, X_i) = |X|^2 |X_i|^2 - \langle X, X_i \rangle^2 + 3\langle X_i, JX \rangle^2 + \langle B_X X, B_{X_i} X_i \rangle - |B_X X_i|^2$$
$$= |X|^2 - \langle X, X_i \rangle^2 + 3\langle X_i, JX \rangle^2 + \langle B_X X, B_{X_i} X_i \rangle - |B_X X_i|^2.$$

$$\langle A_X X_i, V \rangle = -\langle X_i, A_X V \rangle = -\langle X_i, S_{JV} JX \rangle = -\langle B_{JX} X_i, JV \rangle.$$

Hence $|A_X X_i|^2 = |B_{JX} X_i|^2$. It follows from O'Neel's equation

$$K_*(X, X_i) = K(X, X_i) + 3|A_X X_i|^2 = |X|^2 - \langle X, X_i \rangle^2 + 3\langle X_i, JX \rangle^2$$
$$+ \langle B_X X, B_{X_i} X_i \rangle - |B_X X_i|^2 + 3|B_{JX} X_i|^2.$$

$$\mathrm{Ric}_*(X) = \sum_i K_*(X, X_i) = \dim(\mathscr{D})|X|^2 - |X|^2 + 3|JX|^2 + \langle B_X X, H \rangle$$
$$- \sum_i |B_X X_i|^2 + 3 \sum_i |B_{JX} X_i|^2.$$

$$\sum_i |B_X X_i|^2 = \sum_i \sum_j \langle B_X X_i, N_j \rangle^2 = \sum_j \sum_i \langle X_i, S_{N_j} X \rangle^2 = \sum_j |S_{N_j} X|^2$$
$$= \dim(\mathscr{D}^\perp)|X|^2.$$

Hence $\mathrm{Ric}_*(X) = \left(\dim(\mathscr{D}) + 2 + 2\dim(\mathscr{D}^\perp) \right) |X|^2$.

Lemma 23. *Let $M \subset \bar{M}$ be a CR submanifold with the integral foliation of $\mathscr{D}^\perp$ with equidistant leaves. Then the space of leaves of the foliation is Kähler.*

Proof. Define $J_* X \equiv v_*(J \tilde{X})$. To see this is well defined note,

$$V \langle J \tilde{X}, \tilde{Y} \rangle = \langle \nabla_V (J \tilde{X}), \tilde{Y} \rangle + \langle J \tilde{X}, \nabla_V \tilde{Y} \rangle = 0,$$

since the leaves are equidistant. The Kähler condition comes from

$$\langle (\nabla_X^* J_*) Y, Z \rangle$$
$$= \langle \nabla_X^* (J_* Y), Z \rangle - \langle J_*(\nabla_X^* Y), Z \rangle = \langle \nabla_{\tilde{X}}(J \tilde{Y}), \tilde{Z} \rangle - \langle J(\mathscr{D} \nabla_{\tilde{X}} \tilde{Y}), \tilde{Z} \rangle$$
$$= \langle \bar{\nabla}_{\tilde{X}}(J \tilde{Y}), \tilde{Z} \rangle - \langle J(\bar{\nabla}_{\tilde{X}} \tilde{Y}), \tilde{Z} \rangle = \langle (\bar{\nabla}_{\tilde{X}} J) \tilde{Y}, \tilde{Z} \rangle = 0.$$

Definition 19. Let $\pi : M \to B$ be a Riemannian submersion relative to the metric, $\langle, \rangle$, on M. Then the canonical variation metric, $\langle, \rangle_t$, is defined by

$$\langle E, F \rangle_t = \langle X, Y \rangle + t \langle U, V \rangle$$

where t is a positive constant and X, Y and U, V are the respective horizontal and vertical parts of E, F relative to π. π is also a Riemannian submersion relative to $\langle, \rangle_t$.

Exercise 13. Show that the Levi-Civita connection for $\langle, \rangle_t$ is given by

$$\nabla_E^t F = \nabla_E F + (t - 1)\mathscr{H}(\nabla_E F - \nabla_X Y).$$

We adopt the notation, $\theta_E F = \mathcal{H}(\nabla_E F - \nabla_X Y)$.

The author proved the following useful Proposition in [14]:

Proposition 16. *The unnormalised sectional curvature of $\langle,\rangle_t$ is related to that of $\langle,\rangle$ via*

$$K_t(E, F) = tK(E, F) + (1 - t)K_*(X, Y) + t(1 - t)((\theta_E E, \theta_F F) - |\theta_E F|^2).$$

Proof.

$$K_t(E, F) = E\langle \nabla_F^t F, E \rangle_t - \langle \nabla_F^t F, \nabla_E^t E \rangle_t - F\langle \nabla_E F, E \rangle$$
$$+ \langle \nabla_E^t F, \nabla_F^t E \rangle - \langle \nabla_{[E,F]}^t F, E \rangle_t.$$

$$E\langle \nabla_F^t F, E \rangle_t = E\langle \nabla_F F, E \rangle_t + (t - 1)E\langle \theta_F F, X \rangle$$
$$= tE\langle \nabla_F F, E \rangle + (1 - t)E\langle \nabla_F F, X \rangle + (t - 1)E\langle \theta_F F, X \rangle$$
$$= tE\langle \nabla_F F, E \rangle + (1 - t)E\langle \mathcal{H}\nabla_Y Y, X \rangle$$
$$= tE\langle \nabla_F F, E \rangle + (1 - t)X\langle \mathcal{H}\nabla_Y Y, X \rangle$$
$$= tE\langle \nabla_F F, E \rangle + (1 - t)X\langle \nabla_Y^* Y, X \rangle.$$

Similarly

$$F\langle \nabla_E^t F, E \rangle_t = tF\langle \nabla_E F, E \rangle + (1 - t)Y\langle \nabla_X^* Y, X \rangle.$$

$\langle \nabla_E^t E, \nabla_F^t F \rangle_t$
$$= \langle \nabla_E E + (t - 1)\theta_E E, \nabla_F F + \theta_F F \rangle_t$$
$$= \langle \nabla_E E, \nabla_F F \rangle_t + (t - 1)(\langle \nabla_E E, \theta_F F \rangle + \langle \theta_E E, \nabla_F F \rangle) + (t - 1)^2 \langle \theta_E E, \theta_F F \rangle$$
$$= t\langle \nabla_E E, \nabla_F F \rangle + (1 - t)\langle \mathcal{H}\nabla_E E, \mathcal{H}\nabla_F F \rangle) + (t - 1)^2 \langle \theta_E E, \theta_F F \rangle$$
$$+ (t - 1)(\langle \mathcal{H}\nabla_E E, \theta_F F \rangle + \langle \theta_E E, \mathcal{H}\nabla_F F \rangle)$$
$$= t\langle \nabla_E E, \nabla_F F \rangle + (t - 1)\langle \theta_E E, \theta_F F \rangle + (t - 1)^2 \langle \theta_E E, \theta_F F \rangle$$
$$+ (1 - t)\langle \mathcal{H}\nabla_E E - \theta_E E, \mathcal{H}\nabla_F F - \theta_F F \rangle$$
$$= t\langle \nabla_E E, \nabla_F F \rangle + (1 - t)\langle \mathcal{H}\nabla_X X, \mathcal{H}\nabla_Y Y \rangle - t(1 - t)\langle \theta_E E, \theta_F F \rangle$$
$$= t\langle \nabla_E E, \nabla_F F \rangle + (1 - t)\langle \nabla_X^* X, \nabla_Y^* Y \rangle - t(1 - t)\langle \theta_E E, \theta_F F \rangle.$$

Similarly

$$\langle \nabla_E^t F, \nabla_F^t E \rangle_t = t\langle \nabla_E F, \nabla_F E \rangle + (1 - t)\langle \nabla_X^* Y, \nabla_Y^* X \rangle - t(1 - t)|\theta_E F|^2.$$

$$\langle \nabla^t_{[E,F]} F, E \rangle_t = \langle \nabla_{[E,F]} F, E \rangle_t + (t-1)\langle \theta_{[E,F]} F, X \rangle$$
$$= t\langle \nabla_{[E,F]} F, E \rangle + (1-t)\langle \mathscr{H} \nabla_{[E,F]} F, X \rangle + (t-1)\langle \theta_{[E,F]} F, X \rangle$$
$$= t\langle \nabla_{[E,F]} F, E \rangle + (1-t)\langle \mathscr{H} \nabla_{\mathscr{H}[E,F]} Y, X \rangle$$
$$= t\langle \nabla_{[E,F]} F, E \rangle + (1-t)\langle \mathscr{H} \nabla_{\mathscr{H}[X,Y]} Y, X \rangle$$
$$= t\langle \nabla_{[E,F]} F, E \rangle + (1-t)\langle \nabla^*_{[X,Y]} Y, X \rangle.$$

Collecting like terms one arrives at the formula.

Lemma 24. *For a Riemannian submersion*

$$\langle R(X,Y)X, V \rangle = -\langle (\nabla_X A)_X Y, V \rangle - 2\langle A_X Y, T_V X \rangle$$

where X, Y are any horizontal fields and V is any vertical field. Moreover if the fibres are totally geodesic then

$$\langle R(X,Y)X, V \rangle = -\langle (\nabla_X A)_X Y, V \rangle.$$

Proof. We assume X, Y are basic vector fields.

$$\langle R(X,Y)X, V \rangle = X\langle \nabla_Y X, V \rangle - \langle \nabla_Y X, \nabla_X V \rangle - Y\langle \nabla_X X, V \rangle + \langle \nabla_X X, \nabla_Y V \rangle$$
$$- \langle \nabla_{[X,Y]} X, V \rangle$$
$$= X\langle A_Y X, V \rangle - \langle \nabla_Y X, A_X V \rangle - \langle A_Y X, \nabla_X V \rangle + \langle \nabla_X X, A_Y V \rangle$$
$$- \langle A_{\mathscr{H}[X,Y]} X, V \rangle - \langle T_{\mathscr{V}[X,Y]} X, V \rangle$$
$$= \langle \nabla_X (A_Y X), V \rangle - \langle \nabla_Y X, A_X V \rangle + \langle \nabla_X X, A_Y V \rangle$$
$$- \langle A_{[X,Y]} X, V \rangle - 2\langle T_{A_X Y} X, V \rangle.$$

$$\langle T_{A_X Y} X, V \rangle = -\langle T_{A_X Y} V, X \rangle = -\langle T_V A_X Y, X \rangle = \langle A_X Y, T_V X \rangle.$$

$$\langle \nabla_Y X, A_X V \rangle = \langle \mathscr{H} \nabla_Y X, A_X V \rangle = -\langle A_X \mathscr{H} \nabla_Y X, V \rangle$$
$$= \langle A_{\mathscr{H} \nabla_Y X} X, V \rangle = \langle A_{\nabla_Y X} X, V \rangle.$$

$$\langle \nabla_X X, A_Y V \rangle = \langle \mathscr{H} \nabla_X X, A_Y V \rangle = -\langle A_Y \mathscr{H} \nabla_X X, V \rangle = -\langle A_Y \nabla_X X, V \rangle.$$

$$\langle A_{[X,Y]} X, V \rangle = \langle A_{\nabla_X Y - \nabla_Y X} X, V \rangle = \langle A_{\nabla_X Y} X, V \rangle - \langle A_{\nabla_Y X} X, V \rangle.$$

Collecting like terms

$$\langle R(X,Y)X, V\rangle = \langle \nabla_X(A_Y X), V\rangle - \langle A_{\nabla_X Y} X, V\rangle - \langle A_Y \nabla_X X, V\rangle - 2\langle A_X Y, T_V X\rangle$$
$$= \langle (\nabla_X A)_Y X, V\rangle - 2\langle A_X Y, T_V X\rangle.$$

Definition 20. A Riemannian submersion, $\pi : M \to B$, with totally geodesic fibres is said to be Yang-Mills if

$$\sum_i \mathscr{V}(\nabla_{X_i} A)_{X_i} X = 0$$

for each horizontal field, X, where $\{X_i\}$ is an orthonormal basis for the horizontal space.

Proposition 17. *For a Riemannian submersion,*

$$\langle R(U,V)V, X\rangle = \langle (\nabla_U T)_V V, X\rangle - \langle (\nabla_V T)_U V, X\rangle$$

where U, V are any vertical fields and X is any horizontal field. Moreover if the fibres are totally geodesic then $\langle R(U,V)V, X\rangle = 0$.

Proof.

$$\langle R(U,V)V, X\rangle$$
$$= U\langle \nabla_V V, X\rangle - \langle \nabla_V V, \nabla_U X\rangle - V\langle \nabla_U V, X\rangle + \langle \nabla_U V, \nabla_U X\rangle - \langle \nabla_{[U,V]}V, X\rangle$$
$$= U\langle T_V V, X\rangle - \langle T_V V, \nabla_U X\rangle - \langle \mathscr{V}\nabla_V V, T_U X\rangle - V\langle T_U V, X\rangle$$
$$\quad + \langle T_U V, \nabla_U X\rangle + \langle \mathscr{V}\nabla_U V, T_V X\rangle - \langle T_{[U,V]}V, X\rangle$$
$$= \langle \nabla_U(T_V V), X\rangle - \langle \mathscr{V}\nabla_V V, T_U X\rangle - \langle \nabla_V(T_U V), X\rangle + \langle \mathscr{V}\nabla_U V, T_V X\rangle$$
$$\quad - \langle T_{\nabla_U V - T_V U}V, X\rangle$$
$$= \langle \nabla_U(T_V V), X\rangle + \langle T_U \mathscr{V}\nabla_V V, X\rangle - \langle \nabla_V(T_U V), X\rangle - \langle T_V \mathscr{V}\nabla_U V, X\rangle$$
$$\quad - \langle T_{\nabla_U V}, X\rangle + \langle T_{\nabla_V U}V, X\rangle$$
$$= \langle \nabla_U(T_V V), X\rangle - \langle T_{\nabla_U V}V, X\rangle - \langle T_V \nabla_U V, X\rangle$$
$$\quad - (\langle \nabla_V(T_U V), X\rangle - \langle T_{\nabla_V U}V, X\rangle - \langle T_U \nabla_V V, X\rangle)$$
$$= \langle (\nabla_U T)_V V, X\rangle - \langle (\nabla_V T)_U V, X\rangle.$$

Proposition 18. *Let $\pi : M \to B$ be a Riemannian submersion with totally geodesic fibres. Then there are two different Einstein metrics in the canonical variation if and only if*

(i) *π is Yang-Mills, the fibres and base are Einstein with respective constants, $\hat{\lambda} > 0$ and $\check{\lambda} > 0$.*

(ii) *There are constants, μ, ν, such that*

$$\sum_i |A_{X_i} V|^2 = \mu |V|^2, \qquad\qquad \sum_j |A_X V_j|^2 = \nu |X|^2$$

for each vertical field, V, and horizontal field, X, where $\{X_i\}$ and $\{V_j\}$ are respective orthonormal bases for the horizontal and vertical spaces,

(iii) *The inequality, $\check\lambda^2 - 4\hat\lambda(\mu + 2\nu) > 0$, holds.*

Proof.

$$K_t(X + V, X_i) = tK(X + V, X_i) + (1 - t)K_*(X, X_i)$$
$$+ t(1 - t)(\langle \theta_{X+V}(X + V), \theta_{X_i} X_i \rangle - |\theta_{X+V} X_i|^2).$$

$$\theta_{X_i} X_i = \mathscr{H}(\nabla_{X_i} X_i - \nabla_{X_i} X_i) = 0.$$
$$\theta_{X+V} X_i = \mathscr{H}(\nabla_{X+V} X_i - \nabla_X X_i) = \mathscr{H}\nabla_V X = A_X V.$$

$$K(X + V, X_i) = \langle R(X + V, X_i)X_i, X + V \rangle$$
$$= \langle R(X, X_i)X_i, X \rangle + \langle R(X, X_i)X_i, V \rangle + \langle R(V, X_i)X_i, X \rangle$$
$$+ \langle R(V, X_i)X_i, V \rangle$$
$$= K(X, X_i) + 2\langle R(X, X_i)X_i, V \rangle + K(X_i, V)$$
$$= K_*(X, X_i) - 3|A_X X_i|^2 + 2\langle (\nabla_{X_i} A)_{X_i} X, V \rangle + |A_{X_i} V|^2.$$

Collecting like terms,

$$K_t(X + V, X_i) = K_*(X, X_i) - 3t|A_X X_i|^2 + 2t\langle (\nabla_{X_i} A)_{X_i} X, V \rangle + t^2 |A_{X_i} V|^2.$$

$$K_t(X + V, V_j)$$
$$= tK(X + V, V_j) + t(1 - t)(\langle \theta_{X+V}(X + V), \theta_{V_j} V_j \rangle - |\theta_{X+V} V_j|^2).$$

$$\theta_{V_j} V_j = \mathscr{H}(\nabla_{V_j} V_j) = T_{V_j} V_j = 0,$$
$$\theta_{X+V} V_j = \mathscr{H}(\nabla_{X+V} V_j) = A_X V_j + T_V V_j = A_X V_j.$$

$$K(X + V, V_j)$$
$$= \langle R(X + V, V_j)V_j, X + V \rangle$$
$$= \langle R(X, V_j)V_j, X \rangle + \langle R(X, V_j)V_j, V \rangle + \langle R(V, V_j)V_j, X \rangle + \langle R(V, V_j)V_j, V \rangle$$
$$= K(X, V_j) + 2\langle R(V, V_j)V_j, X \rangle + K(V, V_j) = |A_X V_j|^2 + \hat K(V, V_j).$$

Collecting like terms,

$$K_t(X + V, V_j) = t\hat{K}(V, V_j) + t^2|A_X V_j|^2.$$

Note that the union of $\{X_i\}$ and $\{t^{-\frac{1}{2}}V_j\}$ gives an orthonormal basis of tangent vectors.

$$\begin{aligned}
\text{Ric}_t(X + V) &= \sum_i K_t(X + V, X_i) + \sum_j K_t(X + V, t^{-\frac{1}{2}}V_j) \\
&= \sum_i K_t(X + V, X_i) + \sum_j t^{-1} K_t(X + V, V_j) \\
&= \sum_i K_*(X, X_i) - 3t \sum_i |A_X X_i|^2 + 2t \sum_i \langle(\nabla_{X_i} A)_{X_i} X, V\rangle \\
&\quad + t^2 \sum_i |A_{X_i} V|^2 + \sum_j \hat{K}(V, V_j) + t \sum_j |A_X V_j|^2.
\end{aligned}$$

One has

$$\lambda|X + V|_t^2 = \lambda|X|^2 + t\lambda|V|^2.$$

Assume conditions (i)–(iii).

$$\begin{aligned}
\sum_i |A_X X_i|^2 &= \sum_i \sum_j \langle A_X X_i, V_j\rangle^2 = \sum_j \sum_i \langle A_X V_j, X_i\rangle^2 \\
&= \sum_j |A_X V_j|^2 = v|X|^2.
\end{aligned}$$

Hence

$$\begin{aligned}
\text{Ric}_t(X + V) &= \check{\lambda}|X|^2 - 3tv|X|^2 + t^2\mu|V|^2 + \hat{\lambda}|V|^2 + tv|X|^2 \\
&= (\check{\lambda} - 2tv)|X|^2 + (\hat{\lambda} + t^2\mu)|V|^2.
\end{aligned}$$

Thus for Einstein's condition we require

$$\lambda = \check{\lambda} - 2tv = \frac{\hat{\lambda} + t^2\mu}{t}. \tag{22}$$

i.e. $\check{\lambda}t - 2t^2v = \hat{\lambda} + t^2\mu$,   or rather $(\mu + 2v)t^2 - \check{\lambda}t + \hat{\lambda} = 0$.

Since $\Delta = \check{\lambda}^2 - 4\hat{\lambda}(\mu + 2\nu) > 0$ there are two real solutions for t. Since $\mu + 2\nu > 0$, $\hat{\lambda} > 0$ and $-\check{\lambda} < 0$, these two real solutions are positive and hence by (22) give positive Einstein constants.

Conversely assume there are two Einstein metrics in the canonical variation at times, t_1 and t_2, with constants, λ_1 and λ_2, respectively.

$$\lambda_i |X|^2 = \mathrm{Ric}_*(X) - 3t_i \sum_i |A_X X_i|^2 + t_i \sum_j |A_X V_j|^2$$

$$= \mathrm{Ric}_*(X) - 2t_i \sum_j |A_X V_j|^2$$

for $i = 1, 2$.

Hence taking one equation from the other

$$(\lambda_1 - \lambda_2)|X|^2 = 2(t_2 - t_1) \sum_j |A_X V_j|^2.$$

Hence

$$\sum_j |A_X V_j|^2 = \frac{\lambda_1 - \lambda_2}{2(t_2 - t_1)} |X|^2 = \nu |X|^2.$$

Thus

$$\mathrm{Ric}_*(X) = \lambda_i |X|^2 + 2t_i \nu |X|^2 = (\lambda_i + 2t_i \nu)|X|^2 = \check{\lambda}|X|^2.$$

Similarly

$$t_i \lambda_i |V|^2 = \widehat{\mathrm{Ric}}(V) + t_i^2 \sum_i |A_{X_i} V|^2$$

for $i = 1, 2$.

Hence taking one equation from the other

$$(t_1 \lambda_1 - t_2 \lambda_2)|V|^2 = (t_1^2 - t_2^2) \sum_i |A_{X_i} V|^2.$$

$$\sum_i |A_{X_i} V|^2 = \frac{t_1 \lambda_1 - t_2 \lambda_2}{t_1^2 - t_2^2} |V|^2 = \mu |V|^2.$$

Hence

$$\widehat{\mathrm{Ric}}(V) = t_i \lambda_i |V|^2 - t_i^2 \mu |V|^2 = t_i (\lambda_i - t_i \mu)|V|^2 = \hat{\lambda}|V|^2$$

for $i = 1, 2$.

So for $i = 1, 2$ we have

$$\lambda_i = \check{\lambda} - 2t_i \nu = \frac{\hat{\lambda} + t_i^2 \mu}{t_i}.$$

Hence $(\mu + 2\nu)t_i^2 - \check{\lambda}t_i + \hat{\lambda} = 0$ for $i = 1, 2$. Namely t_i are distinct real solutions to $(\mu + 2\nu)t^2 - \check{\lambda}t + \hat{\lambda} = 0$ and thus $\Delta = \check{\lambda}^2 - 4\hat{\lambda}(\mu + 2\nu) > 0$. Note

$$\mathrm{Ric}_{t_i}(X + V) = \mathrm{Ric}_{t_i}(X) + 2t_i \sum_i \langle (\nabla_{X_i} A)_{X_i} X, V \rangle + \mathrm{Ric}_{t_i}(V)$$

$$= \lambda_i |X|^2 + 2t_i \sum_i \langle (\nabla_{X_i} A)_{X_i} X, V \rangle + t_i \lambda_i |V|^2$$

must equal $\lambda_i |X|^2 + t_i \lambda_i |V|^2$. Hence $\sum_i \langle (\nabla_{X_i} A)_{X_i} X, V \rangle = 0$. We have confirmed conditions (i)–(iii).

Proposition 18 is a result of Bérard Bergery that was never published but whose statement appears in Besse's book on Einstein manifolds [5]. Note that the statement in Besse contains a misprint, the 3 in the statement ought to be a 4. This misprint has had consequences that have propagated throughout the Einstein manifold literature [6, 7, 15].

Proposition 19. *Let M be a generic CR submanifold of complex projective space with the integral foliation of $\mathscr{D}^\perp$ with equidistant leaves that are totally geodesic in complex projective space and $\dim(\mathscr{D}^\perp) > 1$. Then the foliation has two Einstein metrics in its canonical variation.*

Proof. Recall from Proposition 10 that M is n-Sasakian where $n = \dim(\mathscr{D}^\perp)$. Recall from Proposition 15 that the space of leaves is Einstein with Einstein constant, $\check{\lambda} = \dim(\mathscr{D}) + 2 + 2\dim(\mathscr{D}^\perp)$. The leaves are n-dimensional space forms and hence are Einstein with Einstein constant, $\hat{\lambda} = \dim(\mathscr{D}^\perp) - 1$.

$$\langle (\nabla_{X_i} A)_{X_i} X, V \rangle = \langle R(X, X_i) X_i, V \rangle = \langle R(X_i, V) X_i, X \rangle$$

$$= \langle \langle X_i, V \rangle X_i - |X_i|^2 V, X \rangle = 0.$$

Hence we have Condition (i).

Recall $|A_X V|^2 = |X|^2 |V|^2$. Hence $\sum_i |A_{X_i} V|^2 = \sum_i |V|^2 = \dim(\mathscr{D})|V|^2$ and $\sum_j |A_X V_j|^2 = \sum_j |X|^2 = \dim(\mathscr{D}^\perp)$. Hence we have Condition (ii) with $\mu = \dim(\mathscr{D})$ and $\nu = \dim(\mathscr{D}^\perp)$.

$$\mu + 2\nu = \dim(\mathscr{D}) + 2\dim(\mathscr{D}^\perp),$$

$$4\hat{\lambda}(\mu + 2\nu) = 4(\dim(\mathscr{D}^\perp) - 1)(\dim(\mathscr{D}) + 2\dim(\mathscr{D}^\perp)),$$

$$\check{\lambda} = \dim(\mathscr{D}) + 2\dim(\mathscr{D}^\perp) + 2.$$

Recall that since M is $\dim(\mathscr{D}^\perp)$-Sasakian that there is an antisymmetric Clifford representation of $\mathscr{D}^\perp$ on $\mathscr{D}$. Hence we have

$$2(\dim(\mathscr{D}^\perp) - 1) \le \dim(\mathscr{D}).$$

Hence since

$$4(\dim(\mathscr{D}^\perp) - 1) \le \dim(\mathscr{D}) + 2(\dim(\mathscr{D}^\perp) - 1) < \dim(\mathscr{D}) + 2\dim(\mathscr{D}^\perp) + 2$$

and

$$\dim(\mathscr{D}) + 2\dim(\mathscr{D}^\perp) < \dim(\mathscr{D}) + 2\dim(\mathscr{D}^\perp) + 2$$

we have $4\hat{\lambda}(\mu + 2\nu) < \check{\lambda}^2$. This yields Condition (iii) and we conclude that M has two Einstein metrics in its canonical variation.

Definition 21. A submanifold, M, of a Kähler manifold, $\bar{M}$, is said to be anti-invariant if $J(T_x M) \subset (T_x M)^\perp$ for each $x \in M$.

Abe showed that the only anti-invariant submanifolds of complex projective space were real projective spaces [1].

Corollary 5. *The leaves of the foliation in Proposition 18 are isometric to* $\mathbb{R}P^n$, *where* $n = \dim(\mathscr{D}^\perp)$.

4 Relation Between Isoparametric Hypersurfaces and n-Sasakian Structures

In the last two lectures we have seen some examples of n-Sasakian manifolds that arise as circle quotients of the focal submanifolds of isoparametric hypersurface families with four principal curvatures via a contact CR structure. This begs the natural question: When is the focal set of an isoparametric hypersurface family with four principal curvatures a contact CR structure?

To begin assume the Cartan polynomial on $V = \mathbb{C}^n$ is invariant under the action of S^1. That is for each $z \in \mathbb{C}$ with $|z| = 1$ that

$$F(zx) = F(x).$$

Lemma 25. *F is S^1-invariant if and only if $\{(zx)(zx)(zx)\} = z\{xxx\}$.*

Proof. Recall $F(x) = 3|x|^4 - \frac{2}{3}\langle\{xxx\}, x\rangle$. Hence F is invariant precisely when

$$3|zx|^4 - \frac{2}{3}\langle\{(zx)(zx)(zx)\}, zx\rangle = 3|x|^4 - \frac{2}{3}\langle\{xxx\}, x\rangle.$$

Since $|zx| = |x|$

$$\langle \{(zx)(zx)(zx)\}, zx \rangle = \langle \{xxx\}, x \rangle. \tag{23}$$

Consider $y(23)$

$$4\langle \{(zx)(zx)(zx)\}, zy \rangle = 4\langle \{xxx\}, y \rangle. \tag{24}$$

Hence $\langle \bar{z}\{(zx)(zx)(zx)\}, y \rangle = \langle \{xxx\}, y \rangle$. Thus $\bar{z}\{(zx)(zx)(zx)\} = \{xxx\}$.
So $\{(zx)(zx)(zx)\} = z\{xxx\}$.
Conversely if $\{(zx)(zx)(zx)\} = z\{xxx\}$, then

$$3|zx|^4 - \frac{2}{3}\langle \{(zx)(zx)(zx)\}, zx \rangle = 3|zx|^4 - \frac{2}{3}\langle z\{xxx\}, zx \rangle$$

$$= 3|x|^4 - \frac{2}{3}\langle \{xxx\}, x \rangle.$$

Lemma 26. *If F is S^1-invariant, then ξ is tangent along M_-.*

Proof. $\xi = ix$ is the fundamental field of the S^1-action. Since $\{(zx)(zx)(zx)\} = z\{xxx\}$ differentiation with respect to θ where $z = e^{i\theta}$ at $\theta = 0$ gives $3\{xx(ix)\} = i\{xxx\} = i(6x)$. Hence $\{xx(ix)\} = 2(ix)$, i.e., $\{xx\xi\}, = 2\xi$, namely $\xi \in V_2(x) = T_x M_-$.

Lemma 27. $V_\kappa(zx) = zV_\kappa(x)$.

Proof. Let $v \in V_\kappa(zx)$. Then $\langle v, zx \rangle = 0$ and $\{(zx)(zx)v\} = \kappa v$. It follows $\langle \bar{z}v, x \rangle = 0$ and $z\{xx(\bar{z}v)\} = \{(zx)(zx)v\} = \kappa v$, i.e., $\{xx(\bar{z}v)\} = \kappa \bar{z}v$. Hence $\bar{z}v \in V_\kappa(x)$ and it follows $v \in zV_\kappa(x)$. The converse follows in much the same fashion.

Corollary 6. *The distribution, $\mathscr{D}$, defined by $\mathscr{D}_x = V_2(x) \cap V_2(ix)$ for each $x \in M_-$ has $\varphi(\mathscr{D}) \subset \mathscr{D}$.*

Proof. Note that for $v \perp \xi$ that $\varphi(v) = iv$.

$$i(V_2(x) \cap V_2(ix)) = iV_2(x) \cap iV_2(ix) = V_2(ix) \cap V_2(i(ix)) = V_2(ix) \cap V_2(x).$$

For an inner product space, V, with subspace, W, we adopt the notation, $V \ominus W$, for the orthogonal complement of W in V.

Corollary 7. *M_- has a contact CR structure if and only if*

$$\mathscr{D}_x^{\perp} = (V_2(x) \cap V_0(ix)) \oplus \mathbb{R}(ix)$$

for each $x \in M_-$.

Proof. Suppose $\mathscr{D}_x^\perp = (V_2(x) \cap V_0(ix)) \oplus \mathbb{R}(ix)$. Then

$$\varphi(\mathscr{D}_x^\perp) = i\,(V_2(x) \cap V_0(ix)) = V_2(ix) \cap V_0(x) \subset V_0(x) = (T_x M_-)^\perp.$$

Conversely assume $\varphi(\mathscr{D}_x^\perp) \subset (T_x M_-)^\perp \subset V_0(x)$. Hence $i\,(\mathscr{D}_x^\perp \ominus \mathbb{R}(ix)) \subset V_0(x)$. Hence $\mathscr{D}_x^\perp \ominus \mathbb{R}(ix) \subset i V_0(x) = V_0(ix)$. Hence $\mathscr{D}_x^\perp \ominus \mathbb{R}(ix) \subset i V_0(x) = V_2(x) \cap V_0(ix)$. That $\mathscr{D}_x^\perp \ominus \mathbb{R}(ix) \supset V_2(x) \cap V_0(ix)$ is manifest.

Proposition 20. *Let $x \in V$, $\langle x, x \rangle = 1$, $\{xxx\} = 6x$ be a minimal tripotent. Then for $u_\kappa, v_\kappa, w_\kappa \in V_\kappa(x)$, where $\kappa = 0, 2$, one has*

$$u_0 \circ v_0 = 0, \quad u_0 \circ v_2 \in V_2(x), \quad u_2 \circ v_2 = 2\langle u_2, v_2 \rangle x + (u_2 \circ v_2)_0, \quad (25)$$

and moreover

$$\{u_0 v_0 w_0\} = 2\,(\langle u_0, v_0 \rangle w_0 + \langle v_0, w_0 \rangle u_0 + \langle w_0, u_0 \rangle v_0), \tag{26}$$

$$\{u_0 v_0 w_2\} = u_0 \circ (v_0 \circ w_2) + v_0 \circ (u_0 \circ w_2), \tag{27}$$

$$\{u_0 v_2 w_2\} = \langle u_0, v_2 \circ w_2 \rangle x + (v_2 \circ (w_2 \circ u_0) + w_2 \circ (v_2 \circ u_0))_0 + \{u_0 v_2 w_2\}_2, \tag{28}$$

$$\{u_2 v_2 w_2\} = 2\,(\langle u_2, v_2 \rangle w_2 + \langle v_2, w_2 \rangle u_2 + \langle w_2, u_2 \rangle v_2) \tag{29}$$

$$\qquad - u_2 \circ (v_2 \circ w_2)_0 - v_2 \circ (w_2 \circ u_2)_0 - w_2 \circ (u_2 \circ v_2)_0 + \{u_2 v_2 w_2\}_0. \tag{30}$$

Exercise 14. Prove Proposition 20 by repeatedly differentiating (13).

Proposition 21. *Suppose that the focal submanifold, M_-, of a isoparametric hypersurface family with four principal curvature is a contact CR submanifold of the sphere. Then the leaves of the integral foliation of $\mathscr{D}^\perp$ are equidistant.*

Proof. Let w lie along $\mathscr{D}$ and v lie along $\mathscr{D}^\perp$ perpendicular to $\xi = ix$. Recall

$$z\{vxw\} = \{(zv)(zx)(zw)\}.$$

Hence differentiation with respect to θ with $z = e^{i\theta}$ at $\theta = 0$ gives

$$i\{vxw\} = \{(iv)xw\} + \{v(ix)w\} + \{vx(iw)\}.$$

Hence

$$i(v \circ w) = (iv) \circ w + i\{(iv)x(iw)\} + v \circ (iw),$$

$$i(v \circ w) = (iv) \circ w + i\,((iv) \circ (iw)) + v \circ (iw).$$

Hence

$$v \circ w = -i\left((iv) \circ w\right) + (iv) \circ (iw) - i\left(v \circ (iw)\right). \tag{31}$$

It follows from (25) that $v \circ w \in V_0(x) \oplus \mathbb{R}x$ and $v \circ (iw) \in V_0(x) \oplus \mathbb{R}x$. Hence $i(v \circ (iw)) \in V_0(ix) \oplus \mathbb{R}(ix)$. Neither of these have a $\mathscr{D}$ component and we thus conclude from (31) that $(iv) \circ (iw)$ and $i((iv) \circ w)$ have the same $\mathscr{D}$ component. It follows

$$\mathscr{D}S_{\varphi(v)}\varphi(w) = \mathscr{D}\varphi(S_{\varphi(v)}w)$$

and that the leaves are equidistant.

Lemma 28. *Let x and v be minimal tripotents of an isoparametric triple, $\{\ldots\}$ on V. Then $v \in V_\kappa(x)$ if and only if $x \in V_\kappa(v)$ where $\kappa = 0, 2$.*

Proof. Suppose $v \in V_\kappa(x)$. Then $\langle v, x \rangle = 0$ and thus $x = x_0 + x_2$ where $x_\mu \in V_\mu(v)$ for $\mu = 0, 2$.

$$\langle \{vvx\}, x \rangle = \langle \{vvx_0\} + \{vvx_2\}, x \rangle = 2\langle x_2, x \rangle = 2|x_2|^2.$$

$$\langle \{vvx\}, x \rangle = \langle \{xxv\}, v \rangle = \langle \kappa v, v \rangle = \kappa.$$

Hence $2|x_2|^2 = \kappa$. If $\kappa = 0$, then $x_2 = 0$, i.e., $x \in V_0(v)$. If $\kappa = 2$, then $|x_2| = 1$ and hence $x_0 = 0$, i.e., $x \in V_2(v)$. The converse follows vertabim.

Proposition 22. *Suppose that the focal submanifold, M_-, of a isoparametric hypersurface family with four principal curvature is a contact CR submanifold of the sphere. Then the leaves of the integral foliation of $\mathscr{D}^\perp$ have $S_{\varphi(V)}W = \langle \varphi(V), \varphi(W) \rangle \xi$ for each pair of anti-holomorphic fields, V, W, if and only if they are totally geodesic in M_-.*

Proof. Let v and w lie along $\mathscr{D}^\perp$ perpendicular to ξ. Suppose the technical condition. Then $(iv) \circ w = \langle iv, iw \rangle ix \in \mathscr{D}^\perp$. Thus the leaves are totally geodesic.

Conversely assume the leaves are totally geodesic. Let v and w lie along $\mathscr{D}^\perp$ perpendicular to ξ. It follows $(iv) \circ w \in (V_2(x) \cap V_0(ix)) \cap \mathbb{R}(ix) = \mathscr{D}_x^\perp$. Let w' be an arbitrary vector lying along $\mathscr{D}^\perp$ and perpendicular to ξ, i.e., $w' \in V_2(x) \cap V_0(ix)$ for each $x \in M_-$.

$$\langle (iv) \circ w, w' \rangle = \langle (iv), w \circ w' \rangle.$$

For $w \in V_0(ix)$ it follows that $\{www\} = 6|w|^2 w$. However since $w \in V_2(x)$ we have $\{www\} = 6|w|^2 w - |(w \circ w)_0|^2$. Hence $(w \circ w)_0 = 0$. Polarisation gives $(w \circ w')_0 = 0$. Thus $\langle (iv) \circ w, w' \rangle = 0$.

$$i\{xxw\} = 2\{(ix)xw\} + \{xx(iw)\}, \qquad i(2w) = 2w \circ (ix).$$

Hence $w \circ ix = iw$.

$$\langle (iv) \circ w, ix \rangle = \langle (iv), w \circ (ix) \rangle = \langle (iv), (iw) \rangle.$$

Hence $(iv) \circ w = \langle (iv), (iw) \rangle (ix)$.

Lemma 29. *Suppose that the focal submanifold, M_-, of a isoparametric hypersurface family with four principal curvature is a contact CR submanifold of the sphere. Then the leaves of the integral foliation of $\mathscr{D}^\perp$ are totally contact geodesic in the sphere if and only if they are totally geodesic in M_-.*

Proof. It is equivalent to show that CR submanifold, M_-/ξ, of complex projective space has totally geodesic leaves in M_-/ξ precisely when the leaves are totally geodesics in the ambient projective space. For this we only need show $B_v v = 0$ for each v lying along $\mathscr{D}^\perp$ perpendicular to ξ.

For $v \in V_2(x) \cap V_0(ix)$ with $|v| = 1$ we have $\{vvv\}v = 6v$ and $v \in V_2(x)$. It follows from Lemma 28 that

$$B_v v = (v \circ v)_0 = \{vvx\}_0 = (2x)_0 = 0.$$

Lemma 30. *Suppose that the focal submanifold, M_-, of a isoparametric hypersurface family with four principal curvatures is a contact CR submanifold of the sphere such that the leaves of the integral foliation of $\mathscr{D}^\perp$ are totally geodesic in M_-. Then $\mathscr{D}_x \subset V_2(q)$ for each $q \in V_2(ix) \cap V_0(x) = \varphi(\mathscr{D}_x^\perp)$ with $|q| = 1$.*

Proof. Let $z \in V_2(x) \cap V_2(ix)$ and $q \in V_2(ix) \cap V_0(x)$. Since the leaves are totally geodesic in M_- they are totally contact geodesic in the sphere. Hence M_-/ξ is n-Sasakian. Recall then that $\varphi(\mathscr{D}^\perp) = V_2(ix) \cap V_0(x)$ induces a symmetric Clifford representation on $\mathscr{D}$. Hence

$$\{qqz\} = 2q \circ (q \circ z) = 2|q|^2 z = 2z.$$

Hence $z \in V_2(q)$. That is $\mathscr{D}_x \subset V_2(q)$ for each $q \in V_2(ix) \cap V_0(x) = \varphi(\mathscr{D}_x^\perp)$ with $|q| = 1$.

Corollary 8. *Suppose that the focal submanifold, M_-, of a isoparametric hypersurface family with four principal curvature is a contact CR submanifold of the sphere such that the leaves of the integral foliation of $\mathscr{D}^\perp$ are totally geodesic in M_-. Then the contact CR structure is generic.*

Proof. The contact CR structure gives the orthogonal splitting,

$$(T_x M)^\perp = (V_0(x) \cap V_2(ix)) \cap (V_0(x) \cap V_0(ix)).$$

Hence to see the structure is generic it suffices to show $V_0(x) \cap V_0(ix) = 0$. Let $iv \in V_0(x) \cap V_0(ix)$ and $iu \in V_0(x) \cap V_2(ix)$ with unit length. Since iv and iu are orthogonal $\{(iu)(iu)(iv)\} = 2(iv)$. We also have $v \in V_0(x) \cap V_0(ix)$. Hence $\{(iu)(iu)v\} = 2v$. Thus $v \in V_2(u) \cap V_2(iu)$. However $V_2(x) \cap V_2(ix) = V_2(u) \cap V_2(iu)$ by Lemma 30. Hence $v = 0$.

That the contact CR structure is generic is the key to the classification of which focal sets give rise to n-Sasakian manifolds. Although the full argument involves rather complicated algebra that make it unsuitable for an introductory lecture course the basic idea is rather simple.

Definition 22. Given symmetric endomorphisms, $P_0, P_1, \ldots, P_n$, of V such that

$$P_i P_j + P_j P_i = 2\delta_{ij} I$$

for each i, j we define an FKM triple to be the isoparametric triple associated with the Cartan polynomial for the FKM system, $F(x) = |x|^4 - 2\sum_k \langle P_k, x \rangle^2$ and a dual FKM triple to be associated with the Cartan polynomial, $-F$.

Proposition 23. *Suppose that M_- is a contact CR submanifold of the sphere with the foliation of the integral leaves of $\mathscr{D}^\perp$ totally geodesic. Moreover assume $\mathscr{D}$ at each point contains at least one minimal tripotent. Then the original triple is a dual Ferus-Karcher-Münzner triple on V with $m_2(V) = \dim(\mathscr{D}^\perp)$.*

The way the above result is argued is that generic contact CR structure implies that the shape operator induces a symmetric Clifford representation of the normal bundle on $\mathscr{D}$ and to use general algebraic structure of the triple along with an old Lemma of Dorfmeister and Neher, see [19], that under such hypotheses the normal bundle, $V_0(x)$, induces an symmetric Clifford representation on V itself via $T(v, x)$ where $v \in V_0(x)$. The representation theory places a very strong restriction in this situation. Let $n = m_2 + 1 = \dim(V_0(x))$. $\dim(\mathscr{D}) = 2\ell\delta(m_2)$ and $\dim(V) = 2k\delta(m_2)$. Since the structure is generic

$$\dim(V) = \dim(\mathscr{D}) + \dim(\mathscr{D}^\perp) + \dim(V_0(x)) + 1$$
$$= 2\ell\delta(m_2) + (m_2 + 2) + (m_2 + 1) + 1 = 2(\ell\delta(m_2) + m_2 + 2).$$

Hence $k\delta(m_2) = \ell\delta(m_2) + m_2 + 2$ or rather $m_2 + 2 = (k - \ell)\delta(m_2)$. Typically $\delta(m_2)$ is much larger than $m_2 + 2$. This restricts us to low multiplicities, namely $m_2 = 0, 2, 6$.

When $\mathscr{D}$ does not contain a minimal tripotent one falls into another interesting situation. Strictly speaking the conclusion of the following is true up to covering.

Proposition 24. *Suppose that M_- is a contact CR submanifold of the sphere with the foliation of the integral leaves of $\mathscr{D}^\perp$ totally geodesic. Moreover assume $\mathscr{D}$ at each point is spanned by maximal tripotents. Then the space of leaves of the foliation is a complex projective space.*

Proof. For $z, y \in \mathscr{D}_x = V_2(x) \cap V_2(ix) \subset V_2(x)$ it follows from (29) that

$$\{zzy\}_2 = 4\langle z, y \rangle z + 2|z|^2 y - 2z \circ (z \circ y)_0 - y \circ (z \circ z)_0.$$

Since $y \in V_2(x)$,

$$\langle \{zzy\}, y \rangle = 4\langle z, y \rangle^2 + 2|z|^2|y|^2 + 2|(z \circ y)_0|^2 - \langle (z \circ z)_0, (y \circ y)_0 \rangle. \qquad (32)$$

Recall $iy \in \mathcal{D}_x$ as well, hence it follows verbatim that

$$\langle \{zz(iy)\}, iy \rangle = 4\langle z, iy \rangle^2 + 2|z|^2|y|^2 + 2|(z \circ (iy))_0|^2 - \langle (z \circ z)_0, ((iy) \circ (iy))_0 \rangle. \qquad (33)$$

For any $n \in V_0(x)$ we have

$$\langle (iy) \circ (iy), n \rangle = \langle iy, n \circ (iy) \rangle = \langle iy, i(n \circ y) \rangle = \langle y, n \circ y \rangle = \langle y \circ y, n \rangle.$$

Hence $((iy) \circ (iy))_0 = (y \circ y)_0$.

As $\mathcal{D}$ consists of scalar multiples of maximal tripotents it follows $\{zzz\} = 3|z|^2 z$ and thus $3\{zzy\} = 6\langle z, y \rangle z + 3|z|^2 y$, i.e., $\{zzy\} = 2\langle z, y \rangle z + |z|^2 y$ by polarisation. Hence

$$\langle \{zzy\}, y \rangle = 2\langle z, y \rangle^2 + |z|^2|y|^2. \qquad (34)$$

Substitute (34) into (32) to find

$$4\langle z, y \rangle^2 + 2|z|^2|y|^2 - 2|(z \circ y)_0|^2 - \langle (z \circ z)_0, (y \circ y)_0 \rangle = \quad 2\langle z, y \rangle^2 + |z|^2|y|^2.$$

Which simplifies to

$$\langle (z \circ z)_0, (y \circ y)_0 \rangle + 2|(z \circ y)_0|^2 = 2\langle z, y \rangle^2 + |z|^2|y|^2. \qquad (35)$$

Analogously it follows from considerations above and (33) that

$$\langle (z \circ z)_0, (y \circ y)_0 \rangle + 2|(z \circ iy)_0|^2 = 2\langle z, iy \rangle^2 + |z|^2|y|^2. \qquad (36)$$

Subtract (36) from (35),

$$2|(z \circ y)_0|^2 - 2|(z \circ iy)_0|^2 = 2\langle z, y \rangle^2 - 2\langle z, iy \rangle^2.$$

Divide by 2 to find

$$|(z \circ y)_0|^2 - |(z \circ iy)_0|^2 = \langle z, y \rangle^2 - \langle z, iy \rangle^2. \qquad (37)$$

Subtract (37) from (35) to find

$$\langle (z \circ z)_0, (y \circ y)_0 \rangle - |(z \circ y)_0|^2 + 3|z \circ (iy))_0|^2 = \quad |z|^2|y|^2 - \langle z, y \rangle^2 + 3\langle z, iy \rangle^2. \qquad (38)$$

Hence substitution of (38) into the unnormalised sectional curvature gives

$$K_*(z, y) = |z|^2|y|^2 - \langle z, y\rangle^2 + 3\langle z, iy\rangle^2 + \langle (z \circ z)_0, (y \circ y)_0\rangle - |(z \circ y)_0|^2$$
$$+ 3|z \circ (iy))_0|^2$$
$$= 2(|z|^2|y|^2 - \langle z, y\rangle^2 + 3\langle z, iy\rangle^2).$$

Thus the space of leaves must be a complex projective space since it is a Kähler manifold with pinching, $\frac{1}{4}$.

Corollary 9. *Suppose that M_+ is a contact CR submanifold of the sphere with the leaves of the integral foliation of $\mathscr{D}^{\perp}$ totally geodesic. Moreover assume $\mathscr{D}$ at each point is spanned by minimal tripotents. Then the space of leaves of the foliation is a complex projective space.*

Example 9. Consider Example 1 with $r = 3$. The triple is homogeneous with isometry group $U(2) \times U(3)$. Hence it suffices to consider

$$x = \frac{1}{\sqrt{2}}\begin{pmatrix} 1 & 0 & 0 \\ 0 & 1 & 0 \end{pmatrix}.$$

To see that $x \in M_+$ we compute

$$xx^* = \frac{1}{2}\begin{pmatrix} 1 & 0 & 0 \\ 0 & 1 & 0 \end{pmatrix}\begin{pmatrix} 1 & 0 \\ 0 & 1 \\ 0 & 0 \end{pmatrix} = \frac{1}{2}\begin{pmatrix} 1 & 0 \\ 0 & 1 \end{pmatrix}.$$

Thus $\{xxx\} = 6xx^*x = 6\left(\frac{1}{2}x\right) = 3x$. Hence $x \in M_+$ as expected.

Now we find the form of $T_xM_+ = V_1(x)$ and $(T_xM_+)^{\perp} = V_3(x)$.
Recall $T(x)z = 2xx^*z + 2zx^*x + 2xz^*x$. We have

$$2xx^*z = \begin{pmatrix} z_1 & z_2 & z_3 \\ w_1 & w_2 & w_3 \end{pmatrix}, \quad 2zx^*x = \begin{pmatrix} z_1 & z_2 & z_3 \\ w_1 & w_2 & w_3 \end{pmatrix}\begin{pmatrix} 1 & 0 & 0 \\ 0 & 1 & 0 \\ 0 & 0 & 0 \end{pmatrix} = \begin{pmatrix} z_1 & z_2 & 0 \\ w_1 & w_2 & 0 \end{pmatrix}$$

and

$$2xz^*x = \begin{pmatrix} 1 & 0 & 0 \\ 0 & 1 & 0 \end{pmatrix}\begin{pmatrix} \bar{z}_1 & \bar{w}_1 \\ \bar{z}_2 & \bar{w}_2 \\ \bar{z}_3 & \bar{w}_3 \end{pmatrix}\begin{pmatrix} 1 & 0 & 0 \\ 0 & 1 & 0 \end{pmatrix} = \begin{pmatrix} \bar{z}_1 & \bar{w}_1 & 0 \\ \bar{z}_2 & \bar{w}_2 & 0 \end{pmatrix}.$$

Adding these three expressions we find

$$T(x)z = \begin{pmatrix} 2z_1 + \bar{z}_1 & 2z_2 + \bar{w}_1 & z_3 \\ 2w_1 + \bar{z}_2 & 2w_2 + \bar{w}_2 & w_3 \end{pmatrix}.$$

Write $z_k = x_k + iy_k$ and $w_k = a_k + ib_k$ to find

$$T(x)z = \begin{pmatrix} 3x_1 + iy_1 & 2z_2 + \bar{w}_1 & z_3 \\ 2w_1 + \bar{z}_2 & 3a_2 + ib_2 & w_3 \end{pmatrix},$$

$$V_1(x) = \left\{ \begin{pmatrix} iy_1 & z_2 & z_3 \\ -\bar{z}_2 & ib_2 & w_3 \end{pmatrix} \right\}, \qquad V_3(x) \oplus \mathbb{R}x = \left\{ \begin{pmatrix} x_1 & z_2 & 0 \\ \bar{z}_2 & a_2 & 0 \end{pmatrix} \right\}.$$

Let

$$\mathscr{D}_x = \left\{ \begin{pmatrix} 0 & 0 & z_3 \\ 0 & 0 & w_3 \end{pmatrix} \right\}, \qquad \mathscr{D}_x^\perp = \left\{ \begin{pmatrix} iy_1 & z_2 & 0 \\ -\bar{z}_2 & ib_2 & 0 \end{pmatrix} \right\}.$$

Note it then follows that $i\mathscr{D}^\perp = V_3(x) \oplus \mathbb{R}x$ hence $\varphi(\mathscr{D}^\perp) = V_3(x)$ and hence we have a generic contact CR structure.

We now check that $\mathscr{D}_x$ consists of scalar multiples of minimal tripotents. Let $z \in \mathscr{D}$ and $|z| = 1$,

$$zz^* = \begin{pmatrix} 0 & 0 & z_3 \\ 0 & 0 & w_3 \end{pmatrix} \begin{pmatrix} 0 & 0 \\ 0 & 0 \\ \bar{z}_3 & \bar{w}_3 \end{pmatrix} = \begin{pmatrix} |z_3|^2 & z_3\bar{w}_3 \\ w_3\bar{z}_3 & |w_3|^2 \end{pmatrix}.$$

$$zz^*z = \begin{pmatrix} |z_3|^2 & z_3\bar{w}_3 \\ w_3\bar{z}_3 & |w_3|^2 \end{pmatrix} \begin{pmatrix} 0 & 0 & z_3 \\ 0 & 0 & w_3 \end{pmatrix} = \begin{pmatrix} 0 & 0 & |z_3|^2z_3 + (z_3\bar{w}_3)w_3 \\ 0 & 0 & (\bar{z}_3w_3)z_3 + |w_3|^2w_3 \end{pmatrix}$$

$$= \begin{pmatrix} 0 & 0 & |z_3|^2z_3 + |w_3|^2z_3 \\ 0 & 0 & |z_3|^2w_3 + |w_3|^2w_3 \end{pmatrix} = \begin{pmatrix} 0 & 0 & z_3 \\ 0 & 0 & w_3 \end{pmatrix}.$$

Hence $\{zzz\} = 6zz^*z = 6z$ as desired.

Example 10. Consider the generic CR submanifold, $M_+ \subset S^{19}$, considered in Example 5. Recall since the example is homogeneous it suffices to consider the specific x discussed there. For this x recall that

$$\mathscr{D}_x = \left\{ \begin{pmatrix} 0 & b \\ -b^T & 0 \end{pmatrix} : b \in \mathbb{C}^4 \right\}.$$

We now check that $\mathscr{D}_x$ consists of scalar multiples of minimal tripotents. For $z \in \mathscr{D}_x$ we have

$$zz^* = \begin{pmatrix} 0 & b \\ -b^T & 0 \end{pmatrix} \begin{pmatrix} 0 & \bar{b} \\ -\bar{b}^T & 0 \end{pmatrix} = \begin{pmatrix} b\bar{b}^T & 0 \\ 0 & b^T\bar{b} \end{pmatrix}.$$

$$zz^*z = \begin{pmatrix} b\bar{b}^T & 0 \\ 0 & b^T\bar{b} \end{pmatrix} \begin{pmatrix} 0 & b \\ -b^T & 0 \end{pmatrix} = \begin{pmatrix} 0 & b\bar{b}^Tb \\ -b^T\bar{b}b^T & 0 \end{pmatrix} = |b|^2 \begin{pmatrix} 0 & b \\ -b^T & 0 \end{pmatrix}.$$

Hence $\{zzz\} = 6zz^*z = 6|z|^2z$ as desired.

In these situations where our spaces of leaves have positive sectional curvature there are particularly strong implications for the existence of metrics of positive sectional curvature on the associated n-Sasakian manifolds. To explore this we will need the following,

Proposition 25. *Let $\pi : M \to B$ be a Riemannian submersion with totally geodesic fibres. Then the sectional curvature of a canonical variation, $\langle, \rangle_t$, has positive sectional curvature for all $0 < t < t_0$ for some t_0 if and only if the fibres have positive sectional curvature and for any linearly independent horizontal fields, X, Y and nonzero vertical field V one has*

$$\langle (\nabla_X A)_X Y, V \rangle^2 < K_*(X, Y)|A_X V|^2.$$

Proof. Suppose the sectional curvature of the canonical variation is positive for all $0 < t < t_0$ for some t_0. Consider $K_t(X, Y + V)$.

$$
\begin{aligned}
K_t(X, Y + V) &= tK(X, Y + V) + (1 - t)K_*(X, Y) - t(1 - t)|A_X V|^2 \\
&= t(K(X, Y) + 2\langle R(Y, X)X, V \rangle + K(X, V)) + (1 - t)K_*(X, Y) \\
&\quad - t(1 - t)|A_X V|^2 \\
&= t(K_*(X, Y) - 3|A_X Y|^2 + 2\langle (\nabla_X A)_X Y, V \rangle + |A_X V|^2) \\
&\quad + (1 - t)K_*(X, Y) - t(1 - t)|A_X V|^2 \\
&= K_*(X, Y) - 3t|A_X Y|^2 + 2t|V|\langle (\nabla_X A)_X Y, \hat{V} \rangle + (t|V|)^2 |A_X \hat{V}|^2.
\end{aligned}
$$

This must be positive for all $t|V|$ thus the discriminant,

$$\Delta = 4\langle (\nabla_X A)_X Y, \hat{V} \rangle^2 - 4|A_X \hat{V}|^2 (K_*(X, Y) - 3t|A_X Y|^2) < 0.$$

Hence

$$\langle (\nabla_X A)_X Y, \hat{V} \rangle^2 < |A_X \hat{V}|^2 (K_*(X, Y) - 3t|A_X Y|^2) \leq |A_X \hat{V}|^2 K_*(X, Y).$$

Multiplication by $|V|^2$ gives

$$\langle (\nabla_X A)_X Y, V \rangle^2 < K_*(X, Y)|A_X V|^2.$$

The converse direction is more difficult and we omit its proof. Details can be found in [16].

Corollary 10. *Let $\pi : S \to \mathcal{O}$ be the foliation of an n-Sasakian manifold, S. Then the sectional curvature of a canonical variation, $\langle, \rangle_t$ has positive sectional curvature for all $0 < t < t_0$ for some t_0 if and only if $\mathcal{O}$ has positive sectional curvature.*

Proof.

$$\langle (\nabla_X A)_X Y, V \rangle = \langle R(Y, X)X, V \rangle = \langle R(X, V)Y, X \rangle$$
$$= \langle \langle Y, V \rangle X - \langle X, Y \rangle V, X \rangle = 0.$$

$$|A_X V|^2 = K(X, V) = \langle R(X, V)V, X \rangle = \langle \langle V, V \rangle X - \langle X, V \rangle V, X \rangle = |X|^2 |V|^2.$$

Hence $\langle (\nabla_X A)_X Y, V \rangle^2 < K_*(X, Y)|A_X V|^2$ for linearly independent X, and V nonzero if and only if $K_*(X, Y) > 0$.

Thus the 3-Sasakian 7-manifold and 5-Sasakian 13-manifold discussed above have associated canonical variation metrics with positive sectional curvature, cf. [3] and [4] respectively.

Acknowledgements Thanks go to Daniele Grandini for his reading over drafts, Quo-Shin Chi for his suggestions.

References

1. K. Abe, Applications of a Riccati type differential equation to Riemannian manifolds with totally geodesic distributions. Tôhoku Math. J. **25**, 425–444 (1973)
2. U. Abresch, Isoparametric hypersurfaces with four or six distinct principal curvatures. Necessary conditions on the multiplicities. Math. Ann. **264**, 283–302 (1983)
3. S. Aloff, N. Wallach, An infinite family of distinct 7-manifolds admitting positively curved Riemannian structures. Bull. Am. Math. Soc. **81**, 93–97 (1975)
4. M. Berger, Les variétés riemanniennes homogènes normales simplement connexes à courbure strictement positive. Ann. Scuola Norm. Sup. Pisa (3) **15**, 179–246 (1961)
5. A.L. Besse, *Einstein Manifolds*. Reprint of the 1987 edition. Classics in Mathematics (Springer, Berlin, 2008)
6. C. Boyer, K. Galicki, 3-Sasakian manifolds, in *Surveys in Differential Geometry VI. Essays on Einstein Manifolds*, ed. by C. LeBrun, M. Wang (International Press, Boston, 1999), pp. 123–184
7. C. Boyer, K. Galicki, *Sasakian Geometry*. Oxford Mathematical Monographs (Oxford University Press, Oxford, 2008)
8. E. Cartan, Sur des familles remarquables d'hypersurfaces isoparamétriques dans les espaces sphériques. Math. Z. **45**, 335–367 (1939)
9. T. Cecil, Q.S. Chi, G. Jensen, Isoparametric hypersurfaces with four principal curvatures. Ann. Math. (2) **166**, 1–76 (2007)
10. Q.S. Chi, A note on the paper, "Isoparametric hypersurfaces with four principal curvatures". Hongyou Wu Memorial Volume. Pac. J. Appl. Math. **3**, 127–134 (2011)
11. Q.S. Chi, Isoparametric hypersurfaces with four principal curvatures, II. Nagoya Math. J. **204**, 1–18 (2011)
12. Q.S. Chi, Isoparametric hypersurfaces with four principal curvatures revisited. Nagoya Math. J. **193**, 129–154 (2009)
13. Q.S. Chi, Isoparametric hypersurfaces with four principal curvatures, III. J. Differential Geom. **94**, 469–540 (2013)

14. O. Dearricott, Positive sectional curvature on 3-Sasakian manifolds. Ann. Global Anal. Geom. **25**, 59–72 (2004)
15. O. Dearricott, n-Sasakian manifolds. Tohoku Math. J. (2) **60**, 329–347 (2008)
16. O. Dearricott, A 7-manifold with positive curvature. Duke Math. J. **158**, 307–346 (2011)
17. J. Dorfmeister, E. Neher, An algebraic approach to isoparametric hypersurfaces in spheres. I, II. Tôhoku Math. J. (2) **35**, 18–224 (1983)
18. J. Dorfmeister, E. Neher, Isoparametric hypersurfaces, case $g = 6$, $m = 1$. Commun. Algebra **13**, 2299–2368 (1985)
19. J. Dorfmeister, E. Neher, Isoparametric triple systems with special $\mathbb{Z}$-structure. Algebras Groups Geom. **7**, 21–94 (1990)
20. D. Ferus, H. Karcher, H.F. Münzner, Cliffordalgebren und neue isoparametrische Hyperflächen. Math. Z. **177**, 479–502 (1981)
21. W. Hsiang, H.B. Lawson, Minimal submanifolds of low cohomogeneity. J. Differ. Geom. **5**, 1–38 (1971)
22. S. Immervoll, On the classification of isoparametric hypersurfaces with four distinct principal curvatures in spheres. Ann. Math. (2) **168**, 1011–1024 (2008)
23. R. Miyaoka, The Dorfmeister-Neher theorem on isoparametric hypersurfaces. Osaka J. Math. **46**, 695–715 (2009)
24. R. Miyaoka, G_2-orbits and isoparametric hypersurfaces. Nagoya Math. J. **203**, 175–189 (2011)
25. P. Molino, *Riemannian foliations* (Translated from the French by Grant Cairns). Progress in Mathematics, vol. 73 (Birkhäuser, Boston, 1988)
26. H.F. Münzner, Isoparametrische Hyperflächen in Sphären. Math. Ann. **251**, 57–71 (1980)
27. B. O'Neill, The fundamental equations of a submersion. Michigan Math. J. **13**, 459–469 (1966)
28. H. Ozeki, M. Takeuchi, On some types of isoparametric hypersurfaces in spheres. I. Tôhoku Math. J. (2) **27**, 515–559 (1975)
29. S. Stolz, Multiplicities of Dupin hypersurfaces. Invent. Math. **138**, 253–279 (1999)
30. K. Yano, M. Kon, *CR Submanifolds of Kaehlerian and Sasakian Manifolds*. Progress in Mathematics, vol. 30 (Birkhäuser, Boston, 1983)

On the Hopf Conjecture with Symmetry

Lee Kennard

The classical Gauss–Bonnet theorem relates the curvature of an orientable surface to its Euler characteristic. For example, if the curvature has fixed sign, the Euler characteristic does as well. In 1925, Hopf proved a generalization of the Gauss–Bonnet theorem for even-dimensional hypersurfaces of Euclidean space (see [11]). In the 1930s, Hopf made the following conjecture that bears his name: An even-dimensional Riemannian manifold with positive (resp. nonnegative) sectional curvature has positive (resp. nonnegative) Euler characteristic. For further background on generalizations of the Gauss–Bonnet theorem and Hopf's conjecture, see [2, 18].

In dimensions two and four, the conjecture follows from the Gauss–Bonnet theorem and its generalization to higher dimensions (see [5]). In dimensions six and greater, there is further evidence in the Kähler case (see [3, 12, 13]) but no general proof. Notably, the naive strategy of proving that the Gauss–Bonnet integrand takes on the desired sign cannot work (see [7, 16]). Finally, we mention that the conjecture is satisfied by all manifolds that are known to admit positive (resp. negative) sectional curvature (see Ziller [24]).

As with many constructions of, and classification problems concerning, positively curved manifolds, much progress on the Hopf conjecture has been made in the presence of a large isometry group. For general surveys on this far-reaching research program, see Grove [8], Wilking [23], or Ziller [25].

Since the isometry group of a compact Riemannian manifold is a compact Lie group, one measure of its size is its rank, i.e., the dimension of a maximal torus. In other words, we may consider the situation where a torus T acts isometrically on M, a closed, even-dimensional, positively curved Riemannian manifold. The results of [19, 20] show that $\chi(M) > 0$ if $\dim(T)$ is bounded from below by

L. Kennard (✉)

Department of Mathematics, South Hall, Room 6607, University of California, Santa Barbara, CA 93106-3080

e-mail: kennard@math.ucsb.edu

O. Dearricott et al., *Geometry of Manifolds with Non-negative Sectional Curvature*, Lecture Notes in Mathematics 2110, DOI 10.1007/978-3-319-06373-7__5,
© Springer International Publishing Switzerland 2014

some linear function in the dimension n. For example, the Hopf conjecture holds if $\dim(T) \geq \frac{n-4}{8}$ (for $n \neq 12$), or if $n \equiv_4 0$ and $\dim(T) \geq \frac{n}{10}$. We are able to improve these linear bounds in dimensions divisible by four (see [14]):

Theorem 1. *Let M^n be a closed, positively curved manifold with $n \equiv 0 \mod 4$. If a torus T acts effectively by isometries on M with $\dim(T) \geq 2\log_2(n) - 2$, then $\chi(M) > 0$.*

In fact, we show in [15] that at least some of the odd Betti numbers vanish. The conjecture that all of the odd Betti numbers vanish for even-dimensional, positively curved manifolds would follow from a combination of the Hopf conjecture and another conjecture of Bott (see [9] and the surveys cited above).

Now let G be a compact Lie group. If G is a torus or a one-connected simple Lie group, then the rank and dimension of G satisfy the following, which can be checked case by case (see, for example, [10]): $\operatorname{rank}(G)^2 \geq \frac{8}{31}\dim(G)$, where equality holds for the exceptional Lie group E_8. Moreover, this inequality is preserved under products, finite covers, and finite extensions, hence this inequality holds for all compact Lie groups G. In particular, G has large rank if it has large dimension. Additionally the cohomogeneity $\operatorname{cohom}(G) = \dim(M/G)$ of the action is bounded above by $n - d$ only if $\dim(G) \geq d$. Theorem 1 therefore immediately implies the following:

Corollary 1. *Let M^n be a closed, positively curved manifold with $n \equiv 0 \mod 4$ and isometry group G. If $\dim(G) \geq 16(\log_2 n)^2$ or $\operatorname{cohom}(M) \leq n - 16(\log_2 n)^2$, then $\chi(M) > 0$.*

Previous results in this direction include those in [22], which imply $\chi(M) > 0$ if $\dim(G) \geq 2n - 6$ or $\operatorname{cohom}(M) \leq \sqrt{n/18} - 1$. In small dimensions, a result in [18] concerning low cohomogeneity remains the best known: $\chi(M) > 0$ if $\operatorname{cohom}(M) < 6$.

1 Proof of Theorem 1

We now sketch the proof of Theorem 1 (see [14] for a complete proof). Since $\pi_1(M)$ is finite for compact, positively curved manifolds, and since the universal cover of M admits an effective, isometric torus action of the same rank, it suffices to prove the theorem in the simply connected case.

The goal is to show that every component F of the fixed point set M^T has singly generated rational cohomology. In particular, this would imply that $\chi(M^T) > 0$, as each component of M^T would have vanishing odd Betti numbers. Since $\chi(M) = \chi(M^T)$ (see [6,17]), and since M^T is non-empty by a theorem of Berger, this would imply the theorem.

To accomplish our goal, we fix a component $F \subseteq M^T$. We show there exists a subtorus $H \subseteq T$ and a component $N \subseteq M^H$ which contains F and whose rational cohomology ring is that of a CROSS. Now the T-action on M induces an effective T'-action on N, where T' is the quotient of T by the kernel of the induced action. Moreover F is a component of the fixed point set $N^{T'}$. Results from Smith theory (see [4], for example) imply that F too has the rational cohomology ring of a CROSS.

The difficult part of the argument, therefore, is finding N and proving that it has the desired rational cohomology ring. The first step here is to study the array of fixed point sets M^H, where H runs over subgroups of involutions in T. The goal is to find components N, N_1, and N_2 of three such fixed point sets M^H, M^{H_1}, and M^{H_2}, respectively, such that F, N_1, and N_2 are contained in N, and such that N_1 and N_2 are submanifolds of N which intersect transversely (in N) and have average codimension (in N) at most $\dim(N)/4$. Once this is done, the second step is to prove the following theorem, which says that N has the rational cohomology ring of a CROSS and hence completes the proof of the theorem:

Theorem 2. *Let N^n be a closed, simply connected Riemannian manifold with positive sectional curvature. If there exists a pair of transversely intersecting, totally geodesic submanifolds $N_1^{n-k_1}$ and $N_2^{n-k_2}$ with $n \equiv_4 0$ and $2k_1 + 2k_2 \leq n$, then N has the rational cohomology ring of a sphere or a complex or quaternionic projective space.*

In the setting of Theorem 2, the connectedness theorem of Wilking implies a certain periodicity in the integral cohomology ring of N. In Sect. 3, we describe how we use the action of the Steenrod algebra to refine this periodicity with mod p or rational coefficients. Theorem 2 is a simple consequence of these periodicity statements and Poincaré duality.

2   A Graph of Involutions

To systematize the search for a subgroup $H \subseteq T$ and a component $N \subseteq M^H$ with the desired properties described in the last section, we make the following definition:

Definition 1. Define an abstract graph Γ by declaring the following:

- An involution $\sigma \in T$ is a vertex in Γ if $\mathrm{cod}\, F(\sigma) \equiv_4 0$ and $\dim \ker(T|_{F(\sigma)}) \leq 1$.
- Distinct vertices σ and σ' in Γ are connected by an edge if $F(\sigma) \cap F(\sigma')$ is not transverse.

Here $F(\sigma)$ denotes the F-component of M^σ, the fixed point set of σ, and $\dim \ker(T|_{F(\sigma)})$ denotes the dimension of the kernel of the induced T-action on $F(\sigma)$. If, for example, Γ has four graph-theoretic components, then we have four involutions σ_i whose fixed point sets have F-components $F(\sigma_i)$ that mutually transversely intersect. Letting $k_i = \mathrm{cod}\, F(\sigma_i)$ and supposing $k_1 \leq \cdots \leq k_4$, we have

$$2k_1 + 2k_2 \leq \sum k_i = \mathrm{cod}(F(\sigma_1) \cap \cdots \cap F(\sigma_4)) \leq n,$$

so Theorem 2 implies M has the rational cohomology ring a CROSS. In this case, the desired N is, in fact, M.

One now considers the cases where Γ has three components, two components, or one component. In each case, we attempt to show the existence of such a submanifold N. In the cases where this attempt fails, we find another submanifold P and a quotient T' of the torus T acting effectively on P with $F \subseteq P \subseteq M$ and $\dim(T') \geq 2 \log_2(\dim P) - 2$. In this case, we can conclude by induction that the result holds for P and, in particular, that F (being a component of $P^{T'}$) has the rational cohomology ring of a CROSS.

3 Proof of Theorem 2

We first recall Wilking's connectedness theorem:

Theorem 3 (Connectedness Theorem, [21]). *Suppose M^n is a closed Riemannian manifold with positive sectional curvature.*

(1) *If N^{n-k} is connected and totally geodesic in M, then $N \hookrightarrow M$ is $(n - 2k + 1)$-connected.*
(2) *If $N_1^{n-k_1}$ and $N_2^{n-k_2}$ are totally geodesic with $k_1 \leq k_2$, then $N_1 \cap N_2 \hookrightarrow N_2$ is $(n - k_1 - k_2)$-connected.*

Recall an inclusion $N \hookrightarrow M$ is called r-connected if $\pi_i(M, N) = 0$ for all $i \leq r$. The following is a topological consequence of highly connected inclusions (see [21]):

Theorem 4. *Let M^n and N^{n-k} be closed, orientable manifolds. If $N \hookrightarrow M$ is $n - k - l$ connected with $n - k - 2l > 0$, then there exists $e \in H^k(M; \mathbb{Z})$ such that the maps $H^i(M; \mathbb{Z}) \to H^{i+k}(M; \mathbb{Z})$ given by $x \mapsto ex$ are surjective for $l \leq i < n - k - l$ and injective for $l < i \leq n - k - l$.*

In the case $l = 0$, we observe that $H^*(M; \mathbb{Z})$ is k-periodic up to degree n, according to the following definition:

Definition 2. For a topological space M and a ring R, we say that $x \in H^k(M; R)$ induces periodicity up to degree n if the maps $H^i(M; R) \to H^{i+k}(M; R)$ given by multiplication by x are surjective for $0 \leq i < n - k$ and injective for $0 < i \leq n - k$. In this case, we also say that $H^*(M; R)$ is k-periodic up to degree n.

We now apply these results to the situation of Theorem 2. We have totally geodesic, transversely intersecting submanifolds $N_1^{n-k_1}$ and $N_2^{n-k_2}$ with $2k_1 + 2k_2 \leq n$. We may assume $k_1 \leq k_2$. It follows that $H^*(N_2; \mathbb{Z})$ is k_1-periodic up to degree $\dim(N_2)$. The Bockstein sequence implies the same property with $\mathbb{Z}_2$- and $\mathbb{Z}_3$-coefficients. One now applies the following lemma, which is a direct generalization of Adem's theorem on singly generated cohomology rings (see [1]):

Theorem 5. *Suppose M is a CW complex, and let p be an odd prime.*

- *If $H^*(M;\mathbb{Z}_2)$ is k-periodic up do degree n with $2k \leq n$, then it is 2^r-periodic up to degree n for some $r \geq 0$ such that $2^r | k$.*
- *If $H^*(M;\mathbb{Z}_p)$ is k-periodic up to degree n with $pk \leq n$, then it is $2\lambda p^s$-periodic up to degree n for some $s \geq 0$ and $\lambda \geq 1$ such that λ divides $p - 1$ and $2\lambda p^s$ divides k.*

The proof uses the Steenrod cohomology operations, following the basic strategy of Adem. Namely, one observes that elements Sq^k of the Steenrod algebra decompose when k is not a power of 2. Since the $\mathbb{Z}_2$-cohomology ring of M is a module over the Steenrod algebra, these decompositions (i.e., these relations in the Steenrod algebra) impose relations in $H^*(M;\mathbb{Z}_2)$. The calculation is involved; however, the upshot is that the minimal degree (homogeneous) element of $H^*(M;\mathbb{Z}_2)$ which induces periodicity has degree equal to a power of 2. The proof for odd primes is similar.

Finally, one translates the conclusion of this lemma into conclusions about the rational cohomology ring of M; namely, one concludes that $H^*(M;\mathbb{Q})$ is periodic with two different periods, one of which is 2^r and the other $4 \cdot 3^s$. A simple argument shows that $H^*(M;\mathbb{Q})$ is therefore periodic with period $\gcd(2^r, 4\cdot 3^s) = 4$. Combining this with Poincaré duality and the fact that $n \equiv_4 0$, we conclude Theorem 2.

References

1. J. Adem, The iteration of the Steenrod squares in algebraic topology. Proc. Natl. Acad. Sci. USA **38**, 720–726 (1952)
2. M. Berger, *Riemannian Geometry During the Second Half of the Twentieth Century*. University Lecture Series, vol. 17 (American Mathematical Society, Providence, 1998)
3. R.L. Bishop, S.I. Goldberg, Some implications of the generalized gauss-bonnet theorem. Trans. Am. Math. Soc. **112**(3), 508–535 (1964)
4. G.E. Bredon, Introduction to compact transformation groups (Academic, New York/London, 1972)
5. S.-S. Chern, On the curvature and characteristic classes of a Riemannian manifold. Abh. Math. Semin. Univ. Hambg. **20**, 117–126 (1956)
6. P.E. Conner, On the action of the circle group. Michigan Math. J. **4**, 241–247 (1957)
7. R. Geroch, Positive sectional curvature does not imply positive Gauss–Bonnet integrand. Proc. Am. Math. Soc. **54**, 267–270 (1976)
8. K. Grove, Developments around positive sectional curvature, in *Geometry, Analysis, and Algebraic Geometry: Forty Years of the Journal of Differential Geometry*. Surveys in Differential Geometry, vol. 13 (International Press, Somerville, MA, 2009), pp. 117–133
9. K. Grove, S. Halperin, Contributions of rational homotopy theory to global problems in geometry. Publ. Math. IHES **56**(1), 171–177 (1982)
10. S. Helgason, *Differential Geometry, Lie Groups, and Symmetric Spaces* (American Mathematical Society, Providence, 2001)
11. H. Hopf, Über die Curvatura integra geschlossener Hyperflächen. Math. Ann. **95**, 340–367 (1925)

12. C.-C. Hsiung, K.M. Shiskowski, Euler–Poincaré characteristic and higher order sectional curvature. I. Trans. Am. Math. Soc. **305**, 113–128 (1988)
13. D.L. Johnson, Curvature nd Euler characteristic for six-dimensional Kähler manifolds. Ill. J. Math. **28**(4), 654–675 (1984)
14. L. Kennard, On the Hopf conjecture with symmetry. Geom. Topol. **17**, 563–593 (2013)
15. L. Kennard, Positively curved Riemannian metrics with logarithmic symmetry rank bounds. Comment. Math. Helv. arXiv:1209.4627v1 [math.DG] (to appear)
16. P.F. Klembeck, On Geroch's counterexample to the algebraic Hopf conjecture. Proc. Am. Math. Soc. **59**(2), 334–336 (1976)
17. S. Kobayashi, Fixed points of isometries. Nagoya Math. J. **13**, 63–68 (1958)
18. T. Püttmann, C. Searle, The Hopf conjecture for manifolds with low cohomogeneity or high symmetry rank. Proc. Am. Math. Soc. **130**(1), 163–166 (2001)
19. X. Rong, X. Su, The Hopf conjecture for manifolds with abelian group actions. Commun. Contemp. Math. **7**, 121–136 (2005)
20. X. Su, Y. Wang, The Hopf conjecture for positively curved manifolds with discrete abelian group actions. Differ. Geom. Appl. **26**(3), 313–322 (2008)
21. B. Wilking, Torus actions on manifolds of positive sectional curvature. Acta Math. **191**(2), 259–297 (2003)
22. B. Wilking, Positively curved manifolds with symmetry. Ann. Math. **163**, 607–668 (2006)
23. B. Wilking, Nonnegatively and positively curved manifolds, in *Metric and Comparison Geometry*. Surveys in Differential Geometry, vol. 11 (International Press, Somerville, MA, 2007), pp. 25–62
24. W. Ziller, Examples of Riemannian manifolds with non-negative sectional curvature, in *Metric and Comparison Geometry*, Surveys in Differential Geometry, vol. 11 (International Press, Somerville, MA, 2007), pp. 63–102
25. W. Ziller, *Riemannian Manifolds with Positive Sectional Curvature*. Lecture Notes in Mathematics, vol. 2110 (2014, in press)

An Introduction to Exterior Differential Systems

Gregor Weingart

1 Introduction

In a rather precise sense the study of exterior differential systems is equivalent to the study of partial differential equations in the language of differential forms. Although the change of language from partial derivatives to differential forms may appear quite surprising nowadays, the concept developed in a time, when differential forms offered the most convenient way to do calculations in differential geometry without reference to local coordinates. Orthodox differential geometry has caught up in the meantime and this initial advantage has been lost to a large extent. Nevertheless exterior differential systems are still an interesting topic to study today, because they unify language, method and results for several different kinds of partial differential equations.

Studying partial differential equations in the unified framework of exterior differential systems allows us to take advantage of the beautiful theory of Cartan–Kähler about analytical solutions to analytical exterior differential systems, which is the central topic of these notes. In essence the theory of Cartan–Kähler replaces actual solutions to a given exterior differential system by formal power series solutions, an idea already used successfully in the predecessor of the Cartan–Kähler theory, the theorem of Cauchy–Kovalevskaya. Calculating the terms of a formal power series solution term by term reduces a complicated partial differential equation effectively to the problem of solving an inhomogeneous linear equation at each order of differentiation. Although this reduction to linear algebra is very appealing, a rather unpleasant problem arises in this approach: Inhomogeneous linear equations need not have a solution in general. An exterior differential system

G. Weingart (✉)

Instituto de Matemáticas (Cuernavaca), Universidad Nacional Autónoma de México, Avenida Universidad s/n, Colonia Lomas de Chamilpa, 62210 Cuernavaca, Morelos, Mexico

e-mail: gw@matcuer.unam.mx

O. Dearricott et al., *Geometry of Manifolds with Non-negative Sectional Curvature*, Lecture Notes in Mathematics 2110, DOI 10.1007/978-3-319-06373-7_6, © Springer International Publishing Switzerland 2014

is called formally integrable, if all the inhomogeneous linear equations encountered at different orders of differentiation in calculating a formal power series solution allow solutions.

A partial answer to the problem of verifying formal integrability is given by the Spencer cohomology $H^{\bullet,\circ}(\mathscr{A})$ associated to an exterior differential system. Spencer cohomology tells us that we can always solve the inhomogeneous linear equation at differentiation order k provided the Spencer cohomology space $H^{k,2}(\mathscr{A}) = \{0\}$ vanishes, on the other hand we may still be able to solve the inhomogeneous linear equation at differentiation order k in case $H^{k,2}(\mathscr{A}) \neq \{0\}$. Despite being only a partial answer Spencer cohomology is a very useful tool in practice, because it is usually much easier to calculate the Spencer cohomology of an exterior differential system than to work with formal power series solutions directly. Moreover the algebraic roots of Spencer cohomology in commutative algebra ensure that only a finite number of problematic differentiation orders k exist with $H^{k,2}(\mathscr{A}) \neq \{0\}$.

Among the several excellent text books on the exterior differential systems let us point out the book [1], which can be seen as an authoritative reference on the topic. In writing these introductory notes I wanted to complement [1] with its numerous examples with a concise exposition of the theory of exterior differential systems and its relationship to Spencer cohomology. Moreover I wanted to discuss some of the points in more detail, which are treated rather superficially in the existing literature, say, for example, the distinction between the reduced symbol comodule $\mathscr{A}$ and the symbol comodule $\mathscr{R}$ and the precise definition of the Cartan character of an exterior differential system. In this way I hope that even the reader well acquainted with the Cartan–Kähler theory of exterior differential systems will find these introductory notes worth reading, the more so a reader looking for a panoramic view on the formal theory of partial differential equations.

Grosso modo these notes on exterior differential systems are structured into three essentially independent parts. In Sect. 2 we will construct the contact systems on three different kinds of jet bundles based on the notion of a canonical contact form. Sections 3 and 4 are dedicated to a detailed study of Spencer cohomology: Sect. 3 focuses on its general algebraic properties, whereas Sect. 4 links Spencer cohomology to three classical statements about partial differential equations. Last but not least the theory of Cartan–Kähler is the topic of Sect. 5, in which we will discuss the general setup of exterior differential system and sketch a proof of the Theorem of Cartan–Kähler about the analytical solutions to partial differential equations with analytical coefficients.

2 Jets and Contact Systems

In essence jets and jet bundles are introduced to geometrize differential operators and/or partial differential equations, splitting their study into algebraic and analytical problems. The resulting hybrid approach is the leitmotif of the formal theory of partial differential equations, which becomes the theory of exterior differential systems when formulated in the language of differential forms and exterior calculus. In this initial section we discuss the geometry of jets and jets bundle focusing on the construction of the contact systems on three different kinds of jets, the jets of smooth maps, the jets of sections of fiber bundles and the jets of submanifolds. In Sects. 3 and 4 we will discuss the algebraic aspects of exterior differential systems, before these two strands of the formal theory of partial differential equations are united into the theory of Cartan–Kähler in Sect. 5.

In order to begin our study of jets let us introduce an equivalence relation on the set of smooth maps from $\mathbb{R}^m$ to $\mathbb{R}^n$ by declaring two maps $f : \mathbb{R}^m \longrightarrow \mathbb{R}^n$ and $g : \mathbb{R}^m \longrightarrow \mathbb{R}^n$ to be in contact $f \sim_{k,x} g$ in a point $x \in \mathbb{R}^m$ to order $k \in \mathbb{N}_0$, if and only if there exists a constant $C > 0$ such that the difference $f - g$ is bounded by the estimate

$$| f(\xi) - g(\xi) | \leq C | \xi - x |^{k+1}$$

for all ξ in a compact neighborhood of x. Apparently this definition depends on the choice of norms $| \cdot |$ on $\mathbb{R}^m$ and $\mathbb{R}^n$ and compact neighborhoods, different choices of norms or neighborhood however only affect the constant, not the existence of the estimate itself. The equivalence class of a given function $f : \mathbb{R}^m \longrightarrow \mathbb{R}^n$ under contact $\sim_{k,x}$ to the order k is called the kth order jet of f in x written $\mathrm{jet}_x^k f$. According to Taylor's Theorem smooth maps f and g are in contact in x to the order k, if and only if all their partial derivatives

$$\frac{\partial^{|A|} f}{\partial x^A}(x) = \frac{\partial^{|A|} g}{\partial x^A}(x) \qquad |A| \leq k \tag{1}$$

up to order k agree in x. In this case there exists a unique polynomial ψ of degree at most k on $\mathbb{R}^m$ with values in $\mathbb{R}^n$, which is in contact to both f and g to the order k in x. Thus the set $\mathrm{Jet}_x^k(\mathbb{R}^m, \mathbb{R}^n)$ of all kth order jets in x of smooth maps from $\mathbb{R}^m$ to $\mathbb{R}^n$ is in bijection

$$\mathrm{Sym}^{\leq k} \mathbb{R}^{m*} \otimes \mathbb{R}^n \xrightarrow{\cong} \mathrm{Jet}_x^k(\mathbb{R}^m, \mathbb{R}^n), \qquad \psi \longmapsto \mathrm{jet}_x^k \psi(\cdot - x) \tag{2}$$

with the vector space $\mathrm{Sym}^{\leq k} \mathbb{R}^{m*} \otimes \mathbb{R}^n$ of polynomials of degree less than or equal to k on $\mathbb{R}^m$ with values in $\mathbb{R}^n$. The proper reason for including the translation $\mathrm{jet}_x^k \psi(\cdot - x)$ instead of the seemingly simpler $\mathrm{jet}_x^k \psi(\cdot)$ is that in this way the jet projections

$$\mathrm{pr}: \quad \mathrm{Jet}^k(\mathbb{R}^m, \mathbb{R}^n) \longrightarrow \mathrm{Jet}^{\tilde{k}}(\mathbb{R}^m, \mathbb{R}^n), \qquad \mathrm{jet}_x^k f \longmapsto \mathrm{jet}_x^{\tilde{k}} f$$

defined for all $k \geq \tilde{k} \geq 0$ become just the standard projections for polynomials

$$\mathrm{pr}: \quad \mathbb{R}^m \times \mathrm{Sym}^{\leq k} \mathbb{R}^{m*} \otimes \mathbb{R}^n \longrightarrow \mathbb{R}^m \times \mathrm{Sym}^{\leq \tilde{k}} \mathbb{R}^{m*} \otimes \mathbb{R}^n,$$

$$(x, \psi) \longmapsto (x, \mathrm{pr}\,\psi)$$

forgetting all homogeneous components of ψ of degree larger than $\tilde{k}$. Although the jet projections pr are defined in this way for all $k \geq \tilde{k} \geq 0$, only the two special cases $\tilde{k} = k - 1$ and $\tilde{k} = 0$ are of any practical importance. Singling out the latter jet projection in notation, which becomes the evaluation map under the identification $\mathrm{Jet}^0(\mathbb{R}^m, \mathbb{R}^n) = \mathbb{R}^m \times \mathbb{R}^n$

$$\mathrm{ev}: \quad \mathrm{Jet}^k(\mathbb{R}^m, \mathbb{R}^n) \longrightarrow \mathbb{R}^m \times \mathbb{R}^n, \qquad \mathrm{jet}^k_x f \longmapsto (x, f(x))$$

we may say that the notation pr virtually always refers to the jet projection with $\tilde{k} = k - 1$.

In order to "globalize" the current definition of jets to smooth maps between manifolds it is very convenient to observe that the second definition (1) of kth order contact together with the general chain rule for iterated partial derivatives of compositions provide us with a well-defined jet composition map on the fibered product of two jet spaces

$$\mathrm{Jet}^k(\mathbb{R}^m, \mathbb{R}^n) \times_{\mathbb{R}^m} \mathrm{Jet}^k(\mathbb{R}^l, \mathbb{R}^m) \longrightarrow \mathrm{Jet}^k(\mathbb{R}^l, \mathbb{R}^n),$$

$$(\mathrm{jet}^k_y f, \mathrm{jet}^k_x g) \longmapsto \mathrm{jet}^k_x(f \circ g)$$

provided the source y of $\mathrm{jet}^k_y f$ agrees $y = g(x)$ with the target $g(x)$ of $\mathrm{jet}^k_x g$, for this reason the composition map is only defined on the fibered product $\times_{\mathbb{R}^m}$. Defining the kth order jet of a smooth map $f : M \longrightarrow N$ between manifolds M and N with respect to local coordinates x and y about $p \in M$ and $f(p) \in N$ simply as $\mathrm{jet}^k_{x(p)}(y \circ f \circ x^{-1})$ we obtain

$$\mathrm{jet}^k_{\tilde{x}(p)}(\tilde{y} \circ f \circ \tilde{x}^{-1}) = \mathrm{jet}^k_{y(f(p))}(\tilde{y} \circ y^{-1}) \circ \mathrm{jet}^k_{x(p)}(y \circ f \circ x^{-1}) \circ \mathrm{jet}^k_{\tilde{x}(p)}(x \circ \tilde{x}^{-1})$$

for every other choice $\tilde{x}$ and $\tilde{y}$ of the local coordinates involved. With the jet composition map being well-defined we conclude that for two smooth maps $f : M \longrightarrow N$ and $g : M \longrightarrow N$ satisfying $f(p) = g(p)$ the validity of an equality of jets of the form

$$\mathrm{jet}^k_{x(p)}(y \circ f \circ x^{-1}) \;=\; \mathrm{jet}^k_{x(p)}(y \circ g \circ x^{-1})$$

is independent of the local coordinates x and y of M and N employed in its formulation:

Definition 2.1 (Jets of Smooth Maps). Two smooth maps $f : M \longrightarrow N$ and $g : M \longrightarrow N$ between manifolds M and N are said to be in contact $f \sim_{k,p} g$ in a point $p \in M$ to the order $k \in \mathbb{N}_0$ provided $f(p) = g(p)$ and

$$\mathrm{jet}^k_{x(p)}(y \circ f \circ x^{-1}) \;=\; \mathrm{jet}^k_{x(p)}(y \circ g \circ x^{-1})$$

for some and hence every local coordinates x of M and y of N about p and $f(p) = g(p)$ respectively. The set of all equivalence classes of smooth maps from M to N in contact

$$\mathrm{Jet}^k(M, N) := \{ \mathrm{jet}^k_p f \mid f : M \longrightarrow N \text{ smooth} \}$$

to order k is a fiber bundle over M with projection $\pi : \mathrm{Jet}^k(M, N) \longrightarrow M$, $\mathrm{jet}^k_p f \longmapsto p$, and over $M \times N$ under the evaluation ev $: \mathrm{Jet}^k(M, N) \longrightarrow M \times N$, $\mathrm{jet}^k_p f \longmapsto (p, f(p))$.

It should be hardly surprising to see that the preceding definition reduces the contact relation for smooth maps between manifolds via a choice of local coordinates to the contact relation for smooth maps between Euclidean spaces, after all the definition of smoothness of maps between manifolds employs local coordinates in exactly the same way. This very observation implies directly that the jet composition map extends to:

$$\mathrm{Jet}^k(M, N) \times_M \mathrm{Jet}^k(L, M) \longrightarrow \mathrm{Jet}^k(L, N),$$

$$(\mathrm{jet}^k_{g(p)} f, \mathrm{jet}^k_p g) \longmapsto \mathrm{jet}^k_p(f \circ g) \tag{3}$$

Using this generalized jet composition map we may define a second kind of jet bundles, the jet bundles of sections of fiber bundles over a manifold M. Consider therefore a smooth fiber bundle $\mathscr{F}M$ over a manifold M with projection $\pi : \mathscr{F}M \longrightarrow M$. The jet composition map induces a well-defined map from the space of jets of smooth maps $M \longrightarrow \mathscr{F}M$ to

$$\mathrm{Jet}^k(M, \mathscr{F}M) \longrightarrow \mathrm{Jet}^k(M, M), \qquad \mathrm{jet}^k_p f \longmapsto \mathrm{jet}^k_{f(p)} \pi \circ \mathrm{jet}^k_p f$$

which sends the kth order jet of a section $f \in \Gamma(\mathscr{F}M)$ to $\mathrm{jet}^k_p(\pi \circ f) = \mathrm{jet}^k_p \mathrm{id}_M$. In turn we define the bundle of kth order jets of sections of $\mathscr{F}M$ as the following submanifold

$$\mathrm{Jet}^k \mathscr{F}M := \{\mathrm{jet}^k_p f \mid \mathrm{jet}^k_{f(p)} \pi \circ \mathrm{jet}^k_p f = \mathrm{jet}^k_p \mathrm{id}_M \} \subset \mathrm{Jet}^k(M, \mathscr{F}M)$$

of $\mathrm{Jet}^k(M, \mathscr{F}M)$ with the induced bundle projection $\pi : \mathrm{Jet}^k \mathscr{F}M \longrightarrow M$, $\mathrm{jet}^k_p f \longmapsto p$. The jet projections pr $: \mathrm{Jet}^k(M, \mathscr{F}M) \longrightarrow \mathrm{Jet}^{\tilde{k}}(M, \mathscr{F}M)$ clearly restrict to jet projections

$$\mathrm{pr} : \quad \mathrm{Jet}^k \mathscr{F}M \longrightarrow \mathrm{Jet}^{\tilde{k}} \mathscr{F}M, \qquad \mathrm{jet}^k_p f \longmapsto \mathrm{jet}^{\tilde{k}}_p f$$

for all $k \geq \tilde{k} \geq 0$ with the special case ev $: \mathrm{Jet}^k \mathscr{F}M \longrightarrow \mathscr{F}M$, $\mathrm{jet}^k_p f \longmapsto f(p)$, singled out in notation to prevent ambiguities as discussed before. The jet projections pr and ev turn the jet bundles $\mathrm{Jet}^k \mathscr{F}M$, $k \geq 0$, associated to $\mathscr{F}M$ into a tower of fiber bundles over M

$$\ldots \xrightarrow{\mathrm{pr}} \mathrm{Jet}^3 \mathscr{F}M \xrightarrow{\mathrm{pr}} \mathrm{Jet}^2 \mathscr{F}M \xrightarrow{\mathrm{pr}} \mathrm{Jet}^1 \mathscr{F}M \xrightarrow{\mathrm{ev}} \mathscr{F}M \xrightarrow{\pi} M \qquad (4)$$

in the sense that every manifold in this tower is a smooth fiber bundle over every manifold further down under the appropriate projection. It should be noted that the two types of jet bundles we have defined so far are very closely related, in fact we could have based all our considerations on the notion of jet bundles of sections. In this approach the jet bundle of smooth maps $M \longrightarrow N$ becomes the bundle $\mathrm{Jet}^k(M, N) := \mathrm{Jet}^k(N \times M)$ of jets of sections of the trivial N-bundle $N \times M$ over M, clearly a section $n : M \longrightarrow N \times M$ is essentially the same thing as the smooth map $\pi_N \circ n : M \longrightarrow N$.

The geometry of the tower of jets bundles (4) associated to a fiber bundle $\mathscr{F}M$ is governed by the structural property that all the fibers of the jet projection pr : $\mathrm{Jet}^k \mathscr{F}M \longrightarrow \mathrm{Jet}^{k-1} \mathscr{F}M$ are naturally affine spaces, more precisely the fiber $\mathrm{pr}^{-1}(\mathrm{jet}_p^{k-1} f) \subset \mathrm{Jet}^k \mathscr{F}M$ over $\mathrm{jet}_p^{k-1} f$ is an affine space modelled on the vector space $\mathrm{Sym}^k T_p^* M \otimes \mathrm{Vert}_{f(p)} \mathscr{F}M$ for all $k \geq 1$. This additional affine structure is of the utmost importance for the formal theory of partial differential equations, because it reduces non-linear partial differential equations effectively to problems concerning affine linear maps, traditionally called symbol maps, which are significantly easier to deal with. In particular the resulting symbolic calculus allows us to climb up the tower (4) recursively one step at a time like we will do in the proof of the Theorem of Cartan–Kähler in Sect. 5.

Ironically enough the construction of the canonical affine structure on the fibers of the jet projection pr invariably involves a non-canonical choice, nevertheless this ambiguity can be reduced significantly by using the concept of anchored coordinate charts. A coordinate chart of a smooth manifold M anchored in a point $p \in M$ is a smooth map $\Phi^M : T_p M \longrightarrow M$ defined at least in some open neighborhood of $0 \in T_p M$ such that $\Phi^M(0) = p$ and such that the differential of Φ^M in the point $0 \in T_p M$ agrees with the identity of $T_p M$:

$$\Phi_{*,p}^M : \quad T_p M \cong T_0(T_p M) \longrightarrow T_p M, \qquad X \longmapsto \frac{d}{dt}\Big|_0 \Phi^M(tX) \overset{!}{=} X \qquad (5)$$

Evidently the concept of coordinate charts of a manifold M anchored in a point $p \in M$ reflects the basic properties of the standard exponential maps studied in affine differential geometry. For every smooth fiber bundle $\pi : \mathscr{F}M \longrightarrow M$ over a manifold M and for every given $f_0 \in \mathscr{F}_{p_0} M$ in the fiber over a point $p_0 \in M$ we can easily find anchored coordinate charts $\Phi^M : T_{p_0} M \longrightarrow M$ and $\Phi^{\mathscr{F}} : T_{f_0} \mathscr{F}M \longrightarrow \mathscr{F}M$ anchored in p_0 and f_0 such that

$$
\begin{array}{ccc}
T_{f_0}\mathscr{F}M & \xrightarrow{\ \Phi^{\mathscr{F}}\ } & \mathscr{F}M \\
{\scriptstyle \pi_{*,f_0}}\downarrow & & \downarrow{\scriptstyle \pi} \\
T_{p_0}M & \xrightarrow{\ \Phi^{M}\ } & M
\end{array}
$$

$$(6)$$

commutes wherever defined. Using such a pair of anchored coordinate charts we define

$$
\mathrm{jet}^k_{p_0} f \ + \ \Delta f \ := \ \mathrm{jet}^k_{p_0}\Big(p \ \longmapsto \ \Phi^{\mathscr{F}}\Big[\, \Phi^{\mathscr{F}-1}(f(p)) \ + \ \Delta f\,(\Phi^{M-1}(p)) \, \Big] \Big)
$$

$$(7)$$

for all $\mathrm{jet}^k_{p_0} f \in \mathrm{Jet}^k_{p_0}\mathscr{F}M$ evaluating to $f(p_0) = f_0$ and all $\Delta f \in \mathrm{Sym}^k T^*_{p_0}M \otimes$ $\mathrm{Vert}_{f_0}\mathscr{F}M$ considered as homogeneous polynomials of degree k on $T_{p_0}M$ with values in the subspace $\mathrm{Vert}_{f_0}\mathscr{F}M \subset T_{f_0}\mathscr{F}M$. The commutativity of the diagram (6) ensures that the expression

$$
M \ \longrightarrow \ \mathscr{F}M, \qquad p \ \longmapsto \ \Phi^{\mathscr{F}}\Big[\, \Phi^{\mathscr{F}-1}(f(p)) \ + \ \Delta f\,(\Phi^{M-1}(p)) \, \Big]
$$

results in a locally defined section of $\mathscr{F}M$. Needless to say this section depends on the pair of anchored coordinate charts $\Phi^{\mathscr{F}}$ and Φ^{M} used in its definition. Different choices for $\Phi^{\mathscr{F}}$ and Φ^{M} however will always lead to local sections in contact in the point $p_0 \in M$ up to order k, because Δf considered as a homogeneous polynomial of degree k on $T_{p_0}M$ has all its partial derivatives of order less than k vanishing in $0 \in T_{p_0}M$.

Taking partial derivatives the first time converts a composition like $\Phi^{\mathscr{F}} \circ \Delta f \circ \Phi^{M-1}$ into a sum of products of partial derivatives of $\Phi^{\mathscr{F}}$, Δf and Φ^{M}, all subsequent partial derivatives are then calculated using the Leibniz rule for products. Hence all partial derivatives of the composition $\Phi^{\mathscr{F}} \circ \Delta f \circ \Phi^{M-1}$ of order less than k in p_0 vanish and the only the non-zero contributions to partial derivatives of order k arise from choosing the critical factor Δf in *all* subsequent applications of the Leibniz rule. The net result is a sum of products of partial derivatives of Δf of order k in $0 \in T_{p_0}M$ with only first order partial derivatives of $\Phi^{\mathscr{F}}$ and Φ^{M} in $0 \in T_{f_0}\mathscr{F}M$ and $0 \in T_{p_0}M$. Exactly these first order derivatives however are fixed by the characteristic property (5) of anchored coordinate charts!

Certainly a lot of work needs to be done to make the argument sketched in the preceding paragraph precise, nevertheless we skip this problem for the time being and conclude that the addition (7) does not dependent on the choice of the pair of anchored coordinate charts $\Phi^{\mathscr{F}}$ and Φ^{M} used in its definition. Moreover the addition satisfies the axioms of a group action for the additive group underlying the vector space $\mathrm{Sym}^k T^*_{p_0}M \otimes \mathrm{Vert}_{f_0}\mathscr{F}M$:

$$
\mathrm{jet}^k_{p_0} f + 0 \ = \ \mathrm{jet}^k_{p_0} f \qquad (\,\mathrm{jet}^k_{p_0} f + \Delta f\,) + \Delta \tilde{f} \ = \ \mathrm{jet}^k_{p_0} f + (\Delta f + \Delta \tilde{f}\,)
$$

Both verifications necessary are essentially trivial, but involve rather bombastic formulas better omitted. Summarizing all our considerations on this topic we have constructed for all $k \geq 1$ a canonical vector group bundle action on $\mathrm{Jet}^k \mathscr{F} M$ fibered over $\mathscr{F} M$

$$+ : \quad \mathrm{Jet}^k \mathscr{F} M \times_{\mathscr{F} M} (\mathrm{Sym}^k T^* M \otimes \mathrm{Vert}\, \mathscr{F} M) \longrightarrow \mathrm{Jet}^k \mathscr{F} M$$

in the sense that the fiber of the vector bundle over $f_0 \in \mathscr{F} M$ acts on subset of jets evaluating to f_0. In the standard jet coordinates on $\mathrm{Jet}^k \mathscr{F} M$ introduced later on it is relatively easy to verify that the addition $+$ is a natural affine space structure on the fibers of the jet projection $\mathrm{pr} : \mathrm{Jet}^k \mathscr{F} M \longrightarrow \mathrm{Jet}^{k-1} \mathscr{F} M$ in the sense that it acts simply transitively on each fiber.

Without doubt the most important use of jets is to provide us with a concise definition of the intuitive notions of (non-linear) differential operators and partial differential equations. In particular the geometrization of partial differential equations brought about by jets can be used to reduce all possible kinds of partial differential equations to a single standard normal form, namely an exterior differential system. Before discussing this point let us point out very briefly that the jet bundles $\mathrm{Jet}^k FM$ of a vector bundle FM over a manifold M are naturally vector bundles again under the obvious choice of scalar multiplication and addition

$$\lambda \, \mathrm{jet}^k_p f \ := \ \mathrm{jet}^k_p(\lambda f) \qquad\qquad \mathrm{jet}^k_p f_1 + \mathrm{jet}^k_p f_2 \ := \ \mathrm{jet}^k_p(f_1 + f_2)$$

moreover the jet bundles $\mathrm{Jet}^k FM$ are also bundles of free modules over the algebra bundle $\mathrm{Jet}^k \mathbb{R} M$ of smooth k-jets of functions. Last but not least the vector bundle structure on $\mathrm{Jet}^k FM$ can be used for an alternative construction of the affine space structure on the fibers of the jet projection $\mathrm{pr} : \mathrm{Jet}^k FM \longrightarrow \mathrm{Jet}^{k-1} FM$. In this alternative construction the addition (7) is mediated by a canonical inclusion of vector bundles

$$\iota : \quad \mathrm{Sym}^k T^* M \otimes FM \longrightarrow \mathrm{Jet}^k FM, \qquad \psi_p \otimes f_p \longrightarrow \mathrm{jet}^k_p(\psi\, f) \quad (8)$$

so that $\mathrm{jet}^k_p f + \Delta f$ can be interpreted simply as the sum of $\mathrm{jet}^k_p f$ and $\iota(\Delta f)$ in the vector space $\mathrm{Jet}^k_p FM$, note that $\mathrm{Vert}_{f_0} FM \cong F_{p_0} M$ are canonically isomorphic for a vector bundle.

Definition 2.2 (Non-linear Partial Differential Equations). A smooth non-linear differential operator of order $k \geq 0$ from sections of a fiber bundle $\mathscr{F} M$ over M to sections of another fiber bundle $\mathscr{E} M$ is a map $D : \Gamma(\mathscr{F} M) \longrightarrow \Gamma(\mathscr{E} M)$ between the sets of locally defined sections such that the value $(Df)(p) \in \mathscr{E}_p M$ of the image of $f \in \Gamma(\mathscr{F} M)$ in a point $p \in M$ depends only on $\mathrm{jet}^k_p f \in \mathrm{Jet}^k_p \mathscr{F} M$. In particular D induces a well-defined smooth map of fiber bundles over M called the total symbol of D:

$$\sigma_D^{\mathrm{total}} : \quad \mathrm{Jet}^k \mathscr{F} M \longrightarrow \mathscr{E} M, \qquad \mathrm{jet}^k_p f \longmapsto (D f)(p)$$

A non-linear partial differential equation for local sections $f \in \Gamma(\mathscr{F}M)$ is an equation of the form $Df = *$ with a distinguished global section $* \in \Gamma(\mathscr{E}M)$ of the target bundle.

This notion of non-linear partial differential equations may not be the most general one, nevertheless it is sufficiently ample to illustrate the use of jets and comprises the important subclass of linear partial differential equations. Naturally enough a differential operator D of order $k \geq 0$ is called a linear differential operator provided both the source and target bundle FM and EM are vector bundles over M and $D : \Gamma(FM) \longrightarrow \Gamma(EM)$ is an $\mathbb{R}$-linear map between the vector spaces of sections, equivalently we may ask for its total symbol $\sigma_D^{\mathrm{total}} : \mathrm{Jet}^k FM \longrightarrow EM$ to be a homomorphism of vector bundles over M. A linear partial differential equation for sections $f \in \Gamma(FM)$ is in turn an equation of the form $Df = 0$ with a linear operator and the distinguished zero section $0 \in \Gamma(EM)$. Given now a partial differential equation $Df = *$ of order $k \geq 1$ we may "solve" the equation algebraically

$$\mathrm{Eq}_p^k M := \{ \mathrm{jet}_p^k f \in \mathrm{Jet}_p^k \mathscr{F}M \mid \sigma_D^{\mathrm{total}}(\mathrm{jet}_p^k f) = *_p \} \subset \mathrm{Jet}_p^k \mathscr{F}M$$

in terms of jets in every point $p \in M$. Evidently a local section $f \in \Gamma(\mathscr{F}M)$ is a solution to the partial differential equation $Df = *$, if and only if its image under the jet operator

$$\mathrm{jet}^k : \quad \Gamma(\mathscr{F}M) \longrightarrow \Gamma(\mathrm{Jet}^k \mathscr{F}M), \qquad f \longmapsto \left(p \longmapsto \mathrm{jet}_p^k f \right) \qquad (9)$$

takes values $\mathrm{jet}_p^k f \in \mathrm{Eq}_p^k M$ in every point $p \in M$. This observation motivates the minimal regularity assumption imposed in the formal theory of partial differential equations, namely we require that the family $\{ \mathrm{Eq}_p^k M \}_{p \in M}$ of subsets of $\mathrm{Jet}^k \mathscr{F}M$ assembles into a subbundle $\mathrm{Eq}^k M \subset \mathrm{Jet}^k \mathscr{F}M$. A partial differential equation failing to satisfy this minimal regularity assumption is outside the scope of the formal theory and has to be treated differently.

Solving a partial differential equation algebraically in every point $p \in M$ introduces the concept of formal solutions into the picture, sections of the fiber bundle $\mathrm{Eq}^k M \subset \mathrm{Jet}^k \mathscr{F}M$. Only those formal solutions $f^k \in \Gamma(\mathrm{Eq}^k M)$ though, which are in the image of the jet operator, correspond to actual solutions $f \in \Gamma(\mathscr{F}M)$. Hence it makes sense to distinguish sections in the image of the jet operator (9) from arbitrary sections of $\mathrm{Eq}^k M$ and call them holonomic sections. In consequence the original partial differential equation $Df = *$ has been reformulated into the problem to find all holonomic sections of $\mathrm{Eq}^k M$. In passing we remark that the question, whether every section of $\mathrm{Eq}^k M$ is fiberwise homotopic to a holonomic section or not, has sparked intensive research on Gromov's h-principle [4].

Interestingly the holonomic sections $\mathrm{jet}^k f \in \Gamma(\mathrm{Jet}^k \mathscr{F}M)$ of a jet bundle are exactly those sections of $\mathrm{Jet}^k \mathscr{F}M$, which satisfy a particular first order partial differential constraint, the contact constraint. In the setup of exterior differential systems this contact constraint is formulated in terms of a canonical 1-form specific to jet bundles, the canonical contact form. Restricting this canonical contact form to the subbundle $\mathrm{Eq}^k M \subset \mathrm{Jet}^k \mathscr{F} M$ of algebraic pointwise solutions of a partial differential equation $Df = *$ induces an exterior differential system on the manifold $\mathrm{Eq}^k M$, whose solutions correspond bijectively to local solutions $f \in \Gamma(\mathscr{F} M)$ of the original partial differential equation $Df = *$. In this way every partial differential equation satisfying the minimal regularity assumption is transformed into an equivalent exterior differential system.

In order to construct the canonical contact form $\gamma^{\mathrm{contact}}$ on the jet bundle $\mathrm{Jet}^k \mathscr{F} M$ of sections of a fiber bundle $\mathscr{F} M$ we remark that every smooth curve $c : \mathbb{R} \longrightarrow \mathrm{Jet}^k \mathscr{F} M$ in the total space of a jet bundle can be written in the form $c(t) = \mathrm{jet}^k_{p_t} f_t$ with smooth curves $t \longmapsto p_t$ in the base M and a curve $t \longrightarrow f_t$ in $\Gamma(\mathscr{F} M)$. Anticipating a Leibniz rule for such combined curves we can decompose every vector tangent to $\mathrm{Jet}^k \mathscr{F} M$ into two parts:

$$\frac{d}{dt}\bigg|_0 \mathrm{jet}^k_{p_t} f_t = \frac{d}{dt}\bigg|_0 \mathrm{jet}^k_{p_t} f_0 + \frac{d}{dt}\bigg|_0 \mathrm{jet}^k_{p_0} f_t \tag{10}$$

Although this formula is entirely correct the decomposition on the right depends on the specific representation of the given tangent vector on the left as a combination of a curve $t \longmapsto p_t$ in the base and a curve $t \longmapsto f_t$ in the local sections of $\mathscr{F} M$. Essentially the problem is that the first summand picks up partial derivatives of order $k + 1$ of f_0 in form of the partial derivatives of $\mathrm{jet}^k f_0$ in the direction of $\frac{d}{dt}\big|_0 p_t$. This problem is easily overcome using the jet projection pr and so we can define the contact form $\gamma^{\mathrm{contact}}$ on $\mathrm{Jet}^k \mathscr{F} M$ via:

$$\gamma^{\mathrm{contact}}\left(\frac{d}{dt}\bigg|_0 \mathrm{jet}^k_{p_t} f_t \right) := \frac{d}{dt}\bigg|_0 \mathrm{jet}^{k-1}_{p_0} f_t \in \mathrm{Vert}_{\mathrm{jet}^{k-1}_{p_0} f_0} \mathrm{Jet}^{k-1} \mathscr{F} M$$

En nuce the contact form tells us, whether or not we are forced to change the local section f_0 in order to reproduce a given vector tangent to $\mathrm{Jet}^k \mathscr{F} M$. Thus every holonomic section $\mathrm{jet}^k f : M \longrightarrow \mathrm{Jet}^k \mathscr{F} M$, $p \longmapsto \mathrm{jet}^k_p f$, with $f \in \Gamma(\mathscr{F} M)$ pulls back the contact form to:

$$\left((\mathrm{jet}^k f)^* \gamma^{\mathrm{contact}} \right) \left(\frac{d}{dt}\bigg|_0 p_t \right) := \gamma^{\mathrm{contact}}\left(\frac{d}{dt}\bigg|_0 \mathrm{jet}^k_{p_t} f \right)$$

$$= \frac{d}{dt}\bigg|_0 \mathrm{jet}^{k-1}_{p_0} f = 0 \tag{11}$$

In order to simplify this characterization of holonomic sections $\mathrm{jet}^k f \in \Gamma(\mathrm{Jet}^k \mathscr{F} M)$ of the jet bundle $\mathrm{Jet}^k \mathscr{F} M$ it is convenient to replace the contact form $\gamma^{\mathrm{contact}}$, which is a 1-form on $\mathrm{Jet}^k \mathscr{F} M$ with values in the slightly unwieldy vector bundle $\mathrm{pr}^*(\mathrm{Vert}\,\mathrm{Jet}^{k-1} \mathscr{F} M)$, by its scalar components aka local sections of the contact subbundle of $T^* \mathrm{Jet}^k \mathscr{F} M$ defined by:

Contact $\mathrm{Jet}^k \mathscr{F} M$

$$:= \mathbf{im}\left(\mathrm{pr}^*(\mathrm{Vert}^*\mathrm{Jet}^{k-1}\mathscr{F} M) \longrightarrow T^*\mathrm{Jet}^k \mathscr{F} M, \ \eta \longmapsto \langle \eta, \gamma^{\mathrm{contact}}\rangle\right)$$

Every contact form $\gamma \in \Gamma(\mathrm{Contact}\,\mathrm{Jet}^k \mathscr{F} M)$ is actually horizontal for the projection pr to $\mathrm{Jet}^{k-1}\mathscr{F} M$, because every pr-vertical tangent vector is π-vertical as well and thus has a presentation $\frac{d}{dt}\big|_0 \mathrm{jet}_p^k f_t$, in which the base point $p \in M$ does not vary. A fortiori we get:

$$\gamma^{\mathrm{contact}}\left(\frac{d}{dt}\bigg|_0 \mathrm{jet}_p^k f_t\right) \ = \ \frac{d}{dt}\bigg|_0 \mathrm{jet}_p^{k-1} f_t \ = \ \mathrm{pr}_*\left(\frac{d}{dt}\bigg|_0 \mathrm{jet}_p^k f_t\right) \ = \ 0$$

The horizontality of contact forms allows us to define the contact system on $\mathrm{Jet}^k \mathscr{F} M$ as the following sequence of vector subbundles of the cotangent bundle of the jet bundle $\mathrm{Jet}^k \mathscr{F} M$

$$\mathrm{Contact}\ \mathrm{Jet}^k \mathscr{F} M \ \subseteq \ \mathrm{Horizontal}\ \mathrm{Jet}^k \mathscr{F} M \ \subseteq \ T^*\mathrm{Jet}^k \mathscr{F} M \qquad (12)$$

where $\mathrm{Horizontal}\,\mathrm{Jet}^k \mathscr{F} M$ denotes the subbundle of horizontal 1-forms with respect to pr.

The preceding calculations offer a good insight into the geometry of the contact system, nevertheless some readers will certainly prefer a more down to earth approach vindicating our findings explicitly in local coordinates on jet bundles. For the time being we will restrict to the jet bundles $\mathrm{Jet}^k(M, \mathscr{F})$ of smooth maps from a manifold M to a manifold $\mathscr{F}$, in any case the difference between $\mathrm{Jet}^k(M, \mathscr{F})$ and $\mathrm{Jet}^k \mathscr{F} M$ virtually disappears in local coordinates for a fiber bundle $\mathscr{F} M$ over M with model fiber $\mathscr{F}$. Choosing local coordinates (x, U) on M and (f, V) on $\mathscr{F}$ we may then define local coordinates on the subset

$$(\mathrm{Jet}^k(M, \mathscr{F}))_{(x,U),\,(f,V)} \ := \ \{\mathrm{jet}_p^k f \mid p \in U \text{ and } f(p) \in V\}$$

of kth order jets of maps $f : M \longrightarrow \mathscr{F}$ with source in U and target in V by setting

$$x^\alpha(\mathrm{jet}_p^k f) \ := \ x^\alpha(p) \qquad f_A^\lambda(\mathrm{jet}_p^k f) \ := \ \frac{\partial^{|A|} f^\lambda}{\partial x^A}(x^1(p), \ldots, x^m(p))$$

for all $\alpha = 1, \ldots, m, \ \lambda = 1, \ldots, n$ and all multi-indices A on $\{1, \ldots, m\}$ of order $|A| \leq k$. The standard jet coordinates constructed in this way from smooth atlases for both M and $\mathscr{F}$ define a smooth atlas for $\mathrm{Jet}^k(M, \mathscr{F})$ turning it into a smooth manifold of dimension:

$$\dim\,\mathrm{Jet}^k(M, \mathscr{F}) \ = \ m + n\binom{m+k}{m}$$

Standard jet coordinates are adapted to the projections pr : $\mathrm{Jet}^k(M, \mathscr{F}) \longrightarrow$ $\mathrm{Jet}^{\tilde{k}}(M, \mathscr{F})$ for all $k \geq \tilde{k} \geq 0$ in the sense that pr simply forgets all the coordinate functions f_A^λ with $|A| > \tilde{k}$, this observation proves explicitly that (4) really is the stipulated tower of smooth fiber bundles over M. Moreover standard jet coordinates on $\mathrm{Jet}^k(M, \mathscr{F})$ allow us to decompose every tangent vector $\frac{d}{dt}\big|_0 \mathrm{jet}_{p_t}^k f_t$ in the way predicted by Leibniz's rule

$$\frac{d}{dt}\bigg|_0 \mathrm{jet}_{p_t}^k f_t$$

$$= \sum_{\alpha=1}^m \left[\frac{d}{dt}\bigg|_0 x^\alpha(\mathrm{jet}_{p_t}^k f_t) \right] \frac{\partial}{\partial x^\alpha} + \sum_{\substack{|A| \leq k \\ \lambda}} \left[\frac{d}{dt}\bigg|_0 f_A^\lambda(\mathrm{jet}_{p_t}^k f_t) \right] \frac{\partial}{\partial f_A^\lambda}$$

$$= \sum_{\alpha=1}^m \delta x^\alpha \left[\frac{\partial}{\partial x^\alpha} + \sum_{\substack{|A| \leq k \\ \lambda}} \left(\frac{\partial}{\partial x^\alpha} \frac{\partial^{|A|} f_0^\lambda}{\partial x^A} \right)(x(p_0)) \frac{\partial}{\partial f_A^\lambda} \right] + \sum_{\substack{|A| \leq k \\ \lambda}} \delta f_A^\lambda \frac{\partial}{\partial f_A^\lambda}$$

where $\delta x^\alpha := \frac{d}{dt}\big|_0 x^\alpha(p_t)$ and $\delta f_A^\lambda := \frac{d}{dt}\big|_0 \frac{\partial^A f_A^\lambda}{\partial x^A}(x(p_0))$. Evaluating $\frac{d}{dt}\big|_0 \frac{\partial^{|A|} f_t^\lambda}{\partial x^A}(x(p_t))$ for a multi-index A of highest order $|A| = k$ we pick up derivatives of f_0^λ of order $k + 1$ as anticipated above, albeit only in the coefficients of the basis vector $\frac{\partial}{\partial f_A^\lambda}$ associated to A. For all multi-indices A of order $|A| < k$ on the other hand the value of the partial derivative $(\frac{\partial}{\partial x^\alpha} \frac{\partial^{|A|+1} f_0^\lambda}{\partial x^{A+\alpha}})(p_0)$ equals $f_{A+\alpha}^\lambda(\mathrm{jet}_{p_0}^k f_0)$ by construction and so we obtain eventually:

$$\mathrm{pr}_*\left(\frac{d}{dt}\bigg|_0 \mathrm{jet}_{p_t}^k f_t \right) = \sum_{\alpha=1}^m \delta x^\alpha \left(\frac{\partial}{\partial x^\alpha} + \sum_{\substack{|A| < k \\ \lambda}} f_{A+\alpha}^\lambda(\mathrm{jet}_{p_0}^k f_0) \frac{\partial}{\partial f_A^\lambda} \right)$$

$$+ \sum_{\substack{|A| < k \\ \lambda}} \delta f_A^\lambda \frac{\partial}{\partial f_A^\lambda}$$

Evidently the first part in this decomposition comes from the variation $\delta p := \frac{d}{dt}\big|_0 p_t$ of the point $p_0 \in M$, while the second part is caused by the variation $\delta f := \frac{d}{dt}\big|_0 \mathrm{jet}_{p_0}^k f_t$ of the kth order jet of the smooth map $f_0 : M \longrightarrow \mathscr{F}$. In this decomposition the canonical contact form $\gamma^{\mathrm{contact}}$ on $\mathrm{Jet}^k(M, \mathscr{F})$ is simply the projection to the second part, so we conclude:

$$\gamma^{\mathrm{contact}}\left(\frac{\partial}{\partial x^\alpha} \right) = - \sum_{\substack{|A| < k \\ \lambda}} f_{A+\alpha}^\lambda \frac{\partial}{\partial f_A^\lambda} \qquad \gamma^{\mathrm{contact}}\left(\frac{\partial}{\partial f_A^\lambda} \right) = + \delta_{|A| < k} \frac{\partial}{\partial f_A^\lambda}$$

More succinctly this explicit version of the canonical contact form γ^{contact} reads:

$$\gamma^{\text{contact}} = \sum_{\substack{|A|<k \\ \lambda}} \underbrace{\left(df_A^\lambda - \sum_{\alpha=1}^m f_{A+\alpha}^\lambda \, dx^\alpha \right)}_{=: \; \gamma_A^\lambda} \otimes \frac{\partial}{\partial f_A^\lambda} = \sum_{\substack{|A|<k \\ \lambda}} \gamma_A^\lambda \otimes \frac{\partial}{\partial f_A^\lambda}$$

In standard jet coordinates the contact system on $\text{Jet}^k(M, \mathscr{F})$ can thus be written

$$\begin{aligned} \text{Contact } \text{Jet}^k(M, \mathscr{F}) &:= \text{span } \{ \quad \gamma_A^\lambda \quad | \text{ for all } \lambda, \, |A| < k \} \\ \text{Horizontal } \text{Jet}^k(M, \mathscr{F}) &:= \text{span } \{ \quad dx^\alpha, df_A^\lambda \quad | \text{ for all } \alpha, \lambda, \, |A| < k \} \end{aligned} \qquad (13)$$

because $\text{Contact Jet}^k(M, \mathscr{F})$ and $\text{Horizontal Jet}^k(M, \mathscr{F})$ are respectively the subbundles of scalar components of γ^{contact} and of horizontal forms with respect to pr. For the calculations to come it is important to observe that the exterior derivative of the scalar contact form γ_A^λ

$$d\gamma_A^\lambda = - \sum_{\alpha=1}^m df_{A+\alpha}^\lambda \wedge dx^\alpha$$

$$= - \sum_{\alpha=1}^m \left(\gamma_{A+\alpha}^\lambda + \sum_{\tilde{\alpha}=1}^m f_{A+\alpha+\tilde{\alpha}}^\lambda \, dx^{\tilde{\alpha}} \right) \wedge dx^\alpha \; \overset{!}{=} \; - \sum_{\alpha=1}^m \gamma_{A+\alpha}^\lambda \wedge dx^\alpha$$

with a multi-index A of order $|A| < k - 1$ lies in the ideal generated by all the scalar contact forms taken together. This is no longer true for multi-indices A of highest order $|A| = k - 1$, but at least $d\gamma_A^\lambda = - \sum_\alpha df_{A+\alpha}^\lambda \wedge dx^\alpha$ is an element of the ideal generated by horizontal forms. In other words the contact system satisfies the characteristic compatibility condition:

$$d : \quad \Gamma(\text{Contact Jet}^k(M, \mathscr{F})) \longrightarrow \Gamma(\text{Horizontal Jet}^k(M, \mathscr{F}) \wedge T^* \text{Jet}^k(M, \mathscr{F})) \qquad (14)$$

An illustrative example for the contact system, whose axiomatization has become a topic of research by itself under the keyword contact manifolds, is the first order jet bundle $\text{Jet}^1 \mathbb{R} M$ of the trivial real line bundle $\mathbb{R} M := \mathbb{R} \times M$ over M, whose sections correspond to smooth functions $f : M \longrightarrow \mathbb{R}$. Rather atypically this bundle splits into the Cartesian product

$$\text{Jet}^1 \mathbb{R} M \; \overset{\cong}{\longrightarrow} \; \mathbb{R} \times T^* M, \qquad \text{jet}_p^1 f \longmapsto (f(p), d_p f) \qquad (15)$$

which identifies the cotangent bundle $T^* M$ of the manifold M with the pointed jet bundle $T^* M :=^* \text{Jet}^1 \mathbb{R} M$ of first order jets of functions evaluating to zero $\text{ev}(\text{jet}_p^1 f) = (0, p)$ in $\text{Jet}^0 \mathbb{R} M = \mathbb{R} \times M$. From this point of view it is natural to define the higher order cotangent bundle as the bundle of pointed kth order jets of functions, compare for example [8]:

$$T^{*k} M \;=\; {}^* \text{Jet}^k \mathbb{R} M \;:=\; \{ \text{jet}_p^k f \mid f : M \longrightarrow \mathbb{R} \text{ smooth and } f(p) = 0 \}$$

More important for our present purpose is that the vertical tangent bundle of $\mathrm{Jet}^0 \mathbb{R}M$ is canonically the trivial line bundle $\mathrm{Vert}\,(\,\mathbb{R} \times M\,) = (\,T\,\mathbb{R}\,) \times M$ over $\mathbb{R} \times M$ due to $T\,\mathbb{R} \cong \mathbb{R} \times \mathbb{R}$, the contact form thus becomes a scalar valued differential form on $\mathrm{Jet}^1 \mathbb{R}M$:

$$\gamma^{\mathrm{contact}} \left(\left. \frac{d}{dt} \right|_0 \mathrm{jet}^1_{p_t} f_t \right) \;=\; \left. \frac{d}{dt} \right|_0 \mathrm{jet}^0_{p_0} f_t \;\hat{=}\; \left. \frac{d}{dt} \right|_0 f_t(\,p_0\,)$$

Comparing this expression with the differential of the tautological function on $\mathrm{Jet}^1 \mathbb{R}M$, which is just the projection $f^{\mathrm{taut}}(\mathrm{jet}^1_p f) := f(\,p\,)$ to the first factor in decomposition (15)

$$df^{\mathrm{taut}} \left(\left. \frac{d}{dt} \right|_0 \mathrm{jet}^1_{p_t} f_t \right) \;=\; \left. \frac{d}{dt} \right|_0 f_t(\,p_t\,) \;=\; \left. \frac{d}{dt} \right|_0 f_t(\,p_0\,) \;+\; d_{p_0} f_0 \left(\left. \frac{d}{dt} \right|_0 p_t \right)$$

we conclude that the contact form comprises $\gamma^{\mathrm{contact}} = df^{\mathrm{taut}} - \mathrm{pr}^*_{T^*M}\,\theta$ both the differential of the tautological function $f^{\mathrm{taut}} \in C^\infty(\mathrm{Jet}^1 \mathbb{R}M)$ and the tautological 1-form θ on T^*M. Correspondingly we get in standard jet coordinates $(\,x^1, \ldots, x^m,\, f,\, f_1, \ldots, f_m\,)$ on $\mathrm{Jet}^1 \mathbb{R}M$ the classical expression for contact forms in Darboux coordinates for contact manifolds:

$$\gamma^{\mathrm{contact}} \;=\; df \;-\; \sum_{\mu=1}^{m} f_\mu \, dx^\mu$$

Besides higher order cotangent bundles we can also define higher order tangent bundles:

Definition 2.3 (Higher Order Tangent Bundles). Recalling the definition of the tangent bundle TM of a manifold M as the set $\mathrm{Jet}^1_0(\,\mathbb{R},\, M\,)$ of equivalence classes of smooth curves $c : \mathbb{R} \longrightarrow M$ under the relation of first order contact in $0 \in \mathbb{R}$ we define the kth order tangent bundle as the set of equivalence classes of curves

$$T^k M \;:=\; \mathrm{Jet}^k_0(\,\mathbb{R},\, M\,)$$

under kth order contact in 0 with projection $\mathrm{Jet}^k_0(\,\mathbb{R},\, M\,) \longrightarrow M,\, \mathrm{jet}^k_0 c \;=:$ $\left. \frac{d^{\leq k}}{dt^{\leq k}} \right|_0 c \longmapsto c(0)$.

In difference to the classical tangent bundle $TM = T^1 M$ the higher order tangent bundles $T^k M$ with $k > 1$ do not carry a natural vector bundle structure. The proper way to think of this problem is to consider the canonical embedding $\Phi : T^k M \longrightarrow \mathrm{Hom}\,(\,^*\mathrm{Jet}^k \mathbb{R}M,\, \mathbb{R}^k M\,)$ of $T^k M$ into the bundle $\mathrm{Hom}\,(\,^*\mathrm{Jet}^k \mathbb{R}M,\, \mathbb{R}^k M\,)$ of linear maps from the kth order cotangent bundle $T^{*k} M =^* \mathrm{Jet}^k \mathbb{R}M$ to the trivial vector bundle $\mathbb{R}^k M$ with fiber $\mathbb{R}^k$ defined by:

$$\Phi \left[\left. \frac{d^{\leq k}}{dt^{\leq k}} \right|_0 c \right] (\mathrm{jet}^k_{c(0)} f) \;:=\; \left(\left. \frac{d^1}{dt^1} \right|_0 (f \circ c),\, \left. \frac{d^2}{dt^2} \right|_0 (f \circ c), \ldots, \left. \frac{d^k}{dt^k} \right|_0 (f \circ c) \right)$$

In the special case $k = 1$ the canonical embedding Φ is of course a version of the canonical pairing $TM \times_M T^*M \longrightarrow \mathbb{R}M$ between the tangent and cotangent bundles TM and T^*M, as such it induces an isomorphism of fiber bundles. In general however the embedding Φ looks in suitable coordinates on $T^k M$ and $\mathrm{Hom}\,(\,{}^*\mathrm{Jet}^k \mathbb{R}M,\ \mathbb{R}^k M\,)$ like the polynomial map

$$(t_1, \ldots, t_k) \longmapsto (t_1,\ t_1^2 + t_2,\ t_1^3 + 2t_1 t_2 + t_3,\ \ldots)$$

from the set $(T_pM)^k$ of k-tuples of vectors in T_pM to the vector space $(\mathrm{Sym}^{1 \le k} T_pM)^k$ of k-tuples of polynomials of degree at most k without constant term in T_pM. Since not every quadratic polynomial is the square of a linear polynomial, the embedding Φ is not surjective for any $k > 1$, nor does it induce a vector space structure on $T^k M$.

Whereas the jets of smooth maps and jets of local sections are very similar and virtually indistinguishable in local coordinates the third kind of jets we want to discuss in this section are slightly different in nature, namely jets of submanifolds. The jet bundles of submanifolds or generalized Graßmannians are introduced to deal with geometrically motivated partial differential equations, which actually ask for a submanifold solution, not a smooth map or local section, consider for example the partial differential equations describing minimal or totally geodesic submanifolds. Generalized Graßmannians can be used as well to describe multivalued solutions to standard partial differential equations as submanifolds of a jet bundle as discussed for example in [6].

In order to define the contact equivalence relation between submanifolds of a given manifold M we recall that the higher order tangent bundles of a manifold M are defined as the set $T_p^k M := \mathrm{Jet}_0^k(\mathbb{R},\ M)$ of equivalence classes $\left.\frac{d^{\le k}}{dt^{\le k}}\right|_0 c$ of curves $c : \mathbb{R} \longrightarrow M$ under contact to order $k \ge 0$ in the point $0 \in \mathbb{R}$. Thinking of the higher order tangent bundle $T_p^k N$ of a submanifold $N \subset M$ in a point $p \in N$ as a subset of the higher order tangent bundle of M

$$T_p^k N := \left\{ \left.\frac{d^{\le k}}{dt^{\le k}}\right|_0 c \ \middle| \ c : \mathbb{R} \longrightarrow N \subset M \text{ smooth curve with } c(0) = p \right\}$$
$$\subset T_p^k M$$

we may say that two submanifolds N and $\tilde{N}$ are in contact up to order $k \ge 0$ in a common point $p \in N \cap \tilde{N}$ provided $T_p^k N = T_p^k \tilde{N} \subset T_p^k M$, equivalently for every curve $c : \mathbb{R} \longrightarrow N$ with $c(0) = p$ there exists a curve $\tilde{c} : \mathbb{R} \longrightarrow \tilde{N}$ in contact to c to order k and vice versa:

Definition 2.4 (Jets of Submanifolds and Graßmannians). Two submanifolds of a manifold M are said to be in contact $N \sim_{k,p} \tilde{N}$ to order $k \ge 0$ in a common point $p \in N \cap \tilde{N}$, if their kth order tangent spaces in p agree $T_p^k N = T_p^k \tilde{N}$ considered as subsets of $T_p^k M$. The equivalence class of a submanifold N under contact to order

$k \geq 0$ in a point $p \in N$ is called the kth order $\mathrm{jet}^k_p N$ of N in p, the set of all kth order jets of submanifolds of dimension n defines the generalized Graßmannian:

$$\mathrm{Gr}^k_n M \ := \ \{ \ \mathrm{jet}^k_p N \ | \ N \text{ an } n\text{–dimensional submanifold of } M \text{ and } p \in N \ \}$$

Two submanifolds N and $\tilde{N}$ sharing a point $p \in N \cap \tilde{N}$ are in contact to order 0 in p irrespective of their dimensions, because $T^0_p M$ is just the manifold M. The 0th order Graßmannian $\mathrm{Gr}^0_n M = M$ is thus not particularly interesting. The first order Graßmannian $\mathrm{Gr}^1_n M$ on the other hand agrees with the fiber bundle of all linear subspaces $\mathrm{Gr}_n(TM)$ of dimension n of the tangent bundle. In consequence two submanifolds N and $\tilde{N}$ of different dimensions $n \neq \tilde{n}$ are never in contact to first and thus never in contact to positive order $k > 0$ due to the existence of the by now familiar tower of fiber bundles over M

$$\ldots \ \xrightarrow{\mathrm{pr}} \ \mathrm{Gr}^3_n M \ \xrightarrow{\mathrm{pr}} \ \mathrm{Gr}^2_n M \ \xrightarrow{\mathrm{pr}} \ \mathrm{Gr}^1_n M \ \xrightarrow{\pi} \ \mathrm{Gr}^0_n M \ = \ M \qquad (16)$$

under the jet projections $\mathrm{pr} : \ \mathrm{Gr}^k_n M \ \longrightarrow \ \mathrm{Gr}^{\tilde{k}}_n M, \mathrm{jet}^k_p N \longmapsto \mathrm{jet}^{\tilde{k}}_p N$. Unlike the towers of jet bundles we have discussed before there is no meaningful evaluation $\mathrm{ev} : \mathrm{Gr}^1_n M \longrightarrow \mathrm{Gr}^0_n M$ defined in this tower other than the fiber bundle projection π.

This minor difference between jet bundles and generalized Graßmannians is reflected faithfully in local standard coordinates on $\mathrm{Gr}^k_n M$. In fact for every $k > 0$ and every choice of local coordinates $(x^1, \ldots, x^m)$ on an open subset $U \subset M$ we may consider the subset

$$(\mathrm{Gr}^k_n M)_{(x,U)}$$
$$:= \{ \mathrm{jet}^k_p N \ | \ p \in U \text{ and } d_p x^1 \big|_{T_p N}, \ \ldots, \ d_p x^n \big|_{T_p N} \text{ linearly independent} \}$$

of the generalized Graßmannian $\mathrm{Gr}^k_n M$ consisting of the kth order jets of n-dimensional submanifolds $\mathrm{jet}^k_p N$ such that the first n coordinate functions $x^1, \ldots, x^n$ restrict to local coordinates $x^1\big|_N, \ldots, x^n\big|_N$ on N in a neighborhood $U_N \subset N \cap U$ of p. Upon restriction to N the other $(m - n)$ coordinate functions $x^{n+1}, \ldots, x^m$ thus become smooth functions of $x^1\big|_N, \ldots, x^n\big|_N$ turning the submanifold N into the graph of the smooth map

$$(x^{n+1}_N, \ldots, x^m_N) : \quad \mathbb{R}^n \longrightarrow \mathbb{R}^{m-n},$$
$$(x^1(q), \ldots, x^n(q)) \longmapsto (x^{n+1}(q), \ldots, x^m(q))$$

defined on $(x^1\big|_N, \ldots, x^n\big|_N)(U_N)$ by:

$$(x^{n+1}_N, \ldots, x^m_N) \ := \ (x^{n+1}, \ldots, x^m) \circ (x^1\big|_N, \ldots, x^n\big|_N)^{-1}$$

In this local description of submanifolds of M the difference between two n-dimensional submanifolds N and $\tilde{N}$ becomes the difference between the associated tuples $(x_N^{n+1}, \ldots, x_N^m)$ and $(x_{\tilde{N}}^{n+1}, \ldots, x_{\tilde{N}}^m)$ of functions of $(x^1, \ldots, x^n)$. Clearly two n-dimensional submanifolds N and $\tilde{N}$ are in contact in a common point $p \in N \cap \tilde{N} \cap U$ to order $k \geq 0$, if and only if

$$\frac{\partial^{|A|} x_N^\beta}{\partial x^A}(x^1(p), \ldots, x^n(p)) = \frac{\partial^{|A|} x_{\tilde{N}}^\beta}{\partial x^A}(x^1(p), \ldots, x^n(p))$$

for all $\beta = n+1, \ldots, m$ and all multi-indices A on $\{1, \ldots, n\}$ of order $|A| \leq k$. With this observation in mind we define the standard jet coordinates on $(\mathrm{Gr}_n^k M)_{(x,U)}$ by setting

$$x^\alpha(\mathrm{jet}_p^k N) := x^\alpha(p) \qquad x_A^\beta(\mathrm{jet}_p^k N) := \frac{\partial^{|A|} x_N^\beta}{\partial x^A}(x^1(p), \ldots, x^n(p))$$

for $\alpha = 1, \ldots, n$, for $\beta = n+1, \ldots, m$ and A a multi-index on $\{1, \ldots, n\}$ of order $|A| \leq k$. Clearly the domains $(\mathrm{Gr}_n^k M)_{(x,U)} \subset \mathrm{Gr}_n^k M$ of these standard jet coordinates associated to local coordinates (x, U) on M cover $\mathrm{Gr}_n^k M$ making it a smooth manifold of dimension

$$\dim \mathrm{Gr}_n^k M = n + (m-n)\binom{n+k}{n}$$

say $\dim \mathrm{Gr}_n^0 M = m$ and $\dim \mathrm{Gr}_n^1 M = m + (m-n)n$ as expected. Moreover these standard jet coordinates are well adapted to the jet projections $\mathrm{pr} : \mathrm{Gr}_n^k M \longrightarrow \mathrm{Gr}_n^{\tilde{k}} M$ and the projection $\pi : \mathrm{Gr}_n^k M \longrightarrow M$ to the base manifold M proving explicitly that the tower (16) of projections specifies a tower of smooth fiber bundles over M.

Among the subtle differences between the jet bundles of smooth maps or sections and the generalized Graßmannians $\mathrm{Gr}_n^k M$ the definition of the canonical contact form is certainly the most significant. In fact we may not simply copy the definition of the canonical contact form $\gamma^{\mathrm{contact}}$ we have used before, because a tangent vector to $\mathrm{Gr}_n^k M$ written in the form

$$\frac{d}{dt}\Big|_0 \mathrm{jet}_{p_t}^k N_t \in T_{\mathrm{jet}_{p_0}^k N_0} \mathrm{Gr}_n^k M$$

implicitly requires $p_t \in N_t$ for all t to be well-defined, so neither the expression $\frac{d}{dt}\big|_0 \mathrm{jet}_{p_0}^k N_t$ nor its counterpart $\frac{d}{dt}\big|_0 \mathrm{jet}_{p_t}^k N_0$ make any sense. For all $k \geq 1$ however we may lift the canonical inclusion $\iota_N : N \longrightarrow M$ of an n-dimensional submanifold $N \subset M$ to the Graßmannian $\mathrm{jet}^{k-1}\iota_N : N \longrightarrow \mathrm{Gr}_n^{k-1} M$, $p \longmapsto \mathrm{jet}_p^{k-1} N$, in such a way that $\pi \circ \mathrm{jet}^{k-1}\iota_N = \iota_N$. The differential of the lifted inclusion $\mathrm{jet}^{k-1}\iota_N$ of the submanifold N in a point $p \in N$

$$(\mathrm{jet}_p^{k-1}\iota_N)_{*,p}:\quad T_pN\;\longmapsto\; T_{\mathrm{jet}_p^{k-1}N}\mathrm{Gr}_n^{k-1}M,\qquad \frac{d}{dt}\Big|_0\,p_t\;\longmapsto\;\frac{d}{dt}\Big|_0\,\mathrm{jet}_{p_t}^{k-1}N$$

is thus an embedding, whose image in the tangent space $T_{\mathrm{jet}_p^{k-1}N}\mathrm{Gr}_n^{k-1}M$ turns out to depend only on the kth order jet of the submanifold $\mathrm{jet}_p^k N\in\mathrm{Gr}_n^k M$. In consequence we can define the canonical contact form on $\mathrm{Gr}_n^k M$ simply by the projection to the corresponding quotient:

$$\gamma^{\mathrm{contact}}\left(\frac{d}{dt}\Big|_0\,\mathrm{jet}_{p_t}^k N_t\right)\;:=\;\frac{d}{dt}\Big|_0\,\mathrm{jet}_{p_t}^{k-1}N_t\;+\;\mathbf{im}\,(\mathrm{jet}_{p_0}^{k-1}\iota_{N_0})_{*,p_0}$$

Although significantly different in definition this contact form serves the same purpose as before, namely it tells us, whether we are forced to vary the submanifold in order to reproduce a given vector tangent to $\mathrm{Gr}_n^k M$. In fact $\mathbf{im}\,(\mathrm{jet}_{p_0}^{k-1}\iota_{N_0})_{*,p_0}$ is precisely the subspace of tangent vectors, which can be realized without a variation of the submanifold N_0!

One advantage of the preceding definition of the canonical contact form on $\mathrm{Gr}_n^k M$ is that it is evidently horizontal for the jet projection $\mathrm{pr}:\mathrm{Gr}_n^k M\longrightarrow\mathrm{Gr}_n^{k-1}M$, because a tangent vector $\frac{d}{dt}\big|_0\,\mathrm{jet}_{p_t}^k N_t$ vertical under pr satisfies $\frac{d}{dt}\big|_0\,\mathrm{jet}_{p_t}^{k-1}N_t=0$ by definition and thus vanishes under $\gamma^{\mathrm{contact}}$. Due to this horizontality we can extend the canonical contact form $\gamma^{\mathrm{contact}}$ to the contact system on the generalized Graßmann bundle $\mathrm{Gr}_n^k M$ of order $k\geq 1$

$$\mathrm{Contact}\,\mathrm{Gr}_n^k M\;\subseteq\;\mathrm{Horizontal}\,\mathrm{Gr}_n^k M\;\subseteq\;T^*\mathrm{Gr}_n^k M \tag{17}$$

where $\mathrm{Horizontal}\,\mathrm{Gr}_n^k M$ denotes the subbundle of horizontal forms with respect to pr and $\mathrm{Contact}\,\mathrm{Gr}_n^k M$ the subbundle of scalar components of the canonical contact form $\gamma^{\mathrm{contact}}$:

$$\mathrm{Contact}_{\mathrm{jet}_p^k N}\,\mathrm{Gr}_n^k M$$

$$:=\mathbf{im}\left(\mathrm{Ann}\,\mathbf{im}\,(\mathrm{jet}^{k-1}\iota_N)_{*,p}\longrightarrow T^*_{\mathrm{jet}_p^k N}\mathrm{Gr}_n^k M,\;\eta\longmapsto\langle\eta,\gamma^{\mathrm{contact}}\rangle\right)$$

In order to find an explicit description of the canonical contact form in standard jet coordinates $(x^\alpha,\,x_A^\beta)$ on $\mathrm{Gr}_n^k M$ let us consider a submanifold $N\subset M$ of dimension n with canonical inclusion $\iota_N:N\longrightarrow M$ written locally as a graph of the smooth map $(x_N^{n+1},\ldots,x_N^m)$:

$$\iota_N:\quad (x^1,\ldots,x^n)\;\longmapsto\;(x^1,\ldots,x^n;\,x_N^{n+1}(x^1,\ldots,x^n),\ldots,x_N^m(x^1,\ldots,x^n))$$

The lift of the inclusion to the Graßmannian $N\longrightarrow\mathrm{Gr}_n^{k-1}M,\;p\longmapsto\mathrm{jet}_p^{k-1}N$, is given by

$$\mathrm{jet}^{k-1}\iota_N:\quad (x^1,\ldots,x^n)\longmapsto\left(x^1,\ldots,x^n;\,\left\{\frac{\partial^{|A|}x_N^\beta}{\partial x^A}(x^1,\ldots,x^n)\right\}_{|A|<k,\,\beta}\right)$$

hence its differential $(\mathrm{jet}^{k-1}\iota_N)_{*,p} : T_pN \longrightarrow T_{\mathrm{jet}_p^{k-1}N}\mathrm{Gr}_n^{k-1}M$ in a point $p \in N$ satisfies:

$$(\mathrm{jet}^{k-1}\iota_N)_{*,p} : \quad \frac{\partial}{\partial x^\alpha} \longmapsto \frac{\partial}{\partial x^\alpha} + \sum_{\substack{|A|<k \\ \beta}} \left(\frac{\partial}{\partial x^\alpha} \frac{\partial^{|A|}x_N^\beta}{\partial x^A} \right)(x^1(p),\dots,x^n(p)) \frac{\partial}{\partial x_A^\beta}$$

On the other hand the definition of the standard jet coordinates (x^α, x_A^β) on $\mathrm{Gr}_n^k M$ becomes

$$x_{A+\alpha}^\beta (\mathrm{jet}_p^k N) \;=\; \left(\frac{\partial}{\partial x^\alpha} \frac{\partial^{|A|}x_N^\beta}{\partial x^A} \right)(x^1(p),\dots,x^n(p))$$

for all multi-indices of order $|A| < k$ less than k, in consequence the image of the differential $(\mathrm{jet}^{k-1}\iota_N)_{*,p}$ depends only on the coordinates (x^α, x_A^β) of the point $\mathrm{jet}_p^k N$ in the generalized Graßmannian $\mathrm{Gr}_n^k M$ as claimed. Specifically we obtain the following congruences

$$\gamma^{\mathrm{contact}}\left(\frac{\partial}{\partial x^\alpha} \right) \equiv - \sum_{\substack{|A|<k \\ \beta}} x_{A+\alpha}^\beta \frac{\partial}{\partial x_A^\beta}$$

$$\gamma^{\mathrm{contact}}\left(\frac{\partial}{\partial x_A^\beta} \right) \equiv + \delta_{|A|<k} \frac{\partial}{\partial x_A^\beta}$$

modulo the would be image of the differential $(\mathrm{jet}^{k-1}\iota_N)_{*,p}$ defined as the subspace:

$$\Sigma(x^\alpha, x_A^\beta) := \text{``}\mathbf{im}\,(\mathrm{jet}^{k-1}\iota_N)_{*,p}\text{''}$$

$$= \mathrm{span}\left\{ \frac{\partial}{\partial x^\alpha} + \sum_{\substack{|A|<k \\ \beta}} x_{A+\alpha}^\beta \frac{\partial}{\partial x_A^\beta} \;\middle|\; \alpha = 1,\dots,n \right\}$$

It is comforting to know that the contact form $\gamma^{\mathrm{contact}}$ thus looks virtually the same as before

$$\gamma^{\mathrm{contact}} \;=\; \underbrace{\sum_{\substack{|A|<k \\ \beta}} \left(dx_A^\beta - \sum_{\alpha=1}^n x_{A+\alpha}^\beta\, dx^\alpha \right)}_{=: \; \gamma_A^\beta} \otimes \left(\frac{\partial}{\partial x_A^\beta} + \Sigma(x^\alpha, x_A^\beta) \right)$$

in standard jets coordinates on $\mathrm{Gr}_n^k M$, in particular the contact system has the familiar form:

$$\text{Contact } \mathrm{Gr}_n^k M := \mathrm{span} \{ \quad \gamma_A^\beta \quad | \text{ for all } \beta, |A| < k \ \}$$
$$\text{Horizontal } \mathrm{Gr}_n^k M := \mathrm{span} \{ \quad dx^\alpha, dx_A^\beta \quad | \text{ for all } \alpha, \beta, |A| < k \ \}$$

With the construction of the canonical contact system we have established an almost complete analogy between the generalized Graßmannian $\mathrm{Gr}_n^k M$ and the jet bundles of maps or sections. What we are still lacking though is an analogue of the addition (7), which turns the fiber of the jet projection pr : $\mathrm{Jet}^k \mathscr{F} M \longrightarrow \mathrm{Jet}^{k-1} \mathscr{F} M$ over a point $\mathrm{jet}_p^{k-1} f$ into an affine space modelled on the vector space $\mathrm{Sym}^k T_p^* M \otimes \mathrm{Vert}_{f(p)} \mathscr{F} M$ for all $k \geq 1$. Much to our chagrin the fiber of the jet projection $\pi : \mathrm{Gr}_n^1 M \longrightarrow M$ is *not* an affine space, rather we may identify it via $\mathrm{jet}_p^1 N \longmapsto T_p N$ with the compact Graßmannian $\mathrm{Gr}_n(T_p M)$. Despite this disappointment we observe that the vertical tangent space of $\mathrm{Gr}_n^1 M$ in a point $\mathrm{jet}_p^1 N = T_p N$

$$\mathrm{Vert}_{\mathrm{jet}_p^1 N} \mathrm{Gr}_n^1 M \ = \ T_{T_p N} \mathrm{Gr}_n(T_p M) \ \cong \ \mathrm{Hom}(T_p N, T_p M / T_p N) \quad (18)$$

can be written in a form $\mathrm{Hom}(T_p N, T_p M / T_p N) = \mathrm{Sym}^1 T_p^* N \otimes (T_p M / T_p N)$ reminiscent of the vector space acting on the first order jet bundle $\mathrm{Jet}^1 \mathscr{F} M$. Somewhat more precisely the identification (18) of the tangent space of the Graßmannian $\mathrm{Gr}_n(T_p M)$ associates to a homomorphism $A \in \mathrm{Hom}(T_p N, T_p M / T_p N)$ the following tangent vector in the point $T_p N$

$$\frac{d}{dt}\Big|_0 \ \mathbf{im}\Big(\mathrm{id} + t A^{\mathrm{lift}} : \ T_p N \ \longrightarrow \ T_p M, \quad X \ \longmapsto \ X + t A^{\mathrm{lift}} X \Big)$$
$$\in T_{T_p N} \mathrm{Gr}_n(T_p M)$$

where $A^{\mathrm{lift}} : T_p N \longrightarrow T_p M$ is a linear lift of A. Of course the curve $t \longmapsto \mathbf{im}(\mathrm{id} + t A^{\mathrm{lift}})$ of n-dimensional subspaces of $T_p M$ defined for t sufficiently small depends on the lift A^{lift} chosen, nevertheless the tangent vector to this curve in $t = 0$ only depends on A.

En nuce the principal idea of the identification (18) of the tangent spaces of the Graßmannian $\mathrm{Gr}_n(T_p M)$ is to replace a subspace $T_p N \subset T_p M$ by its inclusion $T_p N \longrightarrow T_p M$, being an application the latter is more easy to deform. In the context of jets of submanifolds we do not loose information in replacing similarly a submanifold $N \subset M$ by its canonical inclusion $\iota_N : N \longrightarrow M$, because $\mathrm{jet}_p^k \iota_N$ determines $\mathrm{jet}_p^k N$ completely, to wit the inclusion $T_p^k N \longrightarrow T_p^k M$ used to define $\mathrm{jet}_p^k N$ is just the jet composition (3) with $\mathrm{jet}_p^k \iota_N$. In the same vein the addition (7) of jets of smooth maps becomes an addition of jets of submanifolds

$$\mathrm{jet}_p^k N + \Delta N := \mathrm{jet}_p^k \, \mathbf{im}\Big(N \longrightarrow M, q \longmapsto \Phi^M \big[\Phi^{M-1}(q) + \Delta N(\Phi^{N-1}(q)) \big] \Big)$$
$$(19)$$

with homogeneous polynomials $\Delta N \in \mathrm{Sym}^k T_p^* N \otimes T_p M$ of degree k on $T_p N$ with values in $T_p M$. Although this addition is well-defined for all $k \geq 1$ independent of the choice of the anchored coordinate charts Φ^N and Φ^M for N and M, a peculiar problem arises in the case $k = 1$ singled out in our discussion above: The image of the deformed smooth map is not even locally a submanifold of dimension n, because we modify the linear inclusion $T_p N \subset T_p M$ by linear terms. Evidently this problem disappears for jet orders $k \geq 2$ and the equality $\mathrm{jet}_p^{k-1} \, \mathbf{im} \, \iota_N = \mathrm{jet}_p^{k-1} N$ ensures that our addition acts on the fibers of the projection $\mathrm{Gr}_n^k M \longrightarrow \mathrm{Gr}_n^{k-1} M$ in the sense that $\mathrm{jet}_p^k N + \Delta N$ still lies over $\mathrm{jet}_p^{k-1} N$.

Unluckily however the vector space $\mathrm{Sym}^k T_p^* N \otimes T_p M$ is too large to provide us with a simply transitive group action on the fibers of the projection in analogy to the addition (7) on jets of maps or sections. In order to understand this problem let us have another look at the identification (18) of the tangent spaces of Graßmannian $\mathrm{Gr}_n (T_p M)$. The construction of an explicit curve in $\mathrm{Gr}_n (T_p M)$ representing the tangent vector associated to a linear map $A : T_p N \longrightarrow T_p M / T_p N$ required us to lift A to $A^{\mathrm{lift}} : T_p N \longrightarrow T_p M$. The representing curve depended on this lift, but not the tangent vector itself. Changing the homogeneous polynomial $\Delta N \in \mathrm{Sym}^k T_p^* N \otimes T_p M$ used in the addition (19) by a homogeneous polynomial of degree k on $T_p N$ with values in $T_p N$ similarly changes the image submanifold

$$\mathbf{im}\left(N \longrightarrow M, \quad q \longmapsto \Phi^M\left[\Phi^{M-1}(q) + \Delta N(\Phi^{N-1}(q)) \right] \right)$$

but not its equivalence class under contact of submanifolds to order k in p. For example we may always choose the anchored coordinate chart Φ^M in such a way that $\Phi^M(T_p N) \subset N$ holds true. For such a choice and arbitrary $\Delta N \in \mathrm{Sym}^k T_p^* N \otimes T_p N$ the smooth map

$$\varphi : \ N \longrightarrow N, \quad q \longmapsto \Phi^M\left[\Phi^{M-1}(q) + \Delta N(\Phi^{N-1}(q)) \right]$$

is actually a local diffeomorphism (sic!) of N due to $\mathrm{jet}_p^{k-1} \varphi = \mathrm{jet}_p^{k-1} \mathrm{id}_N$ and $k \geq 2$ so that $\mathrm{jet}_p^k N = \mathrm{jet}_p^k N + \Delta N$. Modifying this argument slightly to make it work for changes of $\Delta N \in \mathrm{Sym}^k T_p^* N \otimes T_p M$ by a homogeneous polynomial of degree k with values in $T_p N$ we conclude that the addition (19) descends to a well-defined addition of jets of submanifolds

$$\mathrm{jet}_p^k N + \Delta N$$

$$:= \mathrm{jet}_p^k \, \mathbf{im}\left(N \longrightarrow M, q \longmapsto \Phi^M\left[\Phi^{M-1}(q) + (\Delta N)^{\mathrm{lift}}(\Phi^{N-1}(q)) \right] \right) \quad (20)$$

with $\Delta N \in \mathrm{Sym}^k T_p^* N \otimes (T_p M / T_p N)$ lifted arbitrarily to $(\Delta N)^{\mathrm{lift}} \in \mathrm{Sym}^k T_p^* N \otimes T_p M$. Although it seems difficult to verify the axioms of a group action for the addition $+$ directly due to the ambiguities in choosing Φ^N and Φ^M as well as the lift $(\Delta N)^{\text{lift}}$, this problem disappears in the local standard jet coordinates (x^α, x_A^β) on $\operatorname{Gr}_n^k M$. As an additional bonus this local coordinate presentation makes it rather obvious that $\operatorname{Sym}^k T_p^* N \otimes (T_p M / T_p N)$ acts simply transitive on the fibers of the projection $\operatorname{Gr}_n^k M \longrightarrow \operatorname{Gr}_n^{k-1} M$.

For the purpose of writing the addition (20) as a smooth group bundle action on the Graßmannian $\operatorname{Gr}_n^k M$ of jets of submanifolds we recall that the tautological vector bundle on $\operatorname{Gr}_n^1 M = \operatorname{Gr}_n(TM)$ is defined as the subbundle of the pull back $\pi^* TM$ of the tangent bundle of M via the projection $\pi : \operatorname{Gr}_n^1 M \longrightarrow M$, whose fiber in $\operatorname{jet}_p^1 N \in \operatorname{Gr}_n^1 M$ reads:

$$\operatorname{Taut}_{\operatorname{jet}_p^1 N} \operatorname{Gr}_n^1 M \quad := \quad T_p N$$

Implicitly we have used the tautological vector bundle already in the identification (18)

$$\operatorname{Vert} \operatorname{Gr}_n^1 M \;=\; \operatorname{Taut}^* \operatorname{Gr}_n^1 M \otimes \left(\pi^* TM \big/ \operatorname{Taut} \operatorname{Gr}_n^1 M \right)$$

of the vertical tangent bundle of $\operatorname{Gr}_n^1 M$, in a similar vein the tautological vector bundle appears in the definition of the canonical contact form γ^{contact} on $\operatorname{Gr}_n^1 M$ as the composition

$$T \operatorname{Gr}_n^1 M \xrightarrow{\pi_*} \pi^* TM \xrightarrow{\operatorname{pr}} \pi^* TM \big/ \operatorname{Taut} \operatorname{Gr}_n^1 N$$

of the differential of $\pi : \operatorname{Gr}_n^1 M \longrightarrow M$ with the projection to $\pi^* TM / \operatorname{Taut} \operatorname{Gr}_n^1 M$. The tautological vector bundle pulls back from $\operatorname{Gr}_n^1 M$ to a vector bundle on $\operatorname{Gr}_n^k M$, in turn this pull back bundle allows us to write the addition (20) as a smooth group bundle action

$$+ : \quad \operatorname{Gr}_n^k M \times_{\operatorname{Gr}_n^1 M} \operatorname{Sym}^k \operatorname{Taut}^* \operatorname{Gr}_n^1 M \otimes \left(\pi^* TM \big/ \operatorname{Taut} \operatorname{Gr}_n^1 M \right) \longrightarrow \operatorname{Gr}_n^k M$$

defined on $\operatorname{Gr}_n^k M$ for all $k \geq 2$, which preserves the fibers of the projection to $\operatorname{Gr}_n^{k-1} M$. With the construction of this group bundle action we have established a complete analogy between the three types of jets discussed in this section: Jets of maps, jets of sections of fiber bundles and jets of submanifolds. In particular the contact systems associated to these three types of jets allow us to treat partial differential equations for maps, for sections and for submanifolds in the unified language of exterior differential systems.

3   Comodules and Spencer Cohomology

A comodule over a symmetric coalgebra can be seen as the algebraic analogue
of a jet bundle in differential geometry, in a rather precise sense this analogy
dualizes the better known analogy between differential operators and modules over
polynomial algebras. In the formal theory of partial differential equations the latter
concept is usually studied under the key word D-modules, which is essentially a
proper subtheory of commutative algebra. From our point of view however it is the
former notion of a comodule, which fits nicely into the theory of exterior differential
systems, because the notion can be seen as a straightforward axiomatization of the
commutativity of partial derivatives.

In this section and Sect. 4 we will study the algebraic properties of comodules
over symmetric coalgebras in depth starting from their axiomatic definition in terms
of partial derivatives, introducing the important subclass of tableau comodules on
the way and ending with a detailed discussion of the three most important theorems
about tableau comodules from the point of view of partial differential equations.
Needless to say all the ideas, properties and theorems discussed in this context are
essentially dual to ideas, properties and theorems of commutative algebra. A good
complementary reading to these notes would thus be [2]. Nevertheless we hope that
the reader will find our reformulation of commutative algebra in terms of comodules
helpful for explicit applications in differential geometry:

Definition 3.1 (Comodules over $\mathrm{Sym}\,T^*$**).**   A comodule over the symmetric coal-
gebra $\mathrm{Sym}\,T^*$ is a graded vector space $\mathscr{A}^\bullet$ together with a bilinear map $T \times \mathscr{A}^\bullet \longrightarrow$
$\mathscr{A}^{\bullet-1}$, $(t, a) \longmapsto \frac{\partial a}{\partial t}$, called the directional derivative such that the endomorphism
$\frac{\partial}{\partial t} : \mathscr{A}^\bullet \longrightarrow \mathscr{A}^{\bullet-1}$, $a \longmapsto \frac{\partial a}{\partial t}$, of $\mathscr{A}^\bullet$ with a fixed direction $t \in T$ is homogeneous
of degree -1 and the endomorphisms $\frac{\partial}{\partial t_1}$ and $\frac{\partial}{\partial t_2}$ commute for all $t_1, t_2 \in T$:

$$\frac{\partial}{\partial t_1} \circ \frac{\partial}{\partial t_2} \;=\; \frac{\partial}{\partial t_2} \circ \frac{\partial}{\partial t_1}$$

In consequence we may iterate the axiomatic directional derivatives of a comodule
$\mathscr{A}$ in order to obtain well-defined homogeneous endomorphisms like $\frac{\partial^2}{\partial t_1 \partial t_2}$:
$\mathscr{A}^\bullet \longrightarrow \mathscr{A}^{\bullet-2}$ etc.

Although intimidating in nomenclature the notion of a comodule is nothing but
an axiomatization of a very familiar concept, that of the directional derivatives of
functions on the vector space T. The example motivating this axiomatization is the
vector space $\mathrm{Sym}^\bullet T^* \otimes V$ of polynomials on T with values in a vector space V
graded by homogeneity together with

$$T \;\times\; \mathrm{Sym}^\bullet T^* \otimes V \;\longmapsto\; \mathrm{Sym}^{\bullet-1} T^* \otimes V, \qquad (t, \psi) \longmapsto \frac{\partial \psi}{\partial t}$$

which associates to a polynomial ψ and a direction $t \in T$ the directional derivative:

$$\frac{\partial \psi}{\partial t}(p) \; := \; \frac{d}{d\varepsilon}\Big|_0 \psi(p + \varepsilon t)$$

In this interpretation of comodules as an axiomatization of directional derivatives it is natural to define the Spencer coboundary operator on alternating forms with values in a comodule

$$B: \quad \mathscr{A}^\bullet \otimes \Lambda^\circ T^* \; \longrightarrow \; \mathscr{A}^{\bullet-1} \otimes \Lambda^{\circ+1} T^*, \qquad \omega \longmapsto B\omega$$

in analogy to the de Rham coboundary operator on differential forms by setting

$$(B\omega)(t_0, \ldots, t_r) \; := \; \sum_{\mu=0}^{r} (-1)^\mu \frac{\partial}{\partial t_\mu} \omega(t_0, \ldots, \widehat{t_\mu}, \ldots, t_r)$$

for an alternating r-form $\omega \in \mathscr{A}^k \otimes \Lambda^r T^*$ with values in $\mathscr{A}^k$. Evidently $B\omega$ is then an $(r+1)$-form on T with values in $\mathscr{A}^{k-1}$, in this sense the Spencer coboundary operator B is bihomogeneous of bidegree $(-1, +1)$. The axiomatic commutation of directional derivatives ensures that the Spencer operator B satisfies the coboundary condition, in other words

$$B^2\omega(t_0, \ldots, t_{r+1})$$

$$= \sum_{0 \leq \mu < \nu \leq r+1} (-1)^{\mu+\nu} \Big(+ \frac{\partial^2}{\partial t_\mu \partial t_\nu} \omega(t_0, \ldots, \widehat{t_\mu}, \ldots, \widehat{t_\nu}, \ldots, t_{r+1})$$

$$- \frac{\partial^2}{\partial t_\nu \partial t_\mu} \omega(t_0, \ldots, \widehat{t_\mu}, \ldots, \widehat{t_\nu}, \ldots, t_{r+1}) \Big)$$

vanishes irrespective of ω. In turn B defines a bigraded cohomology theory for comodules:

Definition 3.2 (Spencer Cohomology of a Comodule). The Spencer cohomology of a comodule $\mathscr{A}$ over the symmetric coalgebra $\mathrm{Sym}\, T^*$ of a vector space T is the bigraded cohomology $H^{\bullet, \circ}(\mathscr{A})$ associated to the bigraded Spencer complex

$$\ldots \xrightarrow{B} \mathscr{A}^{\bullet+1} \otimes \Lambda^{\circ-1} T^* \xrightarrow{B} \mathscr{A}^\bullet \otimes \Lambda^\circ T^* \xrightarrow{B} \mathscr{A}^{\bullet-1} \otimes \Lambda^{\circ+1} T^* \xrightarrow{B} \ldots$$

of alternating, multilinear forms on T with values in $\mathscr{A}$:

$$H^{\bullet, \circ}(\mathscr{A}) \; := \; \frac{\ker(\, B: \quad \mathscr{A}^\bullet \quad \otimes \quad \Lambda^\circ T^* \quad \longrightarrow \quad \mathscr{A}^{\bullet-1} \otimes \Lambda^{\circ+1} T^* \,)}{\mathrm{im}(\, B: \quad \mathscr{A}^{\bullet+1} \otimes \Lambda^{\circ-1} T^* \quad \longrightarrow \quad \mathscr{A}^\bullet \quad \otimes \quad \Lambda^\circ T^* \,)}$$

In order to get some idea about Spencer cohomology theory let us calculate it for some examples. Every graded vector space $\mathscr{A}^\bullet$ can be made a comodule $\mathscr{A}^\bullet_{\text{trivial}}$ by declaring all its directional derivatives to vanish $\frac{\partial a}{\partial t} := 0$ for all $a \in \mathscr{A}^k$ and all $t \in T$. The Spencer cohomology of such a comodule aptly called trivial is certainly given by:

$$H^{\bullet,\circ}(\mathscr{A}_{\text{trivial}}) \;=\; \mathscr{A}^\bullet \otimes \Lambda^\circ T^*$$

Somewhat more interesting are the free comodules $\text{Sym}^\bullet T^* \otimes V$ of polynomials on T with values in a vector space V introduced before. The Spencer operator associated to such a free comodule $B : (\text{Sym}^\bullet T^* \otimes V) \otimes \Lambda^\circ T^* \longrightarrow (\text{Sym}^{\bullet-1} T^* \otimes V) \otimes \Lambda^{\circ+1} T^*$ can be written as a sum

$$B \;=\; \sum_{\mu=1}^n \frac{\partial}{\partial t_\mu} \otimes \text{id}_V \otimes dt_\mu \wedge$$

over a dual pair of bases $t_1, \ldots, t_n$ and $dt_1, \ldots, dt_n$ of T and T^*. In order to calculate the cohomology of the Spencer complex we introduce the operator of integration along rays through the origin $B^* : \text{Sym}^\bullet T^* \otimes V \otimes \Lambda^\circ T^* \longrightarrow \text{Sym}^{\bullet+1} T^* \otimes V \otimes \Lambda^{\circ-1} T^*$ as the sum:

$$B^* \;:=\; \sum_{\mu=1}^n dt_\mu \cdot \otimes \text{id}_V \otimes t_\mu \lrcorner$$

After some more or less straightforward calculations we find that the formal Laplace operator

$$\Delta \;:=\; \{ B, B^* \} \;=\; B \circ B^* + B^* \circ B$$

is diagonalizable on $(\text{Sym}^k T^* \otimes V) \otimes \Lambda^r T^*$ with eigenvalue $k + r$. In consequence every closed Spencer cochain $\psi \in (\text{Sym}^k T^* \otimes V) \otimes \Lambda^r T^*$ of bidegree (k, r) satisfying $k + r > 0$ is exact

$$\psi = \frac{1}{k+r} \Delta \psi = \frac{1}{k+r} \left(B(B^*\psi) + B^*(B\psi) \right) = B \left(\frac{1}{k+r} B^* \psi \right)$$

by $B\psi = 0$. Hence the Spencer cohomology of a free comodule $\text{Sym}^\bullet T^* \otimes V$ is concentrated

$$H^{0,0}(\text{Sym } T^* \otimes V) \;=\; V$$

in comodule and form degrees 0. The preceding calculation of the Spencer cohomology of free comodules is an elementary version of Hodge theory and by no means restricted to this special case. Considered as a method to calculate the cohomology of a given coboundary operator B it relies on making a suitable guess for the operator B^* such that the formal Laplace operator $\Delta := \{B, B^*\}$ is diagonalizable. The original complex then decomposes into a direct sum of "eigensubcomplexes" under Δ, because Δ and B commute $[\Delta, B] = 0$, however all these eigensubcomplexes are exact except for the kernel subcomplex!

The limited stock of examples discussed so far can be augmented by simple modifications of the underlying graded vector spaces. For example the shift in grading by an integer $d \in \mathbb{Z}$

$$(\mathscr{A}^{+d})^\bullet := \mathscr{A}^{\bullet+d}$$

certainly results in the shift in grading $H^{\bullet,\circ}(\mathscr{A}^{+d}) = H^{\bullet+d,\circ}(\mathscr{A})$ in Spencer cohomology. A theoretically important variation of the shift is the twist of a comodule $\mathscr{A}$ defined by

$$\mathscr{A}^\bullet(d) := \mathscr{A}^{\bullet+d}$$

for $\bullet \geq 0$ with $\mathscr{A}^\bullet(d) := \{0\}$ for all $\bullet < 0$, here the directional derivatives of $\mathscr{A}^\bullet(d)$ equal the directional derivatives of $\mathscr{A}$ in positive degrees $\bullet > 0$ only. In consequence the Spencer cohomology $H^{\bullet,\circ}(\mathscr{A}(d))$ vanishes in all comodule degrees $\bullet < 0$ and equals

$$H^{0,\circ}(\mathscr{A}(d)) = (\mathscr{A}^d \otimes \Lambda^\circ T^*)\big/ B(\mathscr{A}^{d+1} \otimes \Lambda^{\circ-1} T^*) \qquad (21)$$

in comodule degree $\bullet = 0$, while $H^{\bullet,\circ}(\mathscr{A}(d)) = H^{\bullet+d,\circ}(\mathscr{A})$ as before in degrees $\bullet > 0$. Another interesting variation of the shift is the idea of a free comodule $\mathscr{A}^\bullet = \text{Sym}^\bullet T^* \otimes V^\bullet$ generated by a graded vector space $V^\bullet$, which is essentially a direct sum of shifted free comodules with associated Spencer cohomology $V^\bullet$ concentrated in form degree $\circ = 0$:

$$\text{Sym}^\bullet T^* \otimes V^\bullet = \bigoplus_{k \in \mathbb{Z}} \text{Sym}^{\bullet-k} T^* \otimes V^k$$

Coming back to the general theory we observe that the Spencer operator B commutes with the extended directional derivatives $\frac{\partial}{\partial t} \otimes \text{id}$ on the graded vector space $\mathscr{A}^\bullet \otimes \Lambda^r T^*$ of Spencer cochains of fixed form degree $\circ = r$. In turn the Spencer complex becomes a complex of comodules, the directional derivatives induced on the Spencer cohomology however are all trivial due to the formal version of Cartan's Homotopy Formula

$$\{B, (\text{id} \otimes t \lrcorner)\} := B \circ (\text{id} \otimes t \lrcorner) + (\text{id} \otimes t \lrcorner) \circ B = \frac{\partial}{\partial t} \otimes \text{id} \qquad (22)$$

which implies for every cohomology class $[\omega] \in H^{\bullet,\circ}(\mathscr{A})$ with $B\omega = 0$ and all $t \in T$:

$$\frac{\partial}{\partial t}[\omega] := [(\frac{\partial}{\partial t} \otimes \text{id})\omega] = [B(\text{id} \otimes t \lrcorner)\omega + (\text{id} \otimes t \lrcorner)B\omega] = 0$$

Although the induced comodule structure on Spencer cohomology is thus trivial, the Spencer cohomology of a comodule $H^{\bullet,\circ}(\mathscr{A})$ carries an interesting algebraic structure, namely the right multiplication of Spencer cochains with elements of $\Lambda^\circ T^*$ commutes with the Spencer coboundary operator B and thus descends to a natural graded right $\Lambda^\circ T^*$-module structure. To see this point more clearly we expand the Spencer coboundary operator into the sum

$$B = \sum_{\mu=1}^{n} \frac{\partial}{\partial t_\mu} \otimes dt_\mu \wedge$$

over a dual pair of bases $t_1, \ldots, t_n$ and $dt_1, \ldots, dt_n$ of T and T^* respectively and conclude that right multiplication with $\omega \in \Lambda^\circ T^*$ commutes with left multiplication by dt_μ due to associativity. In the literature the additional module structure on $H^{\bullet,\circ}(\mathscr{A})$ is hardly ever mentioned. Nevertheless it is not only practical in explicit calculations, it is important for the theory as well: In quite precise a sense we can reconstruct a comodule $\mathscr{A}$ from its Spencer cohomology $H^{\bullet,\circ}(\mathscr{A})$ considered as a graded right module over $\Lambda^\circ T^*$.

A pleasant aspect of the very general and abstract Definition 3.1 of comodules we have adopted in these notes is that it very easy to introduce the complementary concept of homomorphisms of comodules. In general a homomorphism of degree $d \in \mathbb{Z}$ from a comodule $\mathscr{A}$ to a comodule $\mathscr{B}$ is a homogeneous linear map $\Phi : \mathscr{A}^\bullet \longrightarrow \mathscr{B}^{\bullet+d}$ between the underlying graded vector spaces, which intertwines the directional derivatives

$$\Phi(\left.\frac{\partial}{\partial t}\right|_{\mathscr{A}} a) = \left.\frac{\partial}{\partial t}\right|_{\mathscr{B}}(\Phi a)$$

for all $t \in T$. The set of all comodule homomorphisms $\Phi : \mathscr{A}^\bullet \longrightarrow \mathscr{B}^{\bullet+d}$ of fixed degree $d \in \mathbb{Z}$ is evidently a vector space $\text{Hom}^d_{\text{Sym}\,T^*}(\mathscr{A}, \mathscr{B})$, in consequence we can talk about the abelian category of comodules over the symmetric coalgebra $\text{Sym}\,T^*$ by defining the vector space of morphisms $\mathscr{A} \longrightarrow \mathscr{B}$ in this category as the direct sum of all these vector spaces:

$$\text{Hom}^\bullet_{\text{Sym}\,T^*}(\mathscr{A}, \mathscr{B}) := \bigoplus_{d \in \mathbb{Z}} \text{Hom}^d_{\text{Sym}\,T^*}(\mathscr{A}, \mathscr{B})$$

This rather complicated definition of morphisms has the advantage of making the following functor from the category of comodules to the category of graded vector spaces representable:

Definition 3.3 (Finitely Generated and Bounded Comodules). The space of generators of a comodule $\mathscr{A}$ over the symmetric coalgebra $\mathrm{Sym}\, T^*$ of a vector space T is the graded vector space of elements of $\mathscr{A}^\bullet$ constant under all partial derivatives:

$$\mathrm{Gen}^\bullet \mathscr{A} := \bigoplus_{k \in \mathbb{Z}} \mathrm{Gen}^k \mathscr{A} \qquad \mathrm{Gen}^k \mathscr{A} := \left\{ a \in \mathscr{A}^k \; \middle| \; \frac{\partial a}{\partial t} = 0 \text{ for all } t \in T \right\}$$

A comodule $\mathscr{A}$ is called finitely generated and bounded below in case $\mathrm{Gen}\,\mathscr{A}$ is a finite-dimensional vector space and $\mathscr{A}^k = \{0\}$ vanishes for all sufficiently small $k \ll 0$.

In passing we observe that the homogeneous subspaces of a finitely generated comodule $\mathscr{A}$ bounded below are finite-dimensional $\dim \mathscr{A}^k < \infty$ for all $k \in \mathbb{Z}$ due to a straightforward induction based on $\mathscr{A}^k = \{0\}$ for $k \ll 0$ and an induction step using the exact sequence:

$$0 \longrightarrow \mathrm{Gen}^\bullet \mathscr{A} \overset{C}{\longrightarrow} \mathscr{A}^\bullet \overset{B}{\longrightarrow} \mathscr{A}^{\bullet-1} \otimes T^*$$

In order to understand the significance of generators let us consider the real numbers as a trivial comodule $\mathbb{R}^\bullet$ concentrated in degree 0 with all directional derivatives necessarily vanishing. The image of $1 \in \mathbb{R}$ under a homomorphism $\Phi : \mathbb{R}^\bullet \longrightarrow \mathscr{A}^{\bullet+k}$ of comodules homogeneous of degree $k \in \mathbb{Z}$ is then a generator $\Phi(1) \in \mathrm{Gen}^k \mathscr{A}$ of $\mathscr{A}$ of degree k due to

$$\frac{\partial}{\partial t}\, \Phi(1) \;=\; \Phi\!\left(\frac{\partial 1}{\partial t}\right) \;=\; 0$$

and vice versa every $a \in \mathrm{Gen}^k \mathscr{A}$ defines the homomorphism $\Phi_a : \mathbb{R}^\bullet \longrightarrow \mathscr{A}^{\bullet+k}$, $x \longmapsto xa$. In other words the functor $\mathrm{Gen}^\bullet$ to the category of graded vector spaces is represented by $\mathbb{R}$:

$$\mathrm{Hom}^\bullet_{\mathrm{Sym}\, T^*}(\mathbb{R}, \mathscr{A}) \overset{\cong}{\longrightarrow} \mathrm{Gen}^\bullet \mathscr{A}, \qquad \Phi \longmapsto \Phi(1)$$

Using a suitable projective resolution of the representing comodule $\mathbb{R}$ it is then easy to prove:

$$H^{\bullet,\circ}(\mathscr{A}) \;\cong\; \mathrm{Ext}^{\circ,\bullet}_{\mathrm{Sym}\, T^*}(\mathbb{R}, \mathscr{A})$$

In consequence the Spencer cohomology calculates the derived functor $\mathrm{Ext}^{\circ,\bullet}_{\mathrm{Sym}\, T^*}(\mathbb{R}, \cdot)$ associated to the functor $\mathrm{Gen}^\bullet = \mathrm{Hom}^\bullet_{\mathrm{Sym}\, T^*}(\mathbb{R}, \cdot)$ from comodules to graded vector spaces!

Remark 3.4 (Interpretation of Spencer Cohomology). In general it seems to be difficult to say directly, what exactly a non-zero Spencer cohomology class tells us about the underlying comodule. Direct interpretations are available however for the Spencer cohomology of a comodule $\mathscr{A}$ over $\mathrm{Sym}\, T^*$ in form degrees 0 and $n := \dim T$, namely $H^{\bullet,0}(\mathscr{A}) = \mathrm{Gen}^\bullet \mathscr{A}$ is true for $\circ = 0$ by our preceding discussion, whereas

$$H^{\bullet,0}(\mathscr{A}) = \mathrm{Gen}^\bullet \mathscr{A} \qquad H^{\bullet,n}(\mathscr{A}) \cong \mathscr{A}^\bullet / _{\mathrm{span}\{\frac{\partial a}{\partial t}\ \mid\ a\, \in\, \mathscr{A}^{\bullet+1}\ \mathrm{and}\ t\, \in\, T\ \}}$$

is satisfied in form degree $\circ = n$ by a straightforward and not too complicated calculation.

Lemma 3.5 (Finiteness of Spencer Cohomology). *Consider a finitely generated comodule $\mathscr{A}$ bounded below. Every subcomodule $\mathscr{B} \subset \mathscr{A}$ and every quotient comodule $\mathscr{A}/\mathscr{B}$ of $\mathscr{A}$ are likewise finitely generated and bounded below. In particular the Spencer cohomology $H^{\bullet,\circ}(\mathscr{A})$ of $\mathscr{A}$ is a finite dimensional vector space:*

$$\dim\, H^{\bullet,\circ}(\mathscr{A}) \ < \ \infty$$

Needless to say the hard part in the proof of this lemma is the assertion that a quotient $\mathscr{A}/\mathscr{B}$ of a finitely generated comodule $\mathscr{A}$ bounded below by a subcomodule $\mathscr{B}$ is finitely generated, all other assertions of the lemma are trivial or direct consequences of this finiteness. For example the rather surprising conclusion about the Spencer cohomology of a finitely generated comodule $\mathscr{A}$ bounded below simply observes that the Spencer complex

$$\cdots \xrightarrow{\;B\;} \mathscr{A}^{\bullet+1} \otimes \Lambda^{\circ-1} T^* \xrightarrow{\;B\;} \mathscr{A}^\bullet \otimes \Lambda^\circ T^* \xrightarrow{\;B\;} \mathscr{A}^{\bullet-1} \otimes \Lambda^{\circ+1} T^* \xrightarrow{\;B\;} \cdots$$

associated to $\mathscr{A}$ is a complex of finitely generated comodules $\mathscr{A} \otimes \Lambda^\circ T^*$ bounded below with generators $\mathrm{Gen}^\bullet(\mathscr{A} \otimes \Lambda^\circ T^*) = (\mathrm{Gen}^\bullet \mathscr{A}) \otimes \Lambda^\circ T^*$. Assuming finite generation of quotients the subquotient comodule $H^\circ(\mathscr{A})$ of the finitely generated comodule $\mathscr{A} \otimes \Lambda^\circ T^*$ bounded below is itself finitely generated and bounded below, on the other hand we have seen that $H^\circ(\mathscr{A})$ is a trivial comodule in the sense that all its directional derivatives vanish. In consequence

$$H^{\bullet,\circ}(\mathscr{A}) \ = \ \mathrm{Gen}^\bullet H^\circ(\mathscr{A})$$

is finite-dimensional as claimed. All in all Lemma 3.5 reduces very easily to the non-trivial statement that quotients of finitely generated comodule bounded below are finitely generated.

In order to give at least a sketch of the principal argument leading to Lemma 3.5 let us consider a quotient $\mathscr{A}/\mathscr{B}$ of a finitely generated comodule $\mathscr{A}$ bounded below. For sufficiently large $d \gg 0$ the spaces of generators $\mathrm{Gen}^{k+d} \mathscr{A} = \{0\}$ vanish for all $k \geq 0$ due to the finite generation of $\mathscr{A}$. Hence for all $k \geq 0$ the following composition of injective maps

$$\mathscr{A}^{k+d} \longrightarrow T^* \otimes \mathscr{A}^{k+d-1} \longrightarrow T^* \otimes T^* \otimes \mathscr{A}^{k+d-2}$$

$$\longrightarrow \quad \ldots \quad \longrightarrow \underbrace{T^* \otimes \ldots \otimes T^*}_{k} \otimes \mathscr{A}^d$$

is injective itself for all $k \geq 0$ and factorizes by coassociativity over the embedding

$$\mathscr{A}^\bullet(d) \xrightarrow{\Delta} \mathrm{Sym}^\bullet T^* \otimes \mathscr{A}^d \tag{23}$$

by means of the comultiplication Δ (sic!), which is defined for $a \in \mathscr{A}^{k+d}$ as the sum

$$\Delta a := \frac{1}{k!} \sum_{\mu_1,\ldots,\mu_k=1}^{n} dt_{\mu_1} \cdot \ldots \cdot dt_{\mu_k} \otimes \frac{\partial^k a}{\partial t_{\mu_1} \ldots \partial t_{\mu_k}} \tag{24}$$

over a dual pair $t_1,\ldots,t_n$ and $dt_1,\ldots,dt_n$ of bases. In consequence the twisted quotient comodule $(\mathscr{A}/\mathscr{B})(d)$ embeds via Δ into a quotient of the free comodule generated by $\mathscr{A}^d$:

$$(\mathscr{A}/\mathscr{B})^\bullet(d) \cong \mathscr{A}^\bullet(d)/\mathscr{B}^\bullet(d) \longrightarrow \mathrm{Sym}^\bullet T^* \otimes \mathscr{A}^d/\Delta(\mathscr{B}^\bullet(d))$$

All generators of $\mathscr{A}/\mathscr{B}$ of degree at least d are thus generators of a quotient of the free comodule $\mathrm{Sym}\, T^* \otimes \mathscr{A}^d$ as well. An upper bound for the dimension of the space of generators of quotients of free comodules however can be calculated quite effectively using the fundamental ideas underlying the construction of Gröbner bases. On the other hand the quotient comodule $\mathscr{A}/\mathscr{B}$ has finite dimensional homogeneous subspaces and thus only a finite dimensional space of generators of degrees less than d, hence we end up with a finite dimensional space $\mathrm{Gen}\,(\mathscr{A}/\mathscr{B})$ of generators of arbitrary degree.

In general the direct calculation of the Spencer cohomology of a comodule can get quite involved. A convenient alternative, at least for a comodule $\mathscr{A}$ with a large symmetry group, is to construct an initial free resolution of length $r \geq 0$ for $\mathscr{A}$ first, this is an exact sequence

$$0 \longrightarrow \mathscr{A}^{\bullet+d_0} \xrightarrow{\Phi_0} \mathrm{Sym}^\bullet T^* \otimes V_0 \xrightarrow{\Phi_1} \mathrm{Sym}^{\bullet-d_1} T^* \otimes V_1$$

$$\xrightarrow{\Phi_2} \quad \ldots \quad \xrightarrow{\Phi_r} \mathrm{Sym}^{\bullet-d_1-\ldots-d_r} T^* \otimes V_r$$

with suitable comodule homomorphisms $\Phi_0,\ldots,\Phi_r$ of degrees $-d_0,\ldots,-d_r$ respectively. The difference here to an actual free resolution of the comodule $\mathscr{A}$ is that we do not ask for Φ_r to be surjective. Comodules allowing an initial free resolution of some length $r \geq 0$ are rather special of course, to the very least they are isomorphic via Φ_0 to subcomodules of free comodules. In practice however it is often easy to guess an initial free resolution and apply the following lemma to obtain information about the Spencer cohomology:

Lemma 3.6 (Initial Free Resolutions and Spencer Cohomology). *Consider a comodule $\mathscr{A}^\bullet$, which allows an initial free resolution of length $r \geq 0$ of the form*

$$0 \longrightarrow \mathscr{A}^{\bullet+d_0} \xrightarrow{\Phi_0} \mathrm{Sym}^\bullet T^* \otimes V_0 \xrightarrow{\Phi_1} \mathrm{Sym}^{\bullet-d_1} T^* \otimes V_1$$

$$\xrightarrow{\Phi_2} \ldots \xrightarrow{\Phi_r} \mathrm{Sym}^{\bullet-d_1-\ldots-d_r} T^* \otimes V_r$$

with comodule homomorphisms $\Phi_0, \ldots, \Phi_r$ of degrees $-d_0, \ldots, -d_r$ respectively. Independent of whether the last comodule homomorphism Φ_r is surjective or not the only non-vanishing Spencer cohomology spaces of $\mathscr{A}$ of form degree $\circ \leq r$ at most equal to r are:

$$H^{d_0,0}(\mathscr{A}) \cong V_0 \quad H^{d_0+d_1-1,1}(\mathscr{A}) \cong V_1 \quad \ldots \quad H^{d_0+\ldots+d_r-r,r}(\mathscr{A}) \cong V_r$$

It is a pity that the only conceptual proof of this lemma I know of requires some knowledge of spectral sequences, which is quite formidable a concept from homological algebra for an introductory text like this one on exterior differential systems. In essence however spectral sequences are just a highly efficient tool to facilitate certain types of diagram chases. The spectral sequence accelerated diagram chases proving the Lemma of Five, the Lemma of Nine and the Snake Lemma for example are almost trivial. Perhaps our use of spectral sequences in this section motivates the reader unacquainted with the concept to study spectral sequences from this point of view to accelerate her or his future diagram chases.

Using spectral sequences the proof of Lemma 3.6 proceeds along the following line of argument. In a first step we extend the given initial free resolution of length $r \geq 0$ to the right by the projection onto the cokernel of Φ_r in order to obtain an exact sequence:

$$0 \longrightarrow \mathscr{A}^{\bullet+d_0} \xrightarrow{\Phi_0} \mathrm{Sym}^\bullet T^* \otimes V_0 \xrightarrow{\Phi_1} \mathrm{Sym}^{\bullet-d_1} T^* \otimes V_1$$

$$\xrightarrow{\Phi_2} \ldots \xrightarrow{\Phi_r} \mathrm{Sym}^{\bullet-d_1-\ldots-d_r} T^* \otimes V_r \xrightarrow{\mathrm{pr}} \mathscr{C}^{\bullet-d_1-\ldots-d_r} \longrightarrow 0$$

Thinking of this exact sequence of comodules as a complex with trivial homology and taking Spencer cochains we obtain a double complex with columns given by the Spencer complexes of the comodules involved, while the rows are all copies of the original exact sequence tensored with $\wedge^\circ T^*$. Of course we would prefer to have the two coboundary operators in this double complex anticommuting instead of commuting, the difference however plays a negligible role in the construction of the two spectral sequences associated to a double complex.

By assumption the initial free resolution extended by the projection to the cokernel comodule $\mathscr{C}$ of Φ_r is exact everywhere, hence the rows first spectral sequence associated to our double complex collapses at its E^1-term, simply because it equals $\{0\}$ everywhere, in consequence the columns first spectral sequence necessarily converges to $\{0\}$ as well. On calculating its E^1-term however we obtain the Spencer cohomology of $\mathscr{A}$ in the first column, the Spencer cohomology of $\mathscr{C}$ in the last column with the vector spaces $V_0, \ldots, V_r$ in between in the first row representing the Spencer cohomology of the free comodules forming the initial free resolution of $\mathscr{A}$. A spectral sequence with such an E^1-term has only one chance left to converge to $\{0\}$, namely the higher order coboundary operators must induce isomorphisms

$$H^{\bullet + d_0 - s, s}(\mathscr{A}) \xrightarrow{\cong} [V_s]_{\bullet = d_1 + \ldots + d_s}$$

of graded vector spaces for all $s = 0, \ldots, r$ as well as for all $s > r$ isomorphisms:

$$H^{\bullet + d_0 - r - 1, s}(\mathscr{A}) \xrightarrow{\cong} H^{\bullet - d_1 - \ldots - d_r, s - r - 1}(\mathscr{C})$$

Apropos spectral sequences by far the most useful spectral sequence in the theory of comodules is not the spectral sequence discussed above, but the spectral sequence arising from a peculiar double Spencer complex. In general the graded tensor product $\operatorname{Sym}^{\bullet} T^* \otimes \mathscr{A}^{\bullet}$ of the free comodule $\operatorname{Sym}^{\bullet} T^*$ with a comodule $\mathscr{A}^{\bullet}$ can be turned into a comodule in two different ways with different directional derivatives. Namely it can be considered as a free comodule $\operatorname{Sym}^{\bullet} T^* \otimes \mathscr{A}^{\bullet}_{\text{trivial}}$ generated by the graded vector space $\mathscr{A}^{\bullet}_{\text{trivial}}$ underlying $\mathscr{A}$ with directional derivatives $\frac{\partial}{\partial t} \otimes \operatorname{id}_{\mathscr{A}}$ or it can be considered as a tensor product $\operatorname{Sym}^{\bullet} T^* \otimes \mathscr{A}^{\bullet}$ of comodules with directional derivatives dictated by the usual Leibniz rule:

$$\left(\frac{\partial}{\partial t}\right)^{\otimes} := \frac{\partial}{\partial t} \otimes \operatorname{id}_{\mathscr{A}} + \operatorname{id}_{\operatorname{Sym}^{\bullet} T^*} \otimes \frac{\partial}{\partial t}$$

For a comodule $\mathscr{A}$ bounded below the resulting two comodules are actually isomorphic via

$$\exp P: \quad \operatorname{Sym}^{\bullet} T^* \otimes \mathscr{A}^{\bullet} \xrightarrow{\cong} \operatorname{Sym}^{\bullet} T^* \otimes \mathscr{A}^{\bullet}_{\text{trivial}}, \qquad \psi \longmapsto \sum_{r \geq 0} \frac{1}{r!} P^r \psi$$

where $P: \operatorname{Sym}^{\bullet} T^* \otimes \mathscr{A}^{\bullet} \longrightarrow \operatorname{Sym}^{\bullet + 1} T^* \otimes \mathscr{A}^{\bullet - 1}$ is defined as the sum over a dual pair

$$P := \sum_{\mu = 1}^{n} dt_{\mu} \cdot \otimes \frac{\partial}{\partial t_{\mu}}$$

of bases $t_1, \ldots, t_n$ and $dt_1, \ldots, dt_n$ for T and T^* respectively. In fact P is at least locally nilpotent for a comodule $\mathscr{A}$ bounded below so that its exponential $\exp P$ is well-defined, moreover the commutator $[\frac{\partial}{\partial t} \otimes \operatorname{id}_{\mathscr{A}}, P] = \operatorname{id}_{\operatorname{Sym}^{\bullet} T^*} \otimes \frac{\partial}{\partial t}$ commutes with P and the identity

$$(\frac{\partial}{\partial t} \otimes \mathrm{id}_{\mathscr{A}}) \circ \exp \ P = \exp \ P \circ (\frac{\partial}{\partial t} \otimes \mathrm{id}_{\mathscr{A}}) + [(\frac{\partial}{\partial t} \otimes \mathrm{id}_{\mathscr{A}}), \exp \ P]$$

$$= \exp \ P \circ \left(\frac{\partial}{\partial t} \otimes \mathrm{id}_{\mathscr{A}} + \mathrm{id}_{\mathrm{Sym}^{\bullet} T^*} \otimes \frac{\partial}{\partial t} \right)$$

shows that $\exp \ P$ is a homomorphism of comodules with inverse $\exp(-P)$. In a sense the resulting isomorphism $\mathrm{Sym}^{\bullet} T^* \otimes \mathscr{A}^{\bullet} \cong \mathrm{Sym}^{\bullet} T^* \otimes \mathscr{A}^{\bullet}_{\mathrm{trivial}}$ of comodules tells us that a general comodule $\mathscr{A}$ bounded below is not too different from a free comodule. A convenient method to make this structural statement about comodules bounded below precise is to consider the two spectral sequences associated to the double Spencer complex

$$
\begin{array}{ccccccc}
& & B \downarrow & & & & B \downarrow \\
\xrightarrow{\ b\ } & \mathrm{Sym}^{\bullet+1}T^* \otimes \mathscr{A}^{\bullet+1} \otimes \Lambda^{\circ}T^* & \xrightarrow{\ b\ } & \mathrm{Sym}^{\bullet}T^* \otimes \mathscr{A}^{\bullet+1} \otimes \Lambda^{\circ+1}T^* & \xrightarrow{\ b\ } & \\
& & B \downarrow & & & & B \downarrow \\
\xrightarrow{\ b\ } & \mathrm{Sym}^{\bullet+1}T^* \otimes \mathscr{A}^{\bullet} \otimes \Lambda^{\circ+1}T^* & \xrightarrow{\ b\ } & \mathrm{Sym}^{\bullet}T^* \otimes \mathscr{A}^{\bullet} \otimes \Lambda^{\circ+2}T^* & \xrightarrow{\ b\ } & \\
& & B \downarrow & & & & B \downarrow \\
\end{array}
$$

$$(25)$$

where B and b are the anticommuting Spencer operators for $\mathscr{A}$ and $\mathrm{Sym}\,T^*$ respectively with the other factor merely serving as additional coefficients. The b-first spectral sequence collapses at its E^1-term, simply because it is concentrated in forms degree $\circ = 0$

$$\delta_{\bullet=0=\circ} \ \mathscr{A}^{\bullet}$$

and so it is impossible that any of the higher coboundary operators are non-trivial. Things are quite different for the B-first spectral sequence however, which turns into an efficient algorithm to reconstruct a comodule $\mathscr{A}$ from its Spencer cohomology:

Lemma 3.7 (Standard Spectral Sequence of a Comodule). *Every finitely generated comodule $\mathscr{A}$ bounded below carries a canonical complete filtration*

$$\mathscr{A}^{\bullet} \supseteq \dots \supseteq (F^{-1}\mathscr{A})^{\bullet} \supseteq (F^{0}\mathscr{A})^{\bullet} \supseteq (F^{+1}\mathscr{A})^{\bullet} \supseteq \dots \supseteq \{0\}$$

by the subcomodules $F^k \mathscr{A}$ generated in degrees greater than or equal to $k \in \mathbb{Z}$ in the sense:

$$(F^{k}\mathscr{A})^{\bullet} := \mathbf{ker}\left(\mathscr{A}^{\bullet} \xrightarrow{\ \Delta\ } \mathrm{Sym}^{\bullet-k+1}T^* \otimes \mathscr{A}^{k-1} \right)$$

Whereas the b-first spectral sequence associated to the double Spencer complex (25) collapses at its E^1-term, the E^1-term of the B-first spectral sequence reflects the Spencer cohomology

$$\mathrm{Sym}^{\bullet}T^* \otimes H^{\bullet,\circ}(\mathscr{A}) \implies \delta_{\circ=0} (F^{\bullet}\mathscr{A} / _{F^{\bullet+1}\mathscr{A}})^{\bullet+\bullet}$$

of $\mathscr{A}$ and the spectral sequence converges to the successive quotients of the filtration subcomodules $F^\bullet \mathscr{A}$. In addition the coboundary operator B_1 for the E^1-term is completely determined by the right $\Lambda^\circ T^*$-module structure on the Spencer cohomology $H^{\bullet,\circ}(\mathscr{A})$.

Perhaps the most striking application of the standard spectral sequence with a very practical appeal is the following explicit formula for the dimensions of the homogeneous subspaces of a finitely generated comodule bounded below, which reflects the equality of the E^1-Euler characteristics of the two spectral sequences associated to the double Spencer complex (25):

Corollary 3.8 (Poincaré Function of a Comodule). *The dimensions of the homogeneous subspaces $\mathscr{A}^k$, $k \in \mathbb{Z}$, of a finitely generated comodule $\mathscr{A}$ bounded below can be calculated from the Betti numbers* dim $H^{\bullet,\circ}(\mathscr{A})$ *of its Spencer cohomology and the dimension $n := $ dim T of the vector space T by means of the formula:*

$$\dim \mathscr{A}^k \;=\; \sum_{\substack{r=0,\ldots,n \\ d \in \mathbb{Z} \\ d+r \leq k}} (-1)^r \binom{k-d-r+n-1}{n-1} \dim H^{d,r}(\mathscr{A})$$

In particular dim $\mathscr{A}^k$ *equals the value of a polynomial in k of degree at most $n-1$ for all $k > d_{\max}$, where $d_{\max} \in \mathbb{Z}$ is chosen so that $H^{d,r}(\mathscr{A}) = \{0\}$ for all $d > d_{\max}$ and all r.*

Proof. The two spectral sequences associated to the double Spencer complex (25) arise from the two anticommuting Spencer coboundary operators B and b, which are trihomogeneous of tridegrees $(0, -1, +1)$ and $(-1, 0, +1)$ respectively with respect to the trigrading on $\mathrm{Sym}^\bullet T^* \otimes \mathscr{A}^\bullet \otimes \Lambda^\circ T^*$. In particular both B and b preserve the total grading so that the both spectral sequences actually decompose into the direct sum of spectral sequences

$$\mathrm{Sym}^\bullet T^* \otimes \mathscr{A}^\bullet \otimes \Lambda^\circ T^* \;=\; \bigoplus_{k \in \mathbb{Z}} \Big(\mathrm{Sym}^\bullet T^* \otimes \mathscr{A}^\bullet \otimes \Lambda^\circ T^* \Big)_{\bullet+\bullet+\circ=k}$$

parametrized by the total degree $k \in \mathbb{Z}$. The total degree k part of the b-first spectral sequence collapses as before at its E^1-term $\delta_{\bullet=0=\circ} \mathscr{A}^k$ of Euler characteristic dim $\mathscr{A}^k$. According to Lemma 3.7 the total degree k-part of the E^1-term of the B-first spectral sequence reads $\mathrm{Sym}^{k-\bullet-\circ} T^* \otimes H^{\bullet,\circ}(\mathscr{A})$, its Euler characteristic is thus finite and given by

$$\sum_{\substack{r=0,\ldots,n \\ d \in \mathbb{Z}}} (-1)^r \dim \mathrm{Sym}^{k-d-r} T^* \otimes H^{d,r}(\mathscr{A})$$

$$=\; \sum_{\substack{r=0,\ldots,n \\ d \in \mathbb{Z} \\ d+r \leq k}} (-1)^r \binom{k-d-r+n-1}{n-1} \dim H^{d,r}(\mathscr{A})$$

because $H(\mathscr{A})$ is a finite-dimensional vector for the finitely generated comodule $\mathscr{A}$ bounded below. The Euler characteristic of every complex on the other hand equals the Euler characteristic of its cohomology, in turn the Euler characteristic is constant all along a spectral sequence, which is in essence a sequence of coboundary operators each defined on the *cohomology* of the previous operator. With the E^∞-terms of the two spectral sequences arising from the double Spencer complex (25) being isomorphic the stipulated formula for dim $\mathscr{A}^k$ simply reflects the equality of the two different E^1-Euler characteristics.

The Spencer cohomology of the finitely generated comodule $\mathscr{A}$ bounded below is a finite dimensional vector space according to Lemma 3.5, hence we may certainly choose $d_{\max} \in \mathbb{Z}$ so that $H^{d,r}(\mathscr{A}) = \{0\}$ for all $d > d_{\max}$ and all $r = 0, \ldots, n$. For all degrees $k > d_{\max}$ the original summation calculating dim $\mathscr{A}^k$ can be simplified to read

$$\dim \mathscr{A}^k \;=\; \sum_{\substack{r=0,\ldots,n \\ d \in \mathbb{Z}}} (-1)^r \binom{k-d-r+n-1}{n-1} \dim H^{d,r}(\mathscr{A}) \qquad (26)$$

because all summands with $d + r > k$ vanish automatically. In fact either $d > d_{\max}$ or $d \le d_{\max} < k$, in the first case dim $H^{d,r}(\mathscr{A}) = 0$, whereas $\binom{k-d-r+n-1}{n-1} = 0$ in the second case due to $n-1 > k-d-r+n-1 \ge 0$. The simplified summation (26) however defines a polynomial of degree at most $n-1$ in k equal to dim $\mathscr{A}^k$ for $k > d_{\max}$. $\qquad\square$

Another direct application of the standard spectral sequence leads to a kind of converse to Lemma 3.6. Consider a comodule $\mathscr{A}$ bounded below satisfying the additional condition that its only non-vanishing Spencer cohomology in form degrees $\circ = 0, \ldots, r$ is concentrated in

$$V_0 := H^{d_0,0}(\mathscr{A}) \quad V_1 := H^{d_0+d_1-1,1}(\mathscr{A}) \quad \ldots \quad V_r := H^{d_0+\ldots+d_r-r,r}(\mathscr{A})$$

for suitable integers $d_0, \ldots, d_r \in \mathbb{Z}$. The integers $d_1, \ldots, d_r \ge 1$ are then actually positive except for d_0 and the comodule $\mathscr{A}$ has an initial free resolution of length $r \ge 0$ by free comodules linked by comodule homomorphisms $\Phi_0, \ldots, \Phi_r$ of degrees $-d_0, \ldots, -d_r$

$$0 \longrightarrow \mathscr{A}^{\bullet+d_0} \xrightarrow{\ \Phi_0\ } \mathrm{Sym}^\bullet T^* \otimes V_0 \xrightarrow{\ \Phi_1\ } \mathrm{Sym}^{\bullet-d_1} T^* \otimes V_1$$

$$\xrightarrow{\ \Phi_2\ } \ \ldots \ \xrightarrow{\ \Phi_r\ } \mathrm{Sym}^{\bullet-d_1-\ldots-d_r} T^* \otimes V_r$$

which are determined by the higher order coboundary operators of the standard spectral sequence of Lemma 3.7. In particular the comodule homomorphism Φ_0 identifies $\mathscr{A}^\bullet$ with a subcomodule of the shifted free comodule $\mathrm{Sym}^{\bullet-d_0} T^* \otimes V_0$ determined by the tableau:

$$\mathscr{A}^{d_0+d_1} \;\cong\; \mathbf{ker}\Big(\, \Phi_1: \; \mathrm{Sym}^{d_1} T^* \otimes V_0 \;\longrightarrow\; V_1 \,\Big)$$

In the following section we will study the structure of such tableau comodules in more detail.

4   Algebraic Properties of Tableau Comodules

In the algebraic analysis of exterior differential systems the comodules of interest are usually tableau comodules, comodules which arise as the kernels of homogeneous homomorphisms between free comodules. Tableau comodules and the partial differential equations they represent are classified, albeit rather superficially, into underdetermined, determined and overdetermined tableau comodules depending on the ranks of the free comodules involved in their definition. Underdetermined partial differential equations can usually be studied successfully with methods from functional analysis, while integrability constraints will likely thwart such an approach for a given overdetermined partial differential equation.

Perhaps the most interesting case of this superficial classification of partial differential equations is the limiting case of both realms: The Euler–Lagrange equations associated to a variational principle and the elliptic differential equations studied in global analysis are always determined partial differential equations. Mathematical physics for example favors determined partial differential equations according to the following metaprinciple: Reasonable field equations should allow for a unique solution for arbitrarily given Cauchy data. In this section we will discuss the three classical statements about under- and overdetermined partial differential equations from the point of view of their associated tableau comodules:

- Formal Integrability of underdetermined differential equations.
- Complex Characterization of finite type differential equations.
- Cartan's Test for Involutivity of first order tableau comodules.

In order to begin our study of tableau comodules let us have a closer look at a non-trivial homogeneous homomorphism $\Phi: \mathrm{Sym}^\bullet T^* \otimes V \longrightarrow \mathrm{Sym}^{\bullet-d} T^* \otimes E$ between free comodules. As a homomorphism of comodules Φ maps the space V of generators of the domain to generators of the codomain $\mathrm{Sym}\, T^* \otimes E$ including 0 so that $d \in \mathbb{N}_0$ is necessarily non-negative. Moreover it is easily seen that Φ is completely determined by its restriction ϕ to the subspace $\mathrm{Sym}^d T^* \otimes V \subset \mathrm{Sym}^\bullet T^* \otimes V$ of elements of degree d. Conversely every linear map $\phi: \mathrm{Sym}^d T^* \otimes V \longrightarrow E$ extends in a unique way to a homomorphism of comodules

$$\Phi: \quad \mathrm{Sym}^\bullet T^* \otimes V \;\longrightarrow\; \mathrm{Sym}^{\bullet-d} T^* \otimes E$$

of degree $-d$, which can be written in terms of directional derivatives as an iterated sum

$$\Phi(\psi \otimes v) := \sum_{\mu_1,\dots,\mu_d=1}^{n} \frac{\partial^d \psi}{\partial t_{\mu_1} \dots \partial t_{\mu_d}} \otimes \phi\left(\frac{1}{d!} dt_{\mu_1} \cdot \ldots \cdot dt_{\mu_d} \otimes v \right) \quad (27)$$

over a basis $t_1,\dots,t_n$ of T and its dual basis $dt_1,\dots,dt_n$ of T^*. In the spirit of partial differential equations we may interpret the original linear map $\phi : \mathrm{Sym}^d T^* \otimes V \longrightarrow E$ as a linear differential operator $D_\phi : C^\infty(T, V) \longrightarrow C^\infty(T, E)$ of order d defined by:

$$(D_\phi \psi)(p) := \phi\left(\sum_{\mu_1,\dots,\mu_d=1}^{n} \frac{1}{d!} dt_{\mu_1} \cdot \ldots \cdot dt_{\mu_d} \otimes \frac{\partial^d \psi}{\partial t_{\mu_1} \dots \partial t_{\mu_d}}(p) \right) \quad (28)$$

The associated partial differential equation $D_\phi \psi = 0$ can be written as a system of dim E scalar differential equations in the dim V unknown scalar components of $\psi \in C^\infty(T, V)$, for this reason the equation is called underdetermined, determined or overdetermined respectively, if there are less, an equal number of or more equations than unknown functions:

$$\begin{array}{lll} \textbf{underdetermined:} & \dim E \leq \dim V & \\ \textbf{determined:} & \dim E = \dim V & (29) \\ \textbf{overdetermined:} & \dim E \geq \dim V & \end{array}$$

The homomorphism $\Phi : \mathrm{Sym}^\bullet T^* \otimes V \longrightarrow \mathrm{Sym}^{\bullet - k} T^* \otimes E$ of free comodules associated to ϕ is nothing else but the restriction of the operator D_ϕ to the subspace $\mathrm{Sym}\, T^* \otimes V \subset C^\infty(T, V)$ of polynomials on T with values in V. In particular its kernel comodule agrees with the space of polynomial solutions $\psi \in \mathrm{Sym}\, T^* \otimes V$ to the partial differential equation $D_\phi \psi = 0$:

Definition 4.1 (Tableaux and Comodules). A tableau of order $d \geq 1$ is by definition a subspace $\mathscr{A}^d \subset \mathrm{Sym}^d T^* \otimes V$ of the vector space $\mathrm{Sym}^d T^* \otimes V$ of homogeneous polynomials of degree d on T with values in V. The tableau comodule $\mathscr{A}^\bullet \subset \mathrm{Sym}^\bullet T^* \otimes V$ associated to a tableau $\mathscr{A}^d$ is the kernel of the homomorphism

$$0 \longrightarrow \mathscr{A}^\bullet \overset{\subset}{\longrightarrow} \mathrm{Sym}^\bullet T^* \otimes V \overset{\Phi}{\longrightarrow} \mathrm{Sym}^{\bullet - d} T^* \otimes E$$

of free comodules induced by some linear map $\phi : \mathrm{Sym}^d T^* \otimes V \longrightarrow E$ with kernel $\mathscr{A}^d$. A tableau comodule $\mathscr{A}$ is called underdetermined, determined or overdetermined provided:

$$\mathrm{codim}\, \mathscr{A}^d \leq \dim V \qquad \mathrm{codim}\, \mathscr{A}^d = \dim V \qquad \mathrm{codim}\, \mathscr{A}^d \geq \dim V$$

Of course one possible choice for the linear map ϕ in the definition is simply the canonical projection pr $:$ $\mathrm{Sym}^d T^* \otimes V \longrightarrow \mathrm{Sym}^d T^* \otimes V/\mathscr{A}^d$, other choices however are convenient to avoid the typographical monster $\mathrm{Sym}^d T^* \otimes V/\mathscr{A}^d$. Whatever the preferred choice the tableau comodule $\mathscr{A}$ does only depend on $\mathscr{A}^d = \mathbf{ker}\,\phi$, for this reason its homogeneous subspaces $\mathscr{A}^{d+1}, \mathscr{A}^{d+2}, \ldots$ are sometimes called the first and the second prolongation of $\mathscr{A}^d$ etc. Of little concern is the equality $\mathscr{A}^k = \mathrm{Sym}^k T^* \otimes V$ for $k < d$, because in general we are interested in the behavior of $\mathscr{A}^k$ at large degrees $k \gg 0$. According to Lemma 3.6 the non-vanishing Spencer cohomology of a tableau comodule in form degrees $\circ = 0, 1$ reads:

$$H^{0,0}(\mathscr{A}) \;=\; V \qquad\qquad H^{d-1,1}(\mathscr{A}) \;=\; \mathrm{Sym}^d T^* \otimes V/_{\mathscr{A}^d} \qquad (30)$$

In order to reduce the complexity it seems like a good idea to replace the linear map ϕ from the complicated and high-dimensional vector space $\mathrm{Sym}^d T^* \otimes V$ with its localizations

$$\phi_\xi : \quad V \longrightarrow E, \qquad v \longmapsto \phi(\tfrac{1}{d!}\xi^d \otimes v)$$

at covectors $\xi \in T^*$. In this way we are interpreting the linear map $\phi : \mathrm{Sym}^d T^* \otimes V \longrightarrow E$ via the vector space isomorphism $\mathrm{Hom}\,(\mathrm{Sym}^d T^* \otimes V, E) \cong \mathrm{Sym}^d T \otimes \mathrm{Hom}\,(V, E)$ as a homogeneous polynomial of degree d on T^* (sic!) with values in $\mathrm{Hom}\,(V, E)$. Motivated by this interpretation of ϕ we define the characteristic (projective) variety of a tableau $\mathscr{A}^d$ by

$$\mathscr{L}(\mathscr{A}^d) \;:=\; \{\,[\xi] \in \mathbb{P}T^* \mid \phi_\xi : V \longrightarrow E \text{ is not surjective}\,\} \qquad (31)$$

this is $\xi \in T^*$ is a characteristic covector, if and only if ϕ_ξ *fails* to be surjective. Evidently the tableau $\mathscr{A}^d$ has to be underdetermined to allow some non-characteristic covector:

Theorem 4.2 (Formal Integrability of Underdetermined Equations). *Consider a linear map $\phi : \mathrm{Sym}^d T^* \otimes V \longrightarrow E$ possessing at least one non-characteristic covector $\xi \in T^*$ in the sense that the localization $\phi_\xi : V \longrightarrow E, v \longmapsto \phi(\tfrac{1}{d!}\xi^d \otimes v)$, of ϕ at ξ is surjective. The homomorphism of free comodules defining the tableau comodule $\mathscr{A}$ associated to the tableau $\mathscr{A}^d := \mathbf{ker}\,\phi$ is surjective, too, with associated short exact sequence:*

$$0 \longrightarrow \mathscr{A}^\bullet \overset{\subset}{\longrightarrow} \mathrm{Sym}^\bullet T^* \otimes V \overset{\Phi}{\longrightarrow} \mathrm{Sym}^{\bullet-d} T^* \otimes E \longrightarrow 0$$

According to Lemma 3.6 the only non-vanishing Spencer cohomology spaces of $\mathscr{A}$ are:

$$H^{0,0}(\mathscr{A}) \;=\; V \qquad\qquad H^{d-1,1}(\mathscr{A}) \;=\; E$$

By far the most important conclusion of this theorem is that underdetermined partial differential equations have no Spencer cohomology of form degree $\circ = 2$, in consequence there are no obstructions at all to the recursive procedure discussed in Sect. 5 to construct infinite order formal power series solutions for arbitrarily specified Cauchy data. In other words Theorem 4.2 is exactly the reason, why the term Spencer cohomology is never even mentioned in text books studying partial differential equations in the language of Functional Analysis: Banach and Sobolev spaces etc.

Despite its importance the proof of Theorem 4.2 is rather straightforward. Fixing a non-characteristic covector $\xi \in T^*$ with surjective localization $\phi_\xi :$ $V \longrightarrow E$, $v \longmapsto \phi(\frac{1}{d!}\xi^d \otimes v)$, we try to construct a preimage of a vector $\frac{1}{k!}\alpha^k \otimes e \in \mathrm{Sym}^k T^* \otimes E$ under the comodule homomorphism $\Phi : \mathrm{Sym}^\bullet T^* \otimes V \longrightarrow$ $\mathrm{Sym}^{\bullet-d} T^* \otimes E$ extending ϕ by making an ansatz

$$\sum_{\mu=0}^{k} \frac{1}{(k-\mu)!}\alpha^{k-\mu}\frac{1}{(d+\mu)!}\xi^{d+\mu} \otimes v_\mu \in \mathrm{Sym}^{k+d} T^* \otimes V$$

with as yet unknown parameter vectors $v_0, \ldots, v_k \in V$. Inserting this ansatz into the definition (27) of the comodule homomorphism Φ we get after some auxiliary calculations:

$$\Phi\left(\sum_{\mu=0}^{k} \frac{1}{(k-\mu)!}\alpha^{k-\mu}\frac{1}{(d+\mu)!}\xi^{d+\mu} \otimes v_\mu\right)$$

$$= \sum_{s=0}^{k} \frac{1}{(k-s)!}\alpha^{k-s}\frac{1}{s!}\xi^s \otimes \phi\left[\sum_{\mu=0\vee(s-d)}^{s} \frac{1}{(s-\mu)!}\alpha^{s-\mu}\frac{1}{(d+\mu-s)!}\xi^{d+\mu-s} \otimes v_\mu\right]$$

$$= \sum_{s=0}^{k} \frac{1}{(k-s)!}\alpha^{k-s}\frac{1}{s!}\xi^s$$

$$\otimes\left(\phi_\xi v_s + \phi\left[\sum_{\mu=0\vee(s-d)}^{s-1} \frac{1}{(s-\mu)!}\alpha^{s-\mu}\frac{1}{(d+\mu-s)!}\xi^{d+\mu-s} \otimes v_\mu\right]\right)$$

Due to the surjectivity of the localization $\phi_\xi : V \longrightarrow E$ at the non-characteristic covector ξ we may thus choose the parameters $v_0, \ldots, v_k \in V$ of our ansatz recursively to satisfy

$$\phi_\xi v_0 = e$$

$$\phi_\xi v_1 = -\phi\left[\frac{1}{1!}\alpha^1\frac{1}{(d-1)!}\xi^{d-1} \otimes v_0\right]$$

$$\phi_\xi v_2 = -\phi\left[\frac{1}{2!}\alpha^2\frac{1}{(d-2)!}\xi^{d-2} \otimes v_0 + \frac{1}{1!}\alpha^1\frac{1}{(d-1)!}\xi^{d-1} \otimes v_1\right]$$

etc. in order to obtain a preimage of $\frac{1}{k!}\alpha^k \otimes e \in \mathrm{Sym}^k T^* \otimes E$ under Φ. In this argument $\alpha \in T^*$ and $e \in E$ as well as $k \in \mathbb{N}_0$ were all arbitrary so that Φ is surjective

$$\mathbf{im}\; \Phi \;\supset\; \mathrm{span}\{\, \tfrac{1}{k!}\alpha^k \otimes e \;\mid\; \alpha \in T^*,\, e \in E,\, k \in \mathbb{N}_0 \,\} \;=\; \mathrm{Sym}\, T^* \otimes E$$

because the polarization formula says that the vectors $\frac{1}{k!}\alpha^k \otimes e$ span $\mathrm{Sym}^k T^* \otimes E$. Unluckily the other two classical statements about tableau comodules discussed in this section are more difficult, in particular the following characterization of partial differential equations of finite type as complex elliptic differential equations requires confidence in multilinear algebra:

Theorem 4.3 (Complex Elliptic Partial Differential Equations). *The homomorphism Φ of free comodules extending a given linear map $\phi : \mathrm{Sym}^d T^* \otimes V \longrightarrow E$ with kernel tableau $\mathscr{A}^d := \ker \phi$ has a finite-dimensional kernel comodule $\mathscr{A}$*

$$0 \longrightarrow \mathscr{A}^\bullet \overset{\subset}{\longrightarrow} \mathrm{Sym}^\bullet T^* \otimes V \overset{\Phi}{\longrightarrow} \mathrm{Sym}^{\bullet-d} T^* \otimes E$$

if and only if only the complex localizations of ϕ at complex valued linear forms $\xi_{\mathbb{C}} \in T^* \otimes_{\mathbb{R}} \mathbb{C}$:

$$\phi_{\xi_{\mathbb{C}}} : V \otimes_{\mathbb{R}} \mathbb{C} \longrightarrow E \otimes_{\mathbb{R}} \mathbb{C}, \qquad v_{\mathbb{C}} \longmapsto \phi(\, \tfrac{1}{d!}\xi_{\mathbb{C}}^d \otimes v_{\mathbb{C}}\,)$$

are injective for every non-zero complex-valued linear form $\xi_{\mathbb{C}} \in T^* \otimes_{\mathbb{R}} \mathbb{C} \setminus \{0\}$.

Essentially this theorem is a consequence of Hilbert's Nullstellensatz in algebraic geometry, although the necessary reformulation (37) can hardly be called obvious. Nevertheless it is well worth the effort to try to understand the main idea of this reformulation, because it provides us with a peculiar kind of upper bound on the growth of a tableau comodule $\mathscr{A}$ in terms of the homogeneous ideal $I^\bullet \subset \mathrm{Sym}^\bullet T$ defining the characteristic variety:

$$\mathscr{Z}(\mathscr{A}^d) \;:=\; \{\, [\xi] \in \mathbb{P}T^* \mid \phi_\xi : V \longrightarrow E \text{ is not injective} \,\} \qquad (32)$$

Needless to say this definition is different to our previous definition (31) of the characteristic variety, although we may reconcile both definitions by asking for covectors $\xi \in T^*$ such that ϕ_ξ fails to be of the maximal possible rank $\min\{\dim V, \dim E\}$. In other words the redefinition (32) is specific to the study of overdetermined partial differential equations.

Let us begin our discussion of Theorem 4.3 with a small side remark about the rank of a linear map $\phi : V \longrightarrow E$. The canonical isomorphism $\mathrm{Hom}(V, E) \cong V^* \otimes E$ allows us to think of ϕ as an element of the algebra $\Lambda V^* \otimes \Lambda E$ with the (untwisted) tensor product multiplication. Pairing powers of ϕ in this algebra with elements of the dual space we get

$$\langle\, \tfrac{1}{r!}\phi^r,\, (v_1 \wedge \ldots \wedge v_r) \otimes \eta\,\rangle \;=\; \eta(\phi v_1, \ldots, \phi v_r) \qquad (33)$$

for all $r \geq 1$ and all $v_1, \ldots, v_r \in V$, $\eta \in \Lambda^r E^*$, in particular $\frac{1}{r!} \phi^r = 0$ is equivalent to ϕ being of rank less than r. Similarly we may interpret the linear map $\phi : \mathrm{Sym}^d T^* \otimes V \longrightarrow E$ defining the tableau $\mathscr{A}^d$ as a homogeneous polynomial of degree d on T^* with values in $\mathrm{Hom}\,(V, E)$ or as an element $\phi \in \mathrm{Sym}^d T \otimes V^* \otimes E$ of the algebra $\mathrm{Sym}\, T \otimes \Lambda\, V^* \otimes \Lambda\, E$. In direct generalization of Eq. (33) the powers of ϕ in this algebra satisfy

$$\langle\, \tfrac{1}{r!} \phi^r,\ \tfrac{1}{(rd)!} \xi^{rd} \otimes (v_1 \wedge \ldots \wedge v_r) \otimes \eta\, \rangle\ =\ \eta\,(\, \phi_\xi v_1, \ldots, \phi_\xi v_r\,) \qquad (34)$$

when paired with elements of the dual space $\mathrm{Sym}^{rd} T^* \otimes \Lambda^r V \otimes \Lambda^r E^*$ with arbitrary $\xi \in T^*$, $v_1, \ldots, v_r \in V$ and $\eta \in \Lambda^r E^*$. Equation (34) implicitly characterizes the covectors $\xi \in T^*$, for which the localization $\phi_\xi : V \longrightarrow E$ fails to have rank at least r, in terms of the power $\frac{1}{r!} \phi^r \in \mathrm{Sym}^{rd} T \otimes \Lambda^r V^* \otimes \Lambda^r E$. In order to make this characterization somewhat more explicit let us consider the following two rearrangements of the factors in (34)

$$\iota_r^\phi :\ \Lambda^r V \otimes \Lambda^r E^* \longrightarrow \mathrm{Sym}^{rd} T$$
$$\mu_r^\phi :\ \Lambda^{r-1} V \otimes \Lambda^r E^* \longrightarrow \mathrm{Sym}^{rd} T \otimes V^*$$

characterized as linear maps by:

$$\begin{aligned}
\eta\,(\, \phi_\xi v_1, \ldots, \phi_\xi v_r\,) &=: \langle\, \iota_r^\phi (v_1 \wedge v_2 \wedge \ldots \wedge v_r \otimes \eta),\ \tfrac{1}{(rd)!} \xi^{rd}\, \rangle \\
&=: \langle\, \mu_r^\phi (v_2 \wedge \ldots \wedge v_r \otimes \eta),\ \tfrac{1}{(rd)!} \xi^{rd} \otimes v_1\, \rangle
\end{aligned} \qquad (35)$$

The localization $\phi_\xi : V \longrightarrow E$ of $\phi : \mathrm{Sym}^d T^* \otimes V \longrightarrow E$ at a covector $\xi \in T^*$ thus has rank less than r, if and only if ξ is a common zero of all polynomials in $\mathrm{im}\, \iota_r^\phi \subset \mathrm{Sym}^{rd} T$ and thus a common zero of all polynomials in the homogeneous ideal generated by $\mathrm{im}\, \iota_r^\phi$:

$$I_r^{\phi\,\bullet}\ :=\ \langle\, \mathrm{im}\, \iota_r^\phi\, \rangle\ \subset\ \mathrm{Sym}^\bullet T$$

In consequence the homogeneous ideal $I_r^{\phi\,\bullet}$, $r \geq 1$, defines the projective algebraic variety

$$\begin{aligned}
\mathscr{Z}_r(\, \mathscr{A}^d\,) &:= \{\, [\xi] \in \mathbb{P}\,T^*\ |\ \psi(\xi) = 0 \text{ for all polynomials } \psi \in I_r^\phi\, \} \\
&= \{\, [\xi] \in \mathbb{P}\,T^*\ |\ \phi_\xi : V \longrightarrow E \text{ has rank less than } r\, \}
\end{aligned}$$

associated to the linear map $\phi :\ \mathrm{Sym}^d T^* \otimes V \longrightarrow E$ or its tableau comodule $\mathscr{A}$, which is called the rth systolic variety in [7]. In particular the characteristic variety (32) is defined by the homogeneous ideal $I_N^{\phi\,\bullet}$ corresponding to $N :=$ dim V.

In the same vein we may consider the graded submodule $M_r^{\phi\,\bullet} := \langle\, \mathrm{im}\, \mu_r^\phi\, \rangle \subset \mathrm{Sym}^\bullet T \otimes V^*$ generated by the image of $\mu_r^\phi : \Lambda^{r-1} V \otimes \Lambda^r E^* \longrightarrow \mathrm{Sym}^{rd} T \otimes V^*$.

Although the submodules $M_r^{\phi\,\bullet} \subset \mathrm{Sym}^\bullet T \otimes V^*$ seem to have no direct geometric interpretation in terms of the characteristic variety, they possess an interesting algebraic property in that they are upper bounds for the tableau comodule $\mathscr{A}$ associated to the tableau $\mathscr{A}^d = \ker \phi$. To see this point clearly let us rewrite the definition of $\mu_r^\phi : \Lambda^{r-1} V \otimes \Lambda^r E^* \longrightarrow \mathrm{Sym}^{rd} T \otimes V^*$ in the form

$$\langle\, \mu_r^\phi(v_2 \wedge \ldots \wedge v_r \otimes \eta),\ \tfrac{1}{(rd)!} \xi^{rd} \otimes v \,\rangle$$

$$:= \eta(\phi(\tfrac{1}{r!} \xi^r \otimes v),\, \phi^{v_2}(\tfrac{1}{r!} \xi^r),\ldots,\phi^{v_r}(\tfrac{1}{r!} \xi^r))$$

$$= \langle\, \eta,\ (\phi \wedge \phi^{v_2} \wedge \ldots \wedge \phi^{v_r})(\Delta[\tfrac{1}{(rd)!} \xi^{rd} \otimes v])\,\rangle$$

where $\phi^v : \mathrm{Sym}^d T^* \longrightarrow E,\ \tfrac{1}{d!}\xi^d \longmapsto \phi(\tfrac{1}{d!}\xi \otimes v)$, denotes the localization of ϕ at some $v \in V$ and $\Delta : \mathrm{Sym}^{rd} T \otimes V^* \longrightarrow (\mathrm{Sym}^d T \otimes V^*) \otimes \mathrm{Sym}^d T \otimes \ldots \otimes \mathrm{Sym}^d T$ the comultiplication:

$$\Delta[\tfrac{1}{(rd)!} \xi^{rd} \otimes v] := (\tfrac{1}{d!}\xi^d \otimes v) \otimes \underbrace{(\tfrac{1}{d!}\xi^d) \otimes \ldots \otimes (\tfrac{1}{d!}\xi^d)}_{r-1 \text{ times}}$$

The decisive observation linking the tableau comodule $\mathscr{A}$ to the submodules $M_r^\phi,\ r \geq 1$, and eventually to the ideals $I_r^\phi,\ r \geq 1$, is that the comultiplication Δ restricts to a map

$$\Delta : \quad \mathscr{A}^{rd} \longrightarrow \mathscr{A}^d \otimes \mathrm{Sym}^d T^* \otimes \ldots \otimes \mathrm{Sym}^d T^*$$

simply because $\mathscr{A}$ is after all a comodule over the symmetric coalgebra $\mathrm{Sym}\, T^*$. Hence

$$\langle\, \mu_r^\phi(v_2 \wedge \ldots \wedge v_r \otimes \eta),\ a \,\rangle \;=\; \langle\, \eta,\ (\phi \wedge \phi^{v_2} \wedge \ldots \wedge \phi^{v_r})(\Delta a)\,\rangle \;=\; 0$$

vanishes for all $a \in \mathscr{A}^{rd}$ and all $v_2,\ldots,v_r \in V,\ \eta \in \Lambda^r E^*$ due to the consequence $(\phi \wedge \phi^{v_2} \wedge \ldots \wedge \phi^{v_r})(\Delta a) = 0$ of the equality $\mathscr{A}^d = \ker \phi$. In turn the canonical pairing between $\mathrm{Sym}^{rd} T \otimes V^*$ and $\mathrm{Sym}^{rd} T^* \otimes V$ vanishes $\langle m,\, a \rangle = 0$ on all pairs $a \in \mathscr{A}^{rd}$ and $m \in \mathrm{im}\, \mu_r^\phi = (M_r^\phi)^{rd}$, and this mutual annihilation property extends immediately

$$\langle m,\, a \rangle \;=\; 0 \tag{36}$$

to all $m \in M_r^{\phi\,\bullet},\ r \geq 1$, and all $a \in \mathscr{A}^\bullet$, because the submodule $M_r^{\phi\,\bullet} \subset \mathrm{Sym}^\bullet T \otimes V^*$ generated by $\mathrm{im}\, \mu_r^\phi$ is spanned by elements of the form $\tfrac{1}{s!} t^s \cdot m$ with $t \in T,\ s \in \mathbb{N}_0$ and $m \in \mathrm{im}\, \mu_r^\phi$, however all these elements satisfy $\langle\, \tfrac{1}{s!} t^s \cdot m, a\,\rangle = \langle m,\, \tfrac{1}{s!} \tfrac{\partial^s}{\partial t^s} a \,\rangle = 0$:

Corollary 4.9 (Upper Bound for Tableau Comodules). *Consider the tableau comodule $\mathscr{A}$ associated to a tableau $\mathscr{A}^d \subset \mathrm{Sym}^d T^* \otimes V$ of order $d \geq 1$ and* *a linear map* $\phi : \text{Sym}^d T^* \otimes V \longrightarrow E$ *realizing* $\mathscr{A}^d$ *in the sense* $\ker \phi = \mathscr{A}^d$. *The powers of the linear map* $\phi \in \text{Sym}^d T \otimes V^* \otimes E$ *in the algebra* $\text{Sym } T \otimes \wedge V^* \otimes \wedge E$ *give rise to a sequence of linear maps* $\mu_r^\phi : \wedge^{r-1} V \otimes \wedge^r E^* \longrightarrow \text{Sym}^{rd} T \otimes V^*$, $r \geq 1$, *with the property*

$$\mathscr{A} \subset \text{Ann } M_r^\phi := \{ a \in \text{Sym } T^* \otimes V \mid \langle m, a \rangle = 0 \text{ for all } m \in M_r^\phi \}$$

where $M_r^\phi := \langle \text{im } \mu_r^\phi \rangle$ *denotes the* $\text{Sym}^\bullet T$-*submodule of* $\text{Sym}^\bullet T \otimes V^*$ *generated by* $\text{im } \mu_r^\phi$.

The preceding lemma is certainly interesting for all $r \geq 1$, nevertheless it has an additional twist for r equal to the dimension $N := \dim V$ of V in that the inclusion $M_r^\phi \subset I_r^\phi \otimes V^*$ becomes an actual equality $M_N^\phi = I_N^\phi \otimes V^*$ for this r. Choosing a dual pair of bases $v_1, \ldots, v_N$ and $dv_1, \ldots, dv_N$ for the vector spaces V and V^* we may in fact reformulate the identity $\langle \mu_r^\phi(\tilde{v}_2 \wedge \ldots \wedge \tilde{v}_r \otimes \eta), \cdot \otimes \tilde{v}_1 \rangle = \langle \iota_r^\phi(\tilde{v}_1 \wedge \tilde{v}_2 \wedge \ldots \wedge \tilde{v}_r \otimes \eta), \cdot \rangle$ derived from the definition (35) of ι_r^ϕ and μ_r^ϕ into an expansion valid for all $\tilde{v}_2, \ldots, \tilde{v}_r \in V$ and all $\eta \in \wedge^r E^*$:

$$\mu_r^\phi(\tilde{v}_2 \wedge \ldots \wedge \tilde{v}_r \otimes \eta) = \sum_{\lambda=1}^N \iota_r^\phi(v_\lambda \wedge \tilde{v}_2 \wedge \ldots \wedge \tilde{v}_r \otimes \eta) \otimes dv_\lambda$$

This expansion tells us $\text{im } \mu_r^\phi \subset \text{im } \iota_r^\phi \otimes V^*$ and so $M_r^\phi \subset I_r^\phi \otimes V^*$ for all $r \geq 1$. The penultimate exterior power $\wedge^{N-1} V$ of V however is spanned by the multivectors obtained by removing a factor v_s from $v_1 \wedge \ldots \wedge v_N \in \wedge^N V$, the preceding equation thus becomes

$$\mu_r^\phi(v_1 \wedge \ldots \wedge \widehat{v_s} \wedge \ldots \wedge v_N \otimes \eta) = (-1)^{s-1} \iota_r^\phi(v_1 \wedge \ldots \wedge v_N \otimes \eta) \otimes dv_s$$

for these multivectors and all $\eta \in \wedge^N E^*$, $s = 1, \ldots, N$ so that $\text{im } \mu_N^\phi = \text{im } \iota_N^\phi \otimes V^*$ and in turn $M_N^\phi = I_N^\phi \otimes V^*$. Combined with Corollary 4.9 this insight establishes the direct link

$$\mathscr{A}^\bullet \subset \text{Ann}(I_N^{\phi \bullet} \otimes V^*) \tag{37}$$

between the tableau comodule $\mathscr{A}$ associated to a tableau $\mathscr{A}^d$ and the homogeneous ideal $I_N^{\phi \bullet}$ defining its characteristic variety $\mathscr{X}(\mathscr{A}^d)$. Before using this direct link in the proof of Theorem 4.3 we want to state an alternative version of (37) in terms of differential operators:

Corollary 4.5 (Scalar Differential Constraints). *Consider a linear map* $\phi : \text{Sym}^d T^* \otimes V \longrightarrow E$ *realizing a tableau* $\mathscr{A}^d \subset \text{Sym}^d T^* \otimes V$ *of order* $d \geq 1$ *in the sense* $\mathscr{A}^d = \ker \phi$. *Equation (28) associates to* ϕ *a linear differential operator* $D_\phi : C^\infty(T, V) \longrightarrow C^\infty(T, E)$ *of order* d, *in complete analogy every homogeneous element* $D \in I_N^k \subset \text{Sym}^k T$ *in the ideal* I_N^ϕ *defining the*

characteristic variety $\mathscr{Z}(\mathscr{A}^d)$ can be interpreted as a scalar differential operator
$D: C^\infty(T) \longrightarrow C^\infty(T)$ of order k. For every solution $\psi \in C^\infty(T, V)$ of the
differential equation $D_\phi \psi = 0$ it holds then true that:

$$(D \otimes \mathrm{id}_V)\,\psi = 0$$

Solutions $\psi \in C^\infty(T, V)$ of the differential equation $D_\phi \psi = 0$ are characterized
by the fact that the homogeneous pieces of their Taylor series $\mathrm{taylor}_p\, \psi \in \overline{\mathrm{Sym}}\, T^* \otimes$
V taken in an arbitrary point $p \in T$ are elements of $\mathscr{A}^k$ for all $k \geq 0$. On the
other hand the value of the scalar differential operator associated to $D \in I_N^k$ on
$\psi \in C^\infty(T, V)$ is given by a sum

$$\left((D \otimes \mathrm{id}_V)\,\psi\right)(p) = \sum_{\lambda=1}^{N}\left(D\langle\, dv_\lambda,\, \psi\,\rangle\right)(p)\, v_\lambda = \sum_{\lambda=1}^{N}\langle\, D \otimes dv_\lambda,\, \mathrm{taylor}_p^k\, \psi\,\rangle\, v_\lambda$$

over a dual pair $v_1, \ldots, v_N$ and $dv_1, \ldots, dv_N$ of bases for V and V^*. Equation (37)
thus tells us that the right hand side vanishes for a solution ψ of the equation
$D_\phi \psi = 0$.

Proof of Theorem 4.3. According to Hilbert's Nullstellensatz from algebraic geometry every homogeneous polynomial $\psi \in \mathbb{C}[x^1, \ldots, x^n]$ of positive degree
vanishing on all points of a projective variety $\mathscr{Z}_\mathbb{C}$, which is the vanishing variety
of some homogeneous ideal I

$$\mathscr{Z}_\mathbb{C} = \{\,[\xi_\mathbb{C}] \in \mathbb{P}\mathbb{C}^n \mid p(\xi_\mathbb{C}) = 0 \text{ for all } p \in I\,\}$$

lies in the radical $\sqrt{I} \subset \mathbb{C}[x^1, \ldots, x^n]$ of I in the sense $\psi^e \in I$ for sufficiently
large exponent $e \in \mathbb{N}$. Among the well-known consequences of this theorem is
that the radical of a homogeneous ideal I with empty vanishing variety $\mathscr{Z}_\mathbb{C} = \emptyset$
equals the "irrelevant" ideal $\sqrt{I} = \mathbb{C}^+[x^1, \ldots, x^n]$ consisting of all polynomials of
positive degree [5]. Exactly this particular consequence of Hilbert's Nullstellensatz
is what Theorem 4.3 is all about.

   Unluckily we have been working over the real numbers as of now and not over
the algebraically closed field $\mathbb{C}$ required by Hilbert's Nullstellensatz. Multilinear
algebra however behaves nicely under complexification inasmuch as we have
canonical identifications

$$(\mathrm{Sym}\, T^*) \otimes_\mathbb{R} \mathbb{C} = \mathrm{Sym}\,(T^* \otimes_\mathbb{R} \mathbb{C}) \qquad (\Lambda\, V^*) \otimes_\mathbb{R} \mathbb{C} = \Lambda\,(V^* \otimes_\mathbb{R} \mathbb{C})$$

of symmetric and exterior powers as well as tensor products etc. Of course we
could go about and repeat all our calculations and constructions for the complexified
linear map

$$\phi \otimes_\mathbb{R} \mathrm{id}: \quad \mathrm{Sym}^d(T^* \otimes_\mathbb{R} \mathbb{C}) \otimes_\mathbb{C} (V \otimes_\mathbb{R} \mathbb{C}) = (\mathrm{Sym}^d\, T^* \otimes V) \otimes_\mathbb{R} \mathbb{C} \longrightarrow E \otimes_\mathbb{R} \mathbb{C}$$

with complex localizations $\phi_{\xi_\mathbb{C}} : V \otimes_\mathbb{R} \mathbb{C} \longrightarrow E \otimes_\mathbb{R} \mathbb{C}, v_\mathbb{C} \longmapsto (\phi \otimes_\mathbb{R}$ id$)(\frac{1}{d!}\xi_\mathbb{C}^d \otimes v_\mathbb{C})$, etc., however the upshot of all these calculations is that the complex characteristic variety

$$\mathcal{Z}_\mathbb{C}(\mathscr{A}^d)$$

$$:= \{[\xi_\mathbb{C}] \in \mathbb{P}(T^* \otimes_\mathbb{R} \mathbb{C}) \mid \phi_{\xi_\mathbb{C}} : (V \otimes_\mathbb{R} \mathbb{C}) \longrightarrow (E \otimes_\mathbb{R} \mathbb{C}) \text{ is not injective}\}$$

can be defined by the complexified ideal $I_N^{\phi \bullet} \otimes_\mathbb{R} \mathbb{C} \subset \mathrm{Sym}^\bullet(T \otimes_\mathbb{R} \mathbb{C})$, while the kernel of the extension $(\Phi \otimes_\mathbb{R}$ id$) : (\mathrm{Sym}^\bullet T^* \otimes V) \otimes_\mathbb{R} \mathbb{C} \longrightarrow (\mathrm{Sym}^{\bullet-d}T^* \otimes E) \otimes_\mathbb{R} \mathbb{C}$ of the complexified linear map $\phi \otimes_\mathbb{R}$ id to comodules over the symmetric coalgebra $\mathrm{Sym}(T^* \otimes_\mathbb{R} \mathbb{C})$ is nothing else but the complexification $\mathscr{A}^\bullet \otimes_\mathbb{R} \mathbb{C}$ of the tableau comodule associated to ϕ.

Let us suppose now that the localization $\phi_{\xi_\mathbb{C}} : V \otimes_\mathbb{R} \mathbb{C} \longrightarrow E \otimes_\mathbb{R} \mathbb{C}$ at some non-zero complex valued form $\xi_\mathbb{C} \in T^* \otimes_\mathbb{R} \mathbb{C}$ is not injective and let $v_\mathbb{C} \in V \otimes_\mathbb{R} \mathbb{C}$ be a non-zero vector in its kernel. With these choices made the product vector $\frac{1}{k!}\xi_\mathbb{C}^k \otimes v_\mathbb{C} \neq 0$ is for all $k \geq 0$ a non-zero element of the kernel $\mathscr{A}^k \otimes_\mathbb{R} \mathbb{C}$ of the comodule extension $\Phi \otimes_\mathbb{R}$ id of $\phi \otimes_\mathbb{R}$ id

$$(\Phi \otimes_\mathbb{R} \text{id})(\tfrac{1}{k!}\xi_\mathbb{C}^k \otimes v_\mathbb{C}) = \tfrac{1}{(k-d)!}\xi_\mathbb{C}^{k-d} \otimes (\phi \otimes_\mathbb{R} \text{id})(\tfrac{1}{d!}\xi_\mathbb{C}^d \otimes v_\mathbb{C}) = 0$$

so that dim $\mathscr{A}^k \geq 1$ for all $k \geq 0$ leading to the comodule $\mathscr{A}$ of infinite dimension. Conversely suppose that all localizations $\phi_{\xi_\mathbb{C}} : V \otimes_\mathbb{R} \mathbb{C} \longrightarrow E \otimes_\mathbb{R} \mathbb{C}$ at non-zero complex valued forms $\xi_\mathbb{C} \in T^* \otimes_\mathbb{R} \mathbb{C}$ are injective. The vectors $t_1, \ldots, t_n$ of a basis for T considered as homogeneous polynomials of degree 1 on $T^* \otimes_\mathbb{R} \mathbb{C}$ trivially vanish on the complex characteristic variety $\mathcal{Z}_\mathbb{C}(\mathscr{A}^d) = \varnothing$ as it is empty, hence Hilbert's Nullstellensatz guarantees the existence of exponents $e_1, \ldots, e_n \in \mathbb{N}$ such that the powers $t_1^{e_1}, t_2^{e_2}, \ldots, t_n^{e_n} \in I_N^\phi$ of the basis vectors are real elements of the homogeneous ideal $I_N^\phi \otimes_\mathbb{R} \mathbb{C}$ describing the complex projective variety $\mathcal{Z}_\mathbb{C}(\mathscr{A}^d)$. In turn the drawers principle asserts that every monomial $t^{k_1} \ldots t^{k_n}$ in the basis vectors $t_1, \ldots, t_n$ of total degree $k_1 + \ldots + k_n > e_1 + \ldots + e_n - n$ is an element of the ideal I_N^ϕ, because at least one of the basis vectors t_μ occurs with an exponent $k_\mu \geq e_\mu$. Since the monomials in basis vectors span the symmetric powers we conclude

$$I_N^{\phi \bullet} = \mathrm{Sym}^\bullet T$$

and so $\mathscr{A}^\bullet = \{0\}$ for all $\bullet > e_1 + \ldots + e_n - n$ due to $\mathscr{A}^\bullet \subset \mathrm{Ann}(I_N^{\phi \bullet} \otimes V^*) = \{0\}$ according to the direct link (37) between $\mathscr{A}$ and I_N^ϕ. With all homogeneous subspaces of $\mathscr{A}^\bullet \subset \mathrm{Sym}^\bullet T^* \otimes V$ being finite dimensional we conclude dim $\mathscr{A} < \infty$.   □

In the second part of this section we want to discuss the main ideas and their ramifications related to Cartan's Involutivity Test for first order tableaux $\mathscr{A}^1 \subset T^* \otimes V$. In contrast to tableaux of higher order tableaux of first order $d = 1$ possess a very interesting discrete invariant, the so-called Cartan character, under the natural action of **GL** $T \times$ **GL** V on the subspaces of $T^* \otimes V = \mathrm{Hom}\,(\,T,\,V\,)$. The nomenclature adopted by Cartan and his collaborators with respect to tableaux alludes directly to the fact that this Cartan character, although usually written as a decreasing sequence of non-negative integers, is actually a Young diagram, a very interesting combinatorial structure with strong ties to the representation theory of the general linear groups: A Young diagram with additional "filling" is traditionally called a Young tableau (sic!) in representation theory.

A Young diagram is by definition a finite set $\mathfrak{Y} \subset \mathbb{N}^2$ of tuples of natural numbers with the property that for every tuple $(r, c) \in \mathfrak{Y}$ all tuples $(\tilde{r}, \tilde{c}) \in \mathbb{N}^2$ of natural numbers satisfying both inequalities $\tilde{r} \leq r$ and $\tilde{c} \leq c$ are elements $(\tilde{r}, \tilde{c}) \in \mathfrak{Y}$, too. In the parlance of partially ordered sets and lattices we may equivalently define a Young diagram as a finite lower subset $\mathfrak{Y} \subset \mathbb{N}^2$ with respect to the componentwise partial order $\geq$ on $\mathbb{N}^2$. It is much more appropriate though to think of a Young diagram as a picture of little squares neatly aligned in rows and columns in an arrangement similar to matrices:

$$\mathfrak{Y} = \{\, (1,1),\, (1,2),\, (1,3),\, (1,4),\, (1,5),$$
$$(2,1),\, (2,2),\, (2,3),\, (2,4),\, (3,1),$$
$$(3,2),\, (4,1),\, (4,2),\, (5,1),\, (6,1)\,\}$$

Due to this interpretation the elements of a Young diagram are called its boxes, the number $\sharp\mathfrak{Y}$ of boxes of a Young diagram $\mathfrak{Y}$ is called its order. Say the Young diagram depicted above has order 15 with boxes arranged in columns of lengths $6 \geq 4 \geq 2 \geq 2 \geq 1$ and rows of lengths $5 \geq 4 \geq 2 \geq 2 \geq 1 \geq 1$. Similarly every Young diagram $\mathfrak{Y}$ is completely determined by the lengths $c_1 \geq c_2 \geq c_3 \geq \ldots$ of its columns or the lengths $r_1 \geq r_2 \geq r_3 \geq \ldots$ of its rows. The image of a Young diagram $\mathfrak{Y} \subset \mathbb{N}^2$ under the reflection along the main diagonal $(r, c) \longmapsto (c, r)$ interchanging rows and columns is again a Young diagram of the same order called the diagram $\mathfrak{Y}^*$ conjugated to $\mathfrak{Y}$. Moreover the finite set

$$\mathbf{YD}(\,D\,) \quad := \quad \{\, \mathfrak{Y} \subset \mathbb{N}^2 \mid \mathfrak{Y} \text{ is a Young diagram of order } \sharp\mathfrak{Y} = D \,\}$$

of all Young diagrams of fixed order $D \in \mathbb{N}_0$ comes along with a partial order $\geq$ defined by:

$$\mathfrak{Y} \geq \tilde{\mathfrak{Y}} \qquad \Leftrightarrow \qquad \sum_{\mu=1}^{s} c_\mu \geq \sum_{\mu=1}^{s} \tilde{c}_\mu \quad \text{for all } s \geq 1 \qquad (38)$$

In other words $\mathfrak{Y} \geq \tilde{\mathfrak{Y}}$, if and only if $\mathfrak{Y}$ has at least as many boxes in the first column as $\tilde{\mathfrak{Y}}$, at least as many boxes in the first two columns together as $\tilde{\mathfrak{Y}}$ and so on. Under this partial order the set $\mathbf{YD}(\,D\,)$ of Young diagrams of order $D \geq 0$ is actually a self dual lattice with antimonotone involution $* : \mathbf{YD}(\,D\,) \longrightarrow \mathbf{YD}(\,D\,),\, \mathfrak{Y} \longmapsto \mathfrak{Y}^*$. Lacking a pretext we will not discuss these beautiful examples of self-dual lattices in more detail, because only the partial order $\geq$ enters into the definition of the Cartan character of a first order tableau $\mathscr{A}^1 \subset \mathrm{Hom}\,(\,T,\,V\,)$. Perhaps the reader will enjoy studying the following Hasse diagram of $\mathbf{YD}(\,11\,)$ though, in which the Young diagrams are ordered descendingly from left to right:

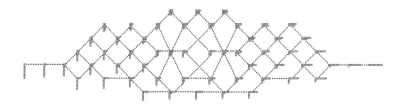

The similarity between Young diagrams and matrices mentioned before provides the fundamental link between Young diagrams and tableaux. Recall from your first semesters at university that the choice of bases $t_1, \ldots, t_n$ for T and $v_1, \ldots, v_N$ for V turns the vector space $\mathrm{Hom}\,(\,T,\,V\,)$ into the vector space of all $N \times n$-matrices via the following linear map

$$\mathrm{mat}: \quad \mathrm{Hom}\,(\,T,\,V\,) \;\overset{\cong}{\longrightarrow}\; \mathrm{Mat}_{N \times n}\mathbb{R}, \qquad A \longmapsto \begin{pmatrix} dv_1(\,At_1\,) & \ldots & dv_1(\,At_n\,) \\ \vdots & & \vdots \\ dv_N(\,At_1\,) & \ldots & dv_N(\,At_n\,) \end{pmatrix}$$

where $dv_1, \ldots, dv_N \in V^*$ is the basis dual to $v_1, \ldots, v_N$. Modulo the choice of bases for T and V every tableau $\mathscr{A}^1 \subset \mathrm{Hom}\,(\,T,\,V\,)$ may thus be thought of as a subspace of matrices, in turn the coefficient of the image matrix in row r and column c becomes the linear functional:

$$\mathrm{mat}_{rc}: \quad \mathscr{A}^1 \longrightarrow \mathbb{R}, \qquad A \longmapsto \mathrm{mat}_{rc}(A) := dv_r(\,At_c\,)$$

It should be noted that the matrix coefficients $\mathrm{mat}_{rc} \in \mathscr{A}^{1*}$ span the space $\mathscr{A}^{1*}$ of linear functionals on $\mathscr{A}^1$ since $\mathrm{mat}: \mathscr{A}^1 \longrightarrow \mathrm{Mat}_{N \times n}\mathbb{R}$ is injective. Hence we may choose a basis $\{\,\mathrm{mat}_{rc}\,\}_{(r,c)\in\mathfrak{Y}}$ of $\mathscr{A}^{1*}$ consisting entirely of the matrix coefficients indexed by a suitable subset $\mathfrak{Y} \subset \{1, \ldots, N\} \times \{1, \ldots, n\}$. The remaining matrix coefficients are then *fixed* linear combinations of the matrix coefficients in $\mathfrak{Y}$.

It may be somewhat surprising, but the preceding rather esoteric discussion about the linear independence of matrix coefficients captures exactly what we do automatically, whenever we specify a subspace of matrices. Consider for example the following subspace

$$\left\{ \begin{pmatrix} a & b \\ 0 & 2a-b \end{pmatrix} \;\middle|\; a, b \in \mathbb{R} \right\} \;=\; \left\{ \begin{pmatrix} \tfrac{1}{2}b + \tfrac{1}{2}d & b \\ 0 & d \end{pmatrix} \;\middle|\; b, d \in \mathbb{R} \right\}$$

of $\mathrm{Mat}_{2\times 2}\mathbb{R}$. Although defining the same subspace the left and right hand side definitions differ significantly in form, implicitly we have chosen the basis $\{\,a,\ b\,\}$ of matrix coefficients corresponding to $\{\,(1,1),\ (1,2)\,\}$ on the left and the basis $\{\,b,\ d\,\}$ corresponding to $\{\,(1,2),\ (2,2)\,\}$ on the right hand side. The subset $\{\,(1,1),\ (2,2)\,\}$ of matrix coefficients corresponds to a third alternative definition of the same subspace of 2×2-matrices, whereas subsets containing $(2,1)$ certainly do not correspond to bases of matrix coefficients:

Definition 4.6 (Young Diagrams Presenting a Tableau). A Young diagram $\mathfrak{Y}$ of order D is said to present a first order tableau $\mathscr{A}^1 \subset \mathrm{Hom}\,(\,T,\ V\,)$ of dimension $D\ :=\ \dim\,\mathscr{A}^1$, if there exist some bases $t_1,\dots,t_n$ and $v_1,\dots,v_N$ of T and V respectively such that the associated matrix coefficients $\mathrm{mat}_{rc} \in \mathscr{A}^{1*}$ indexed by $(r,c) \in \mathfrak{Y}$

$$\mathrm{mat}_{rc}:\quad \mathscr{A}^1 \longrightarrow \mathbb{R},\qquad A \longmapsto dv_r(\,At_c\,)$$

are a basis of $\mathscr{A}^{1*}$. Schematically we may then write $\mathrm{mat}(\,\mathscr{A}^1\,) \subset \mathrm{Mat}_{N\times n}\mathbb{R}$ in the form

$$\mathrm{mat}(\,\mathscr{A}^1\,)\ =\ \left\{\left(\ \vcenter{\hbox{}}\ \right)\right\}$$

where the coefficients in $\mathfrak{Y}$ can be assigned arbitrary values, the fixed linear combinations of these values calculating the other coefficients characterize the subspace $\mathrm{mat}(\,\mathscr{A}^1\,)$.

In saying that a Young diagram $\mathfrak{Y}$ of order D presents a tableau $\mathscr{A}^1 \subset \mathrm{Hom}\,(\,T,\ V\,)$ of dimension $D\ =\ \dim\,\mathscr{A}^1$ we deliberately draw attention away from the bases of T and V realizing this presentation. In this way the set of Young diagrams presenting a given tableau $\mathscr{A}^1$ of dimension D becomes an invariant of the tableau under the natural action $\mathbf{GL}\,T \times \mathbf{GL}\,V$ on the Graßmannian of D-dimensional subspaces of $\mathrm{Hom}\,(\,T,\ V\,)$. It should not pass by unnoticed that this invariant with values in the subsets of $\mathbf{YD}(\,D\,)$ has a compelling interpretation in terms of the Plücker embedding $\mathrm{Gr}_D\,\mathrm{Hom}\,(\,T,\ V\,) \longrightarrow \mathbb{P}(\,\Lambda^D\mathrm{Hom}\,(\,T,\ V\,)\,)$. According to the representation theory of general linear groups [3] the domain of the Plücker embedding decomposes under $\mathbf{GL}\,T \times \mathbf{GL}\,V$ into a direct sum of irreducible subrepresentations

$$\Lambda^D\mathrm{Hom}\,(\,T,\ V\,)\ =\ \bigoplus_{\substack{\mathfrak{Y}\subset\mathbf{YD}(\,D\,)\\ \sharp\,\mathrm{rows}\leq N\\ \sharp\,\mathrm{columns}\leq n}}\ \mathrm{Schur}^{\mathfrak{Y}*}\,T^* \otimes \mathrm{Schur}^{\mathfrak{Y}}\,V$$

parametrized by all Young diagrams of order D with at most $n := \dim T$ columns and $N := \dim V$ rows. The subset of Young diagrams $\mathfrak{Y}$ presenting a tableau $\mathscr{A}^1$ agrees with the subset of the irreducible subrepresentations in this decomposition, to which the Plücker line in $\Lambda^D \operatorname{Hom}(T, V)$ associated to $\mathscr{A}^1$ projects non-trivially.

Leaving the construction of invariants aside there is another good reason not to spend too much importance on the bases of T and V used to write a given tableau $\mathscr{A}^1 \subset \operatorname{Hom}(T, V)$ as a subspace of matrices of size $N \times n$ in $\mathfrak{Y}$-schematical form: This characteristic property does not pertain to the bases themselves, but actually to their associated flags. Recall at this point that a complete flag on T is an increasing sequence $F_\bullet$ of subspaces

$$\{0\} \;=:\; F_0 \;\subsetneqq\; F_1 \;\subsetneqq\; \cdots \;\subsetneqq\; F_{n-1} \;\subsetneqq\; F_n \;:=\; T$$

satisfying $\dim F_s = s$ for all $s = 1, \ldots, n$. A basis $t_1, \ldots, t_n$ of T is called adapted to a complete flag $F_\bullet$ provided $t_s \in F_s \setminus F_{s-1}$ for all $s = 1, \ldots, n$. Evidently every basis $t_1, \ldots, t_n$ of T is adapted to exactly one complete flag defined by $F_s := \operatorname{span}\{t_1, \ldots, t_s\}$ for all s, on the other hand there are certainly many different bases adapted to a given flag. Nevertheless two bases $t_1, \ldots, t_n$ and $t'_1, \ldots, t'_n$ adapted to the same complete flag $F_\bullet$ on T are necessarily related by an invertible lower triangular matrix $B \in \operatorname{Mat}_{n \times n} \mathbb{R}$ via:

$$t'_c \;=\; \sum_{s=1}^{c} B_{cs}\, t_s$$

For the matrix coefficients $\operatorname{mat}'_{rc} \in \mathscr{A}^{1*}$ associated to the basis $t'_1, \ldots, t'_n$ this becomes

$$\operatorname{mat}'_{rc}(A) \;=\; dv_r\Big(A\big(\sum_{s=1}^{c} B_{cs} t_s \big) \Big) \;=\; \sum_{s=1}^{c} B_{cs}\, \operatorname{mat}_{rs}(A)$$

so that the matrix coefficients $\operatorname{mat}'_{rc} \in \mathscr{A}^{1*}$ for $(r, c) \in \mathfrak{Y}$ are invertible linear combinations of the matrix coefficients $\operatorname{mat}_{rs} \in \mathscr{A}^{1*}$ with $(r, s) \in \mathfrak{Y}$. A very similar argument applies to changing the basis of V, while keeping the associated complete flag on V unchanged. This dependence on flags is precisely the reason, why we are *not* interested in arbitrary subsets of matrix coefficients, but in subsets specified by Young diagrams.

Another way to understand the relationship between complete flags on T and Young diagram presentations of a given first order tableau $\mathscr{A}^1$ is to study its restrictions to subspaces $F \subset T$. The linear restriction map $\operatorname{res}_F : \operatorname{Hom}(T, V) \longrightarrow \operatorname{Hom}(F, V)$, $A \longmapsto A|_F$, associated to a subspace $F \subset T$ gives rise in fact to a short exact sequence of tableaux

$$0 \;\longrightarrow\; \mathscr{A}^1_F \;\overset{\subset}{\longrightarrow}\; \mathscr{A}^1 \;\overset{\operatorname{res}_F}{\longrightarrow}\; \operatorname{res}_F \mathscr{A}^1 \;\longrightarrow\; 0 \tag{39}$$

where $\mathrm{res}_F \mathscr{A}^1 \subset \mathrm{Hom}\,(\,F,\,V\,)$ is the image of $\mathscr{A}^1 \subset \mathrm{Hom}\,(\,T,\,V\,)$ under res_F and $\mathscr{A}_F^1 \subset \mathscr{A}^1$ is the subspace of all $A \in \mathscr{A}^1$ satisfying $At = 0$ for all $t \in F$. In general the dimensions of the two derived tableaux $\mathscr{A}_F^1$ and $\mathrm{res}_F \mathscr{A}^1$ depend delicately on the chosen subspace $F \subset T$, hence it makes sense to call a subspace $F \subset T$ of dimension s a regular subspace provided:

$$\dim \mathrm{res}_F \mathscr{A}^1 \;=\; \max \{\; \dim \mathrm{res}_{\hat{F}} \mathscr{A}^1 \;\mid\; \hat{F} \subset T \text{ is a subspace of dimension } s \;\}$$

In the opposite case F is a singular subspace with respect to the tableau $\mathscr{A}^1 \subset \mathrm{Hom}\,(\,T,\,V\,)$ in the sense that the dimension of the intersection $\mathscr{A}_F^1 = \mathscr{A}^1 \cap \mathrm{Hom}\,(\,T/F,\,V\,)$ is larger than it needs to be. After a little bit of multilinear algebra of the kind we used to establish Eq. (36) above the latter characterization of singular subspaces turns into an explicit space of polynomials on the Graßmannian $\mathrm{Gr}_s T$ considered as an algebraic variety such that a subspace $F \subset T$ of dimension s is singular with respect to $\mathscr{A}^1$, if and only if $F \in \mathrm{Gr}_s T$ is a common zero of all these polynomials. The complementary subset of regular subspaces of T in dimension s is thus a non-empty Zariski dense subset:

$$\mathrm{Gr}_s^{\mathrm{reg}} T := \{\; F \;\mid\; F \subset T \text{ is an } \mathscr{A}^1\text{-regular subspace of dimension } s \;\} \subseteq \mathrm{Gr}_s T$$

Coming back to complete flags we conclude that the set of all complete flags $F_\bullet$ on T, which feature a regular subspace $F_s \in \mathrm{Gr}_s^{\mathrm{reg}} T$ in a given dimension s, is a Zariski dense subset of the algebraic variety $\mathrm{Flag}\,T$ of all complete flags on T. Finite intersections of Zariski dense subsets however are still Zariski dense, in consequence the subset of regular flags on T

$$\mathrm{Flag}^{\mathrm{reg}} T \;:=\; \{\; F_\bullet \in \mathrm{Flag}\,T \;\mid\; \text{every } F_s \text{ is } \mathscr{A}^1\text{-regular in its dimension } s \;\}$$

is a Zariski dense subset of the algebraic variety $\mathrm{Flag}\,T$ of all complete flags on T, in particular $\mathrm{Flag}^{\mathrm{reg}} T$ is a non-empty, dense subset of $\mathrm{Flag}\,T$ with respect to the manifold topology as well. The existence of regular complete flags for arbitrary first order tableaux allows us to define the Cartan character $\mathfrak{Y}^{\mathscr{A}}$ of a tableau $\mathscr{A}^1 \subset \mathrm{Hom}\,(\,T,\,V\,)$ in the following way:

Lemma 4.7 (Cartan Character of First Order Tableaux). *Associated to every first order tableau $\mathscr{A}^1 \subset T^* \otimes V$ of dimension $D := \dim \mathscr{A}^1$ is the set of all Young diagrams of order D presenting $\mathscr{A}^1$. With respect to the partial order $\geq$ this subset of $\mathbf{YD}(\,D\,)$ has a unique maximal element called the Cartan character $\mathfrak{Y}^{\mathscr{A}}$ of the tableau $\mathscr{A}^1$, its column lengths $c_1^{\mathscr{A}} \geq c_2^{\mathscr{A}} \geq \ldots \geq c_n^{\mathscr{A}} \geq 0$ satisfy for all $s = 1, \ldots, n$:*

$$c_1^{\mathscr{A}} + c_2^{\mathscr{A}} + \ldots + c_s^{\mathscr{A}} \;:=\; \max \{\; \dim \mathrm{res}_F \mathscr{A}^1 \;\mid\; F \in \mathrm{Gr}_s T \;\}$$

Proof. For the purpose of proof let us assume that the column lengths of the Cartan character $\mathfrak{Y}^{\mathscr{A}}$ in spe of a fixed first order tableau $\mathscr{A}^1 \subset \mathrm{Hom}\,(\,T,\,V\,)$

of dimension $D \geq 0$ are defined simply as a sequence of non-negative numbers $c_1^{\mathscr{A}}, c_2^{\mathscr{A}}, \ldots, c_n^{\mathscr{A}} \geq 0$ via:

$$c_1^{\mathscr{A}} + c_2^{\mathscr{A}} + \ldots + c_s^{\mathscr{A}} := \max \{ \dim \operatorname{res}_F \mathscr{A}^1 \mid F \in \operatorname{Gr}_s T \} \qquad (40)$$

For every given Young diagram $\mathfrak{Y} \in \mathbf{YD}(D)$ presenting $\mathscr{A}^1$ we may choose bases $t_1, \ldots, t_n$ and $v_1, \ldots, v_N$ of T and V respectively such that the matrix coefficients $A \longmapsto dv_r(At_c)$ indexed by $(r, c) \in \mathfrak{Y}$ are a basis of $\mathscr{A}^{1*}$. The specific matrix coefficients $\operatorname{mat}_{rc} \in \mathscr{A}^{1*}$ indexed by boxes $(r, c) \in \mathfrak{Y}$ in the first $s = 1, \ldots, n$ columns with $c \leq s$ actually come from the restriction of $\mathscr{A}^1$ to the subspace $F_s \in \operatorname{Gr}_s T$ spanned by $\{t_1, \ldots, t_s\}$ in the sense $\operatorname{mat}_{rc}(A) := dv_r([A|_{F_s}]t_c)$. Hence the image of the adjoint $(\operatorname{res}_{F_s} \mathscr{A}^1)^* \longrightarrow \mathscr{A}^{1*}$ of the restriction $\mathscr{A}^1 \longrightarrow \operatorname{res}_{F_s} \mathscr{A}^1$, $A \longmapsto A|_{F_s}$, to F_s contains the $c_1 + \ldots + c_s$ linearly independent matrix coefficients indexed by boxes $(r, c) \in \mathfrak{Y}$ in the first s columns and so:

$$c_1 + c_2 + \ldots + c_s \leq \dim \operatorname{res}_{F_s} \mathscr{A}^1$$

$$\leq \max \{ \dim \operatorname{res}_F \mathscr{A}^1 \mid F \in \operatorname{Gr}_s T \}$$

$$= c_1^{\mathscr{A}} + c_2^{\mathscr{A}} + \ldots + c_s^{\mathscr{A}}$$

Since this inequality is true for all $s = 1, \ldots, n$, we conclude that $\mathfrak{Y} \leq \mathfrak{Y}^{\mathscr{A}}$ provided we can show that the non-negative numbers $c_1^{\mathscr{A}}, \ldots, c_n^{\mathscr{A}} \geq 0$ are actually the column lengths of a Young diagram $\mathfrak{Y}^{\mathscr{A}}$ presenting the tableau $\mathscr{A}^1$.

For this purpose let us fix a regular flag $F_\bullet \in \operatorname{Flag}^{\mathrm{reg}} T$ for the tableau $\mathscr{A}^1$ and an adapted basis $t_1, \ldots, t_n$ for T satisfying $F_s = \operatorname{span}\{t_1, \ldots, t_s\}$ for all $s = 1, \ldots, n$. Evidently the kernel $\mathscr{A}^1_{F_s} \subset \mathscr{A}^1$ of the restriction to F_s consists of those $A \in \mathscr{A}^1 \subset \operatorname{Hom}(T, V)$ satisfying $A t_\mu = 0$ for all $1 \leq \mu \leq s$, this simple observation gives rise to the short exact sequences

$$0 \longrightarrow \mathscr{A}^1_{F_s} \overset{\subset}{\longrightarrow} \mathscr{A}^1_{F_{s-1}} \overset{\frac{\partial}{\partial t_s}}{\longrightarrow} \mathscr{A}^1_{F_{s-1}} t_s \longrightarrow 0 \qquad (41)$$

for all $s = 1, \ldots, n$, where $\frac{\partial}{\partial t_s} A := A t_s$ and $\mathscr{A}^1_{F_0} := \mathscr{A}^1$ in case of doubt. In consequence

$$\dim \operatorname{res}_{F_s} \mathscr{A}^1 = \dim \mathscr{A}^1 - \dim \mathscr{A}^1_{F_s} = c_1^{\mathscr{A}} + \ldots + c_s^{\mathscr{A}}$$

$$\dim \mathscr{A}^1_{F_{s-1}} t_s = \dim \mathscr{A}^1_{F_{s-1}} - \dim \mathscr{A}^1_{F_s} = c_s^{\mathscr{A}}$$

where the first equation simply reflects the regularity of the chosen flag $F_\bullet \in \operatorname{Flag}^{\mathrm{reg}} T$ with respect to the tableau $\mathscr{A}^1$ and the second the short exact sequence (41).

The crucial observation to be made at this point is that the sequence of subspaces $\mathscr{A}^1_{F_{s-1}} t_s, s = 1, \ldots, n$, of dimensions $c_s^{\mathscr{A}}$ is actually a monotonely decreasing filtration

$$V \supseteq \mathscr{A}^1_{F_0} t_1 \supseteq \mathscr{A}^1_{F_1} t_2 \supseteq \mathscr{A}^1_{F_2} t_3 \supseteq \ldots \supseteq \mathscr{A}^1_{F_{n-1}} t_n \supseteq \{0\} \quad (42)$$

so that the non-negative integers $c^{\mathscr{A}}_1, \ldots, c^{\mathscr{A}}_n \geq 0$ we have been using up to now are monotonely decreasing $c^{\mathscr{A}}_1 \geq c^{\mathscr{A}}_2 \geq \ldots \geq c^{\mathscr{A}}_n \geq 0$ as appropriate for the column lengths of a Young diagram $\mathfrak{Y}^{\mathscr{A}}$ of order $c^{\mathscr{A}}_1 + \ldots + c^{\mathscr{A}}_n = \dim \mathscr{A}^1$. By the our choice of a regular flag $F_{\bullet} \in \mathrm{Flag}^{\mathrm{reg}} T$ all the subspaces $F_s \in \mathrm{Gr}^{\mathrm{reg}}_s T$ are regular in their dimension s with respect to the tableau $\mathscr{A}^1$. With regularity being a Zariski open condition we conclude that for an arbitrary deformation vector $t \in T$ the deformation $F^\varepsilon_s := F_{s-1} \oplus \mathbb{R}(t_s + \varepsilon t)$ with ε sufficiently close to 0 is still a regular subspace $F^\varepsilon_s \in \mathrm{Gr}^{\mathrm{reg}}_s T$. Comparing the short exact sequence (41) for F_s with the short exact sequence constructed similarly for F^ε_s

$$0 \longrightarrow \mathscr{A}^1_{F^\varepsilon_s} \overset{\subset}{\longrightarrow} \mathscr{A}^1_{F_{s-1}} \longrightarrow \mathscr{A}^1_{F_{s-1}}(t_s + \varepsilon t) \longrightarrow 0$$

we observe that the regularity of F^ε_s is equivalent to $\dim \mathscr{A}^1_{F_{s-1}}(t_s + \varepsilon t) = \dim \mathscr{A}^1_{F_{s-1}} t_s$, hence sufficiently close to 0 the curve $\varepsilon \longmapsto \mathscr{A}^1_{F_{s-1}}(t_s + \varepsilon t)$ is a smooth curve in the Graßmannian of subspaces of V of dimension $c^{\mathscr{A}}_s$. In particular the trivial inclusion of subspaces

$$\mathscr{A}^1_{F_s} t = \mathscr{A}^1_{F_s}(t_s + \varepsilon t) \subseteq \mathscr{A}^1_{F_{s-1}}(t_s + \varepsilon t)$$

valid for all $\varepsilon \neq 0$ continues to hold true for $\varepsilon = 0$ by the way the topology is defined on the Graßmannians. In consequence $\mathscr{A}^1_{F_{s-1}} t_s \supseteq \mathscr{A}^1_{F_s} t$ for all $s = 1, \ldots, n$ and an arbitrary deformation vector $t \in T$, in particular $\mathscr{A}^1_{F_{s-1}} t_s \supseteq \mathscr{A}^1_{F_s} t_{s+1}$ in filtration (42).

Last but not least we complement the chosen basis $t_1, \ldots, t_n$ for T adapted to the regular flag $F_{\bullet} \in \mathrm{Flag}^{\mathrm{reg}} T$ by a basis $v_1, \ldots, v_N$ of V adapted to the decreasing filtration (42) in the sense that for all $s = 1, \ldots, n$ the filtration subspace $\mathscr{A}^1_{F_{s-1}} t_s$ is spanned by the first $c^{\mathscr{A}}_s$ basis vectors $v_1, \ldots, v_{c^{\mathscr{A}}_s}$. In order to show that the special matrix coefficients $\mathrm{mat}_{rc} \in \mathscr{A}^{1*}$ indexed by boxes $(r, c) \in \mathfrak{Y}^{\mathscr{A}}$ with respect to these bases are actually a basis of $\mathscr{A}^{1*}$ it is sufficient to verify that they generate $\mathscr{A}^{1*}$, in other words we need to prove that every $A \in \mathscr{A}^1$ satisfying $\mathrm{mat}_{rc}(A) = 0$ for all $(r, c) \in \mathfrak{Y}^{\mathscr{A}}$ necessarily vanishes $A = 0$.

Due to $A \in \mathscr{A}^1 = \mathscr{A}^1_{F_0}$ the vector $At_1 \in \mathscr{A}^1_{F_0} t_1$ is a linear combination of the first $c^{\mathscr{A}}_1$ basis vectors $v_1, \ldots, v_{c^{\mathscr{A}}_1} \in V$, hence the assumption $\mathrm{mat}_{r1}(A) = 0$ for all $(r, 1) \in \mathfrak{Y}^{\mathscr{A}}$ implies $At_1 = 0$ or equivalently $A \in \mathscr{A}^1_{F_1}$. Iterating this argument we find that $At_2 \in \mathscr{A}^1_{F_1} t_2$ is a linear combination of the first $c^{\mathscr{A}}_2$ basis vectors of V and conclude $At_2 = 0$ or $A \in \mathscr{A}^1_{F_2}$ as before from the assumption $\mathrm{mat}_{r2}(A) = 0$ for all $(r, 2) \in \mathfrak{Y}^{\mathscr{A}}$. Continuing in this way we eventually arrive at the conclusion $A \in \mathscr{A}^1_{F_n} = \{0\}$ or equivalently $A = 0$. Summing up this argument we conclude that the matrix coefficients $\mathrm{mat}_{rc} \in \mathfrak{Y}^{\mathscr{A}}$ indexed by the boxes $(r, c) \in \mathfrak{Y}^{\mathscr{A}}$ are a basis of $\mathscr{A}^{1*}$ so that $\mathfrak{Y}^{\mathscr{A}}$ presents the tableau $\mathscr{A}^1$. $\square$

Theorem 4.8 (Cartan's Involutivity Test for Tableaux). *Every Young diagram* $\mathfrak{Y}$ *presenting a first order tableau* $\mathscr{A}^1 \subset \mathrm{Hom}\,(T,\,V)$ *with associated tableau comodule* $\mathscr{A}$ *over the symmetric coalgebra* $\mathrm{Sym}\,T^*$ *of a vector space* T *of dimension* $n := \dim T$ *provides us with an a priori estimate on the dimension of the first prolongation* $\mathscr{A}^2 \subset \mathrm{Sym}^2 T^* \otimes V$ *of the tableau in terms of its column lengths* $c_1 \geq c_2 \geq \ldots \geq c_n \geq 0$

$$\dim \mathscr{A}^2 \;\leq\; c_1 + 2\,c_2 + 3\,c_3 + 4\,c_4 + \ldots + n\,c_n$$

which may only be sharp for the Cartan character $\mathfrak{Y}^{\mathscr{A}}$ *of* $\mathscr{A}^1$. *If this estimate is in fact sharp for the Cartan character, then the Spencer cohomology of the comodule* $\mathscr{A}$ *is concentrated in comodule degree zero with* $H^{\bullet,\circ}(\mathscr{A}) = \{0\}$ *for* $\bullet \neq 0$. *Moreover the dimensions of* $\mathscr{A}^k$ *and* $H^{0,r}(\mathscr{A})$ *can be calculated for all* $k,\,r > 0$ *from the column lengths of* $\mathfrak{Y}^{\mathscr{A}}$ *via:*

$$\dim \mathscr{A}^k = + \binom{k-1}{0} c_1^{\mathscr{A}} + \binom{k}{1} c_2^{\mathscr{A}} + \ldots + \binom{k+n-2}{n-1} c_n^{\mathscr{A}}$$

$$\dim H^{0,r}(\mathscr{A}) = \binom{n}{r} \dim V - \binom{n-1}{r-1} c_1^{\mathscr{A}} - \binom{n-2}{r-1} c_2^{\mathscr{A}} + \ldots - \binom{0}{r-1} c_n^{\mathscr{A}}$$

Last but not least the comodule $\mathscr{A}$ *has a canonical resolution by free comodules of the form:*

$$0 \longrightarrow \mathscr{A}^{\bullet} \overset{\subset}{\longrightarrow} \mathrm{Sym}^{\bullet} T^* \otimes H^{0,0}(\mathscr{A}) \longrightarrow \mathrm{Sym}^{\bullet-1} T^* \otimes H^{0,1}(\mathscr{A})$$

$$\longrightarrow \mathrm{Sym}^{\bullet-2} T^* \otimes H^{0,2}(\mathscr{A}) \longrightarrow \ldots \longrightarrow \mathrm{Sym}^{\bullet-n} T^* \otimes H^{0,n}(\mathscr{A}) \longrightarrow 0$$

Without doubt Cartan's Involutivity Test is the most beautiful gem in the theory of exterior differential systems, although or perhaps because it is in essence a theorem of commutative algebra. Involutivity of a tableau is a notion actually defined by the theorem: A first order tableau $\mathscr{A}^1 \subset \mathrm{Hom}\,(T,\,V)$ is called an involutive tableau provided it passes Cartan's Test positively with dim $\mathscr{A}^2 = c_1^{\mathscr{A}} + 2\,c_2^{\mathscr{A}} + \ldots + n\,c_n^{\mathscr{A}}$, in consequence the associated Spencer cohomology is concentrated in comodule degree $\bullet = 0$ with $H^{\bullet,\circ}(\mathscr{A}) = \{0\}$ for all $\bullet \neq 0$. The converse of is statement is true as well, although we will not prove this fact: A first order tableau $\mathscr{A}^1$, whose Spencer cohomology is concentrated in comodule degree zero, necessarily passes Cartan's Test dim $\mathscr{A}^2 = c_1^{\mathscr{A}} + 2\,c_2^{\mathscr{A}} + \ldots + n\,c_n^{\mathscr{A}}$ positively. The proof presented below of the direct implication of Theorem 4.8 relies heavily on the following technical lemma:

Lemma 4.9 (Technical Lemma for Cartan's Involutivity Test). *Consider for a given comodule* $\mathscr{A}$ *over the symmetric coalgebra* $\mathrm{Sym}\,T^*$ *of a vector space* T *the subcomodule* $\mathscr{A}^{\bullet}_{\mathbb{R}t} \subset \mathscr{A}^{\bullet}$ *of elements constant in the direction of a fixed vector* $t \in T$:

$$\mathscr{A}_{\mathbb{R}t}^{\bullet} := \mathbf{ker}\left(\frac{\partial}{\partial t} : \mathscr{A}^{\bullet} \longrightarrow \mathscr{A}^{\bullet-1}, \quad a \longmapsto \frac{\partial a}{\partial t} \right)$$

In case the directional derivative $\frac{\partial}{\partial t} : \mathscr{A}^{k+1} \longrightarrow \mathscr{A}^k$ in the direction t is surjective for some $k \in \mathbb{Z}$ the inclusion $\mathscr{A}_{\mathbb{R}t}^{\bullet} \longrightarrow \mathscr{A}^{\bullet}$ induces a surjection on the level of Spencer cohomology:

$$H^{k,\circ}(\mathscr{A}_{\mathbb{R}t}) \longrightarrow H^{k,\circ}(\mathscr{A}), \qquad [\omega] \longmapsto [\omega]$$

If in addition to $\frac{\partial}{\partial t} : \mathscr{A}^{k+1} \longrightarrow \mathscr{A}^k$ being surjective the following Spencer cohomology spaces

$$H^{k+1,0}(\mathscr{A}) = 0 \qquad H^{k+1,1}(\mathscr{A}) = 0 \qquad H^{k,2}(\mathscr{A}_{\mathbb{R}t}) = 0$$

of $\mathscr{A}$ and $\mathscr{A}_{\mathbb{R}t}$ vanish, then the directional derivative $\frac{\partial}{\partial t} : \mathscr{A}^{k+2} \longrightarrow \mathscr{A}^{k+1}$ is surjective again.

The first statement is an almost trivial consequence of Cartan's Homotopy Formula. Starting with an arbitrary representative $\omega \in \mathscr{A}^k \otimes \Lambda^r T^*$ of a cohomology class $[\omega] \in H^{k,r}(\mathscr{A})$ we use the surjectivity of the directional derivative $\frac{\partial}{\partial t} : \mathscr{A}^{k+1} \longrightarrow \mathscr{A}^k$ and the algebraic analogue of Cartan's Homotopy Formula (22) to find a cochain $\omega^{\mathrm{pre}} \in \mathscr{A}^{k+1} \otimes \Lambda^r T^*$ satisfying:

$$\omega = (\frac{\partial}{\partial t} \otimes \mathrm{id}) \, \omega^{\mathrm{pre}} = \{ B, (\mathrm{id} \otimes t_{\lrcorner}) \} \, \omega^{\mathrm{pre}}$$

$$:= (\mathrm{id} \otimes t_{\lrcorner}) \, B \, \omega^{\mathrm{pre}} + B \, (\mathrm{id} \otimes t_{\lrcorner}) \, \omega^{\mathrm{pre}}$$

In consequence $(\mathrm{id} \otimes t_{\lrcorner}) \, B \, \omega^{\mathrm{pre}} \equiv \omega$ modulo $\mathbf{im}\, B$ is still closed and represents the same cohomology class $[(\mathrm{id} \otimes t_{\lrcorner}) \, B \, \omega^{\mathrm{pre}}] = [\omega] \in H^{k,r}(\mathscr{A})$, however its directional derivative

$$(\frac{\partial}{\partial t} \otimes \mathrm{id}) \left((\mathrm{id} \otimes t_{\lrcorner}) \, B \, \omega^{\mathrm{pre}} \right) = (\mathrm{id} \otimes t_{\lrcorner}) \, B \left((\frac{\partial}{\partial t} \otimes \mathrm{id}) \, \omega^{\mathrm{pre}} \right)$$

$$= (\mathrm{id} \otimes t_{\lrcorner}) \, B \, \omega = 0$$

in the direction of t vanishes, recall that ω is assumed to represent a cohomology class and so it is necessarily closed $B \, \omega = 0$. The modified representative $(\mathrm{id} \otimes t_{\lrcorner}) \, B \, \omega^{\mathrm{pre}} \in \mathscr{A}_{\mathbb{R}t}^k \otimes \Lambda^r T^*$ is thus constant in the direction of t and provides us with the looked for preimage of $[\omega]$ under the map $H^{k,r}(\mathscr{A}_{\mathbb{R}t}) \longrightarrow H^{k,r}(\mathscr{A})$ induced by the inclusion $\mathscr{A}_{\mathbb{R}t}^{\bullet} \longrightarrow \mathscr{A}^{\bullet}$. Turning to the proof of the second statement we study the commutative diagram

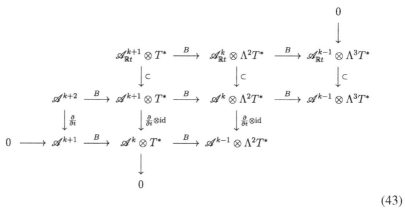

$$(43)$$

whose rows and columns are all complexes. By assumption the row complexes are exact on the diagonal $\mathscr{A}^{k+1}$, $\mathscr{A}^{k+1} \otimes T^*$ and $\mathscr{A}^k_{\mathbb{R}t} \otimes \Lambda^2 T^*$, while the column complexes are exact on the parallel diagonal $\mathscr{A}^k \otimes T^*$, $\mathscr{A}^k \otimes \Lambda^2 T^*$ and $\mathscr{A}^{k-1}_{\mathbb{R}t} \otimes \Lambda^3 T^*$ due to the definition of the comodule $\mathscr{A}_{\mathbb{R}t}$ and our assumption that the directional derivative $\frac{\partial}{\partial t} : \mathscr{A}^{k+1} \longrightarrow \mathscr{A}^k$ is surjective. A very delightful diagram chase all over this diagram proves that the directional derivative $\frac{\partial}{\partial t} : \mathscr{A}^{k+2} \longrightarrow \mathscr{A}^{k+1}$ on the left is surjective as well.

Proof of Theorem 4.8: Once and for all let us fix a first order tableau $\mathscr{A}^1 \subset \text{Hom}(T, V)$ of dimension $D \geq 0$ on vector spaces T and V of dimensions n and N respectively. On the set of Young diagrams of order D with at most n columns we define the weighted sum $\|\cdot\| : \mathbf{YD}_n(D) \longrightarrow \mathbb{N}_0$ of the column lengths $c_1 \geq c_2 \geq \ldots \geq c_n \geq 0$ of the argument by:

$$\|\mathfrak{Y}\| := c_1 + 2c_2 + \ldots + n c_n$$
$$= (n+1)D - (c_1) - (c_1 + c_2) - \ldots - (c_1 + c_2 + \ldots + c_n)$$

Using the second expression for the weighted sum $\|\cdot\|$ and the definition (38) of the partial order $\geq$ on the lattice $\mathbf{YD}(D)$ we conclude directly that $\mathfrak{Y} \geq \tilde{\mathfrak{Y}}$ implies $\|\mathfrak{Y}\| \leq \|\tilde{\mathfrak{Y}}\|$ with equality, if and only if $\mathfrak{Y} = \tilde{\mathfrak{Y}}$. Every Young diagram presenting $\mathscr{A}^1$ has at most n columns of course and the maximality of the Cartan character $\mathfrak{Y}^{\mathscr{A}}$ among the diagrams presenting $\mathscr{A}^1$ implies that $\|\cdot\|$ attains its minimum in and only in the Cartan character:

$$\|\mathfrak{Y}^{\mathscr{A}}\| = \min\{\|\mathfrak{Y}\| \mid \mathfrak{Y} \in \mathbf{YD}_n(D) \text{ presents the tableau } \mathscr{A}^1\} \quad (44)$$

In the following paragraphs we will provide two different arguments to establish the *a priori* estimate $\dim \mathscr{A}^2 \leq \|\mathfrak{Y}\|$ for the dimension of the first prolongation $\mathscr{A}^2$ in terms of a Young diagram $\mathfrak{Y}$ presenting $\mathscr{A}^1$. Whereas the first argument is rather explicit and is intended to provide the reader with a meaningful interpretation (46) for the weighted sum $\|\cdot\|$, the second argument is more elegant and provides us with the a useful description of the equality case dim $\mathscr{A}^2 = \|\mathfrak{Y}\|$, on which the inductive proof of the main statement is based.

Consider an arbitrary Young diagram $\mathfrak{Y}$ presenting the tableau $\mathscr{A}^1$ and corresponding bases $t_1, \ldots, t_n$ and $v_1, \ldots, v_N$ for T and V respectively so that the matrix coefficients $\mathrm{mat}_{rc}(A) := dv_r(At_c)$ indexed by the boxes $(r, c) \in \mathfrak{Y}$ are a basis of $\mathscr{A}^{1*}$. Recall that the directional derivative in the tableau comodule $\mathscr{A}^\bullet$ associated to $\mathscr{A}^1$ agrees with insertion $\frac{\partial}{\partial t} : \mathscr{A}^1 \longrightarrow V$, $A \longmapsto At$, in comodule degree $\bullet = 1$, this observation allows us to generalize the matrix coefficients $\mathrm{mat}_{rc} \in \mathscr{A}^{1*}$ to matrix coefficients defined on $\mathscr{A}^2$ by:

$$\mathrm{mat}_{r;\bar{c}c} : \quad \mathscr{A}^2 \longrightarrow \mathbb{R}, \qquad a \longmapsto dv_r\left(\frac{\partial^2}{\partial t_{\bar{c}} \, \partial t_c} a \right)$$

It is well known that these generalized matrix coefficients $\mathrm{mat}_{r;\bar{c}c}$ with their symmetry $\mathrm{mat}_{r;\bar{c}c} = \mathrm{mat}_{r;c\bar{c}}$ taken into account are a basis of the vector space dual to $\mathrm{Sym}^2 T^* \otimes V$, hence they certainly generate $\mathscr{A}^{2*}$ due to $\mathscr{A}^2 \subset \mathrm{Sym}^2 T^* \otimes V$. On the other hand we know that the matrix coefficients $\mathrm{mat}_{\bar{r}\bar{c}} \in \mathscr{A}^{1*}$ indexed by $(\bar{r}, \bar{c}) \notin \mathfrak{Y}$ are fixed linear combinations

$$\mathrm{mat}_{\bar{r}\bar{c}} = \sum_{(r,c)\in\mathfrak{Y}} C^{rc}_{\bar{r}\bar{c}} \, \mathrm{mat}_{rc} \tag{45}$$

of the matrix coefficients indexed by $(r, c) \in \mathfrak{Y}$, where the constants $C^{rc}_{\bar{r}\bar{c}} \in \mathbb{R}$ are characteristic for the tableau $\mathscr{A}^1$, the trivial identity $\mathrm{mat}_{\bar{r};s\bar{c}}(a) = \mathrm{mat}_{\bar{r}\bar{c}}(\frac{\partial a}{\partial t_s})$ thus implies $\mathrm{mat}_{\bar{r};s\bar{c}} = \sum_{(r,c)\in\mathfrak{Y}} C^{rc}_{\bar{r}\bar{c}} \, \mathrm{mat}_{r;sc}$ for every $s = 1, \ldots, n$ as well. In the light of the symmetry $\mathrm{mat}_{r;c\bar{c}} = \mathrm{mat}_{r;\bar{c}c}$ we conclude that the generalized matrix coefficients $\mathrm{mat}_{r;\bar{c}c}$ indexed by triples $(r; \bar{c}, c)$ satisfying $c \geq \bar{c} \geq 1$ and $(r, c) \in \mathfrak{Y}$ already generate all of $\mathscr{A}^{2*}$, hence:

$$\dim \mathscr{A}^2 \leq \|\mathfrak{Y}\| = \sharp\{ (r; \bar{c}, c) \mid r \geq 1, \ c \geq \bar{c} \geq 1 \text{ and } (r, c) \in \mathfrak{Y} \} \tag{46}$$

Somewhat more elegantly the *a priori* estimate $\dim \mathscr{A}^2 \leq \|\mathfrak{Y}\|$ can be established using the complete flag $F_\bullet$ associated to the chosen basis $t_1, \ldots, t_n \in T$ with $F_s := \mathrm{span}\{ t_1, \ldots, t_s \}$. Associated to this complete flag is a descending filtration of the tableau $\mathscr{A}^1$ by subtableaux

$$\mathscr{A}^1 = \mathscr{A}^1_{F_0} \supseteq \mathscr{A}^1_{F_1} \supseteq \mathscr{A}^1_{F_2} \supseteq \cdots \supseteq \mathscr{A}^1_{F_{n-1}} \supseteq \mathscr{A}^1_{F_n} = \{0\}$$

which in turn define tableau comodules $\mathscr{A}^\bullet_{F_s}$ for all $s = 1, \ldots, n$. The most important observation to be made at this point is that these tableau comodules are interrelated by

$$\mathscr{A}_{F_s}^\bullet = \mathbf{ker}\left(\frac{\partial}{\partial t_s} : \mathscr{A}_{F_{s-1}}^\bullet \longrightarrow \mathscr{A}_{F_{s-1}}^{\bullet-1}, \ a \longmapsto \frac{\partial a}{\partial t_s} \right) =: (\mathscr{A}_{F_{s-1}}^\bullet)_{\mathbb{R}\, t_s} \tag{47}$$

valid for all $s = 1, \ldots, n$. In fact the tableau comodule $\mathscr{A}_{F_{s-1}}^\bullet \subset \operatorname{Sym}^\bullet T^* \otimes V$ is more or less by definition the subspace of polynomials ψ on T with values in V, whose differential $B_p \psi : T \longrightarrow V, t \longmapsto \frac{\partial \psi}{\partial t}(p)$, is an element of the subspace $\mathscr{A}_{F_{s-1}}^1 \subset \operatorname{Hom}(T, V)$ in every point $p \in T$. Clearly such a polynomial is constant in the direction of t_s, if and only if its differential in every point kills the vector t_s and thus lies in the subspace $\mathscr{A}_{F_s}^1$. The resulting equality (47) between the tableau comodule $\mathscr{A}_{F_s}^\bullet$ and the kernel of $\frac{\partial}{\partial t_s}$ in the tableau comodule $\mathscr{A}_{F_{s-1}}^\bullet$ for all $s = 1, \ldots, n$ for all $s = 1, \ldots, n$ implies the exactness of the sequences:

$$0 \longrightarrow \mathscr{A}_{F_s}^2 \xrightarrow{\ \subset\ } \mathscr{A}_{F_{s-1}}^2 \xrightarrow{\ \frac{\partial}{\partial t_s}\ } \mathscr{A}_{F_{s-1}}^1 \tag{48}$$

Combining this exactness with the estimate $\dim \mathscr{A}_{F_s}^1 = D - \dim \operatorname{res}_{F_s} \mathscr{A}^1 \leq c_{s+1} + \ldots + c_n$ established in the proof of Lemma 4.7 to obtain the *a priori* estimate as a telescope sum

$$\dim \mathscr{A}^2 = \sum_{s=1}^n \left(\dim \mathscr{A}_{F_{s-1}}^2 - \dim \mathscr{A}_{F_s}^2 \right) \leq \sum_{s=1}^n \dim \mathscr{A}_{F_{s-1}}^1 \leq \| \mathfrak{Y} \|$$

telescoping to $\dim \mathscr{A}_{F_0}^2 - \dim \mathscr{A}_{F_n}^2 = \dim \mathscr{A}^2$ according to $\mathscr{A}_{F_0}^1 = \mathscr{A}^1$ and $\mathscr{A}_{F_n}^1 = \{0\}$.

For the second part of the proof let us assume that the *a priori* estimate is actually sharp $\dim \mathscr{A}^2 = \| \mathfrak{Y}^{\mathscr{A}} \|$ for the Cartan character $\mathfrak{Y}^{\mathscr{A}}$. Under this assumption all inequalities in the preceding argument must be equalities, in particular $\dim \mathscr{A}_{F_s}^1 = c_{s+1}^{\mathscr{A}} + \ldots + c_n^{\mathscr{A}}$ holds true for all $s = 1, \ldots, n$ and all the exact sequences (48) are surjective on the right and thus short exact. This simple observation provides the basis in $k = 1$ for an inductive argument to the end that we have for all $k \geq 1$ and all $s = 1, \ldots, n$ short exact sequences:

$$0 \longrightarrow \mathscr{A}_{F_s}^{k+1} \xrightarrow{\ \subset\ } \mathscr{A}_{F_{s-1}}^{k+1} \xrightarrow{\ \frac{\partial}{\partial t_s}\ } \mathscr{A}_{F_{s-1}}^k \longrightarrow 0 \tag{49}$$

In the induction step from k to $k+1$ we apply the first statement of the technical Lemma 4.9 to all tableau comodules $\mathscr{A}_{F_s}^\bullet = (\mathscr{A}_{F_{s-1}}^\bullet)_{\mathbb{R}\, t_s}$ in turn to obtain a chain of surjections

$$H^{k,\circ}(\mathscr{A}_{F_n}) \longrightarrow H^{k,\circ}(\mathscr{A}_{F_{n-1}}) \longrightarrow \ldots \longrightarrow H^{k,\circ}(\mathscr{A}_{F_1}) \longrightarrow H^{k,\circ}(\mathscr{A}_{F_0})$$

in Spencer cohomology, in which the first term $H^{k,\circ}(\mathscr{A}_{F_n}) = \{0\}$ vanishes due to $k > 0$, after all the tableau comodule $\mathscr{A}_{F_n}^\bullet = \delta_{\bullet=0} V$ is concentrated in degree zero. In consequence

$$H^{k+1,0}(\mathscr{A}_{F_s}) \;=\; \{0\} \qquad H^{k+1,1}(\mathscr{A}_{F_s}) \;=\; \{0\} \qquad H^{k,2}(\mathscr{A}_{F_s}) \;=\; \{0\}$$

vanish for all $s = 0, \ldots, n$, in fact we just have finished proving the third assertion, whereas the first two are true for every first order tableau comodule according to Eq. (30) and the assumption $k > 0$. All requirements of the second statement of Lemma 4.9 are thus met, and we conclude that the short exact sequences one degree higher up are exact

$$0 \;\longrightarrow\; \mathscr{A}_{F_s}^{k+2} \;\overset{\subset}{\longrightarrow}\; \mathscr{A}_{F_{s-1}}^{k+2} \;\overset{\frac{\partial}{\partial t_s}}{\longrightarrow}\; \mathscr{A}_{F_{s-1}}^{k+1} \;\longrightarrow\; 0$$

on the right for all $s = 1, \ldots, n$ completing thus the induction. More or less as a by-product we have proved that the Spencer cohomology $H^{\bullet,\circ}(\mathscr{A}_{F_s}) = \{0\}$ vanishes in comodule degrees $\bullet \neq 0$ for all comodules $\mathscr{A}_{F_s}$, $s = 0, \ldots, n$ and thus for $\mathscr{A} = \mathscr{A}_{F_0}$ as well. With such a Spencer cohomology the E^1-term of the standard spectral sequence of Lemma 3.7

$$\mathrm{Sym}^{\bullet} T^* \otimes H^{\bullet,\circ}(\mathscr{A}) \;\Longrightarrow\; \delta_{\bullet=0=\circ} \mathscr{A}^{\bullet}$$

is concentrated in degrees $\bullet = 0$. Hence this spectral sequence has only one chance left to converge to the original comodule $\mathscr{A}$: The coboundary operators leading to the E^2-term have to link up to form a resolution of $\mathscr{A}$ by the free comodules with basis $H^{0,\circ}(\mathscr{A})$, in particular the standard spectral sequence collapses at its E^2-term equal to $\mathscr{A}$.

The short exact sequences (49) valid for all $k \geq 1$ and $s = 1, \ldots, n$ allow us to calculate the dimension of the homogeneous subspaces $\mathscr{A}_{F_s}^k$ of the comodules $\mathscr{A}_{F_s}$ as telescope sums

$$\dim \mathscr{A}_{F_s}^k \;=\; \sum_{\mu \geq 1} \left(\dim \mathscr{A}_{F_s+\mu-1}^k - \dim \mathscr{A}_{F_s+\mu}^k \right) \;=\; \sum_{\mu \geq 1} \dim \mathscr{A}_{F_s+\mu-1}^{k-1}$$

for all $k \geq 2$. Straightforward induction on $k \geq 1$ using this equation in the induction step and the equality $\dim \mathscr{A}_{F_s}^1 = c_{s+1}^{\mathscr{A}} + \ldots + c_n^{\mathscr{A}}$ established as a direct consequence of the assumption $\dim \mathscr{A}^2 = \| \mathfrak{Y}^{\mathscr{A}} \|$ as induction base thus proves the proves the explicit formula

$$\dim \mathscr{A}_{F_s}^k \;=\; \sum_{\mu=1}^{n-s} \binom{k-2+\mu}{\mu-1} c_{s+\mu}^{\mathscr{A}}$$

for all $k \geq 1$ and $s = 0, \ldots, n$, which becomes the stipulated formula for $\dim \mathscr{A}^k$ in the special case $s = 0$. Eventually the Betti numbers $\dim H^{0,r}(\mathscr{A})$ of the comodule $\mathscr{A}$ can be calculated by binomial inversion from the following identity, which is obtained by equating the preceding formula for $\dim \mathscr{A}^k$ with the formula from Corollary 3.8:

$$\sum_{s=1}^{n} \binom{k-2+s}{s-1} c_s^{\mathscr{A}} = \sum_{r=0,\dots,n} (-1)^r \binom{k-r+n-1}{n-1} \dim H^{0,r}(\mathscr{A})$$

$\square$

Perhaps the reader may have wondered, why we took the time to prove the *a priori* estimate $\dim \mathscr{A}^2 \le \|\mathfrak{Y}\|$ for the dimension of the first prolongation of a tableau $\mathscr{A}^1 \subset \mathrm{Hom}\,(T, V)$ in terms of a Young diagram $\mathfrak{Y}$ presenting $\mathscr{A}^1$. The point is that the argument using matrix coefficients generalizes to all degrees $k \ge 1$ in the formulation that the matrix coefficients

$$\mathrm{mat}_{r;\,c^1\dots c^k}(a) := dv_r\left(\frac{\partial^k}{\partial t_{c^1}\dots\partial t_{c^k}}\,a\right)$$

indexed by tuples $(r;\,c^1,\dots,c^k)$ satisfying $c^k \ge \dots \ge c^1 \ge 1$ and $(r, c^k) \in \mathfrak{Y}$ generate $\mathscr{A}^{k*}$. For the Cartan character $\mathfrak{Y} = \mathfrak{Y}^{\mathscr{A}}$ of an involutive tableau $\mathscr{A}^1 \subset \mathrm{Hom}\,(T, V)$ however the formulas for $\dim \mathscr{A}^k$ in terms of the column lengths of $\mathfrak{Y}^{\mathscr{A}}$ imply

$$\dim \mathscr{A}^k = \sharp\{\,(r;\,c^1,\dots,c^k)\mid r \ge 1,\ c^k \ge \dots \ge c^1 \ge 1 \text{ and } (r, c^k) \in \mathfrak{Y}^{\mathscr{A}}\,\}$$

so that these special matrix coefficient are a basis of $\mathscr{A}^{k*}$. In a sense this statement can be seen as an interpolation formula, because it implies that for every choice of real constants $a_{r;\,c^1\dots c^k} \in \mathbb{R}$ for every tuple with $r \ge 1$ and $c^k \ge \dots \ge c^1 \ge 1$ as well as $(r, c^k) \in \mathfrak{Y}$ there exists a unique element $a \in \mathscr{A}^k$ of the comodule satisfying:

$$\mathrm{mat}_{r;\,c^1\dots c^k}(a) = a_{r;\,c^1\dots c^k}$$

In other words there exists an essentially algorithmic way to calculate all elements of the tableau comodule $\mathscr{A}$ corresponding to an involutive tableau $\mathscr{A}^1 \subset \mathrm{Hom}\,(T, V)$ in terms of its structure constants $C_{r\tilde{c}}^{rc}$ of Eq. (45), although it seems difficult to write an actual computer program to implement this algorithm. In light of this interpolation property of involutive tableaux, it is very interesting to know that every finitely generated comodule $\mathscr{A}$ bounded below becomes eventually an involutive tableau comodule:

Theorem 4.10 (Twisted Comodules and Involutivity). *Consider a finitely generated comodule $\mathscr{A}$ bounded below and let $d_{\max} < \infty$ be the maximal comodule degree $\bullet$ realized by a non-trivial Spencer cohomology space $H^{\bullet,0}(\mathscr{A}) \ne \{0\}$ in its finite dimensional Spencer cohomology, this is $H^{d,r}(\mathscr{A}) = \{0\}$ for all $d > d_{\max}$ and all r. For all $d \ge d_{\max}$ the twist $\mathscr{A}^{\bullet}(d) \subset \mathrm{Sym}^{\bullet}T^* \otimes \mathscr{A}^d$ associated to $\mathscr{A}$ is an involutive tableau comodule associated to the "prolonged" tableau $B(\mathscr{A}^{d+1}) \subset \mathrm{Hom}\,(T, \mathscr{A}^d)$ with:*

$$B(\,\mathscr{A}^{d+1}\,) \;\; := \;\; \{\, B\,a : T \longrightarrow \mathscr{A}^d, \;\; t \longmapsto \tfrac{\partial a}{\partial t} \;\mid\; a \in \mathscr{A}^{d+1} \,\}$$

Its Cartan character $\mathfrak{Y}^{\mathscr{A}(d)}$ *has columns of length* $c_1^{\mathscr{A}(d)} \geq c_2^{\mathscr{A}(d)} \geq \ldots \geq c_n^{\mathscr{A}(d)} \geq$
0 *given by:*

$$c_s^{\mathscr{A}(d)} \;\; = \;\; \sum_{\substack{r=0,\ldots,n \\ l \in \mathbb{Z}}} (-1)^r \binom{n+d-s-l-r}{n-s} \dim H^{l,r}(\mathscr{A})$$

where $\binom{x}{m}$ *denotes the binomial polynomial* $\frac{1}{m!}\, x\,(x-1)\cdots(x-m+1)$ *for all*
$m \in \mathbb{N}_0$. *In passing we note the identity* $\dim \mathscr{A}^d = c_1^{\mathscr{A}(d)} + \dim H^{d,n}(\mathscr{A})$ *valid*
for all $d \geq d_{\max}$.

The Prolongation Theorem is actually a recompilation of all the properties we have discussed in the last two sections, for this reason we will not go into the details of its proof. Perhaps the strangest conclusion of Theorem 4.10 is that the Betti numbers of every finitely generated comodule $\mathscr{A}$ bounded below satisfy the following *a priori* inequalities for all $s = 1, \ldots, n$

$$\sum_{\substack{r=0,\ldots,n \\ l \in \mathbb{Z}}} (-1)^r \binom{n+d_{\max}-s-l-r-1}{n-s} \dim H^{l,r}(\mathscr{A}) \;\; \geq \;\; 0$$

which reflect the standard column length inequalities for the Cartan character $\mathfrak{Y}^{\mathscr{A}(d_{\max})}$.

5 Cartan–Kähler Theory

In essence the notion of an exterior differential system studied in this section can be seen as an axiomatization of the contact systems on the jet bundles of maps or section and the similar contact system on the generalized Graßmannians constructed in Sect. 2. Exaggerating somewhat we may say that exterior differential systems axiomatize the very concept of partial differential equations itself. En nuce the Cartan–Kähler theory of exterior differential systems is based on the simple idea to replace the submanifold solutions passing through a point by their infinite order Taylor series in this point, an idea already present in the beautiful theorem of Cauchy–Kovalevskaya for underdetermined partial differential equations. The purpose of this section is to sketch a proof of the formal version of the theorem of Cartan–Kähler, which generalizes the theorem of Cauchy–Kovalevskaya to other partial differential equations, while linking the topic to the Spencer cohomology of comodules discussed in Sects. 3 and 4.

Certainly the most striking feature common to both the contact system (12) on the bundle of jets of maps or sections and the contact system (17) on the generalized Graßmannians is the existence of a filtration of the cotangent bundle of the total space M by subbundles

$$0 \subseteq CM \subseteq HM \subseteq T^*M$$

such that the characteristic compatibility condition $d\,\Gamma(CM) \subset \Gamma(HM \wedge T^*M)$ holds true:

Definition 5.1 (Exterior Differential Systems). An exterior differential system on a manifold M is a filtration of the cotangent bundle T^*M by subbundles CM and HM called the bundles of contact and horizontal forms respectively

$$0 \subseteq CM \subseteq HM \subseteq T^*M$$

such that the exterior derivative of every contact form $\gamma \in \Gamma(CM)$ is a section of the ideal bundle $d\gamma \in \Gamma(HM \wedge T^*M)$ generated by HM. The annihilator subbundles

$$C^\perp M := \operatorname{Ann} CM = \{\, X_p \in TM \mid \gamma(X_p) = 0 \text{ for all } \gamma \in C_p M \,\} \subset TM$$

and $H^\perp M := \operatorname{Ann} HM$ defining the reciprocal filtration of the tangent bundle of M

$$TM \supseteq C^\perp M \supseteq H^\perp M \supseteq \{0\}$$

are called the vector bundles of admissible and vertical vectors on M respectively.

The reader may well wonder how such a definition may be used to treat partial differential equations in the language of differential forms, this question however is as futile as asking for the proper meaning of an answer 42 without knowing the question exactly. In other words the preceding definition is pretty useless without being accompanied by the complementary notion of a solution to a given exterior differential system $CM \subseteq HM \subseteq T^*M$:

Definition 5.2 (Solutions to Exterior Differential Systems). A solution to an exterior differential system $CM \subset HM$ on a manifold M is a submanifold $N \subset M$ of dimension $n := \dim HM - \dim CM$ such that every vector tangent to N is both admissible $T_p N \subset C_p^\perp M$ and non-vertical $T_p N \cap H_p^\perp M = \{0\}$. In every point $p \in N$ the tangent space $T_p N$ is thus a linear complement to the vertical in the admissible vectors:

$$C_p^\perp M = T_p N \oplus H_p^\perp M$$

Whatever else exterior differential systems and their solutions may be good for, their *raison d'être* is to unify different types of partial differential into a common framework formulated in the language of differential forms. For this reason let us postpone the development of the general theory for the moment in order to verify that the solutions to the contact systems discussed in Sect. 2 faithfully represent our intuitive understanding of what a solution to a partial differential equation should be. For convenience we will only consider the contact system on the generalized Graßmannian $\mathrm{Gr}_n^k M$, the reader is invited to repeat this analysis with the contact systems on $\mathrm{Jet}^k(N, M)$ and/or $\mathrm{Jet}^k \mathscr{F} M$.

Recall to begin with that the standard jet coordinates on the generalized Graßmannian $\mathrm{Gr}_n^k M$ associated to local coordinates $(x^1, \ldots, x^m)$ on M take the form (x^α, x_A^β) with indices $\alpha = 1, \ldots, n$ and $\beta = n + 1, \ldots, m$ as well as multi-indices A on $\{1, \ldots, n\}$ of order $|A| \leq k$. Moreover the scalar components of the canonical contact form γ^{contact} on $\mathrm{Gr}_n^k M$ are indexed by $\beta = n + 1, \ldots, m$ and multi-indices A of order $|A| < k$ and read:

$$\gamma_A^\beta := dx_A^\beta - \sum_{\alpha=1}^n x_{A+\alpha}^\beta \, dx^\alpha$$

Augmented by horizontal forms the contact system (17) on $\mathrm{Gr}_n^k M$ can thus be written:

$$
\begin{aligned}
C(\mathrm{Gr}_n^k M) &:= \mathrm{span} \{ \quad \gamma_A^\beta \quad \mid \quad \text{for all } \beta,\ |A| < k \ \} \\
H(\mathrm{Gr}_n^k M) &:= \mathrm{span} \{ \ dx^\alpha, dx_A^\beta \ \mid \quad \text{for all } \alpha, \beta,\ |A| < k \ \}
\end{aligned}
$$

In particular the annihilator subbundles of the reciprocal filtration of $T\mathrm{Gr}_n^k M$ are given by

$$
\begin{aligned}
H^\perp(\mathrm{Gr}_n^k M) &:= \mathrm{span} \{ \quad \frac{\partial}{\partial x_A^\beta} \quad \mid \quad \text{for all } \beta,\ |A| = k \ \} \\
C^\perp(\mathrm{Gr}_n^k M) &:= \mathrm{span} \{ \ \frac{\partial}{\partial x_A^\beta}, \frac{d}{dx^\alpha} \ \mid \quad \text{for all } \alpha, \beta,\ |A| = k \ \}
\end{aligned}
$$

where the total derivatives $\frac{d}{dx^\alpha}$ associated to the jet coordinates (x^α, x_A^β) are defined by:

$$\frac{d}{dx^\alpha} := \frac{\partial}{\partial x^\alpha} + \sum_{\substack{|A|<k \\ \beta}} x_{A+\alpha}^\beta \, \frac{\partial}{\partial x_A^\beta}$$

With a view on the calculations to come we remark that in this special exterior differential system the dual quotient bundles $H(\mathrm{Gr}_n^k M)/C(\mathrm{Gr}_n^k M)$ and $C^\perp(\mathrm{Gr}_n^k M)/H^\perp(\mathrm{Gr}_n^k M)$ are spanned by the dual classes represented by $dx^1, \ldots, dx^n$ and $\frac{d}{dx^1}, \ldots, \frac{d}{dx^n}$. Every linear complement to the vertical in the admissible vectors in a point $p \in \mathrm{Gr}_n^k M$ is of the form

$$\text{span} \left\{ \left. \frac{d}{dx^\alpha} \right|_p + \sum_{\substack{|A|=k \\ \beta}} x^\beta_{A,\alpha} \left. \frac{\partial}{\partial x^\beta_A} \right|_p \right\} \subset C^\perp_p(\text{Gr}^k_n M) \tag{50}$$

with suitably chosen constants $x^\beta_{A,\alpha} \in \mathbb{R}$ defined for all α, β and multi-indices A of order $|A| = k$. According to this description of all linear complements possible in $p \in \text{Gr}^k_n M$ the differentials $d_p x^1, \ldots, d_p x^n$ of the coordinate functions $x^1, \ldots, x^n$ stay linearly independent upon restriction to the tangent space $T_p N$ of a solution submanifold $N \subset \text{Gr}^k_n M$ passing through p, hence N can be written at least locally as the graph of a smooth map

$$(x^1, \ldots, x^n) \longmapsto (x^1, \ldots, x^n, x^\beta_A(x^1, \ldots, x^n))$$

with parameter functions $x^\beta_A(x^1, \ldots, x^n)$ to be specified for all β and all multi-indices A of order $|A| \leq k$. In terms of these parameter functions the tangent space $T_p N$ can be written

$$T_p N \;=\; \text{span} \left\{ \left. \frac{\partial}{\partial x^\alpha} \right|_p + \sum_{\substack{|A| \leq k \\ \beta}} \frac{\partial x^\beta_A}{\partial x^\alpha}(x^1, \ldots, x^n) \left. \frac{\partial}{\partial x^\beta_A} \right|_p \right\}$$

and comparing coefficients with the general form (50) of linear complements to the vertical in the admissible vectors we obtain the following constraints on the functions $x^\beta_A(x^1, \ldots, x^n)$

$$x^\beta_{A+\alpha}(x^1, \ldots, x^n) \;=\; \frac{\partial x^\beta_A}{\partial x^\alpha}(x^1, \ldots, x^n) \qquad\qquad x^\beta_{A,\alpha} \;=\; \frac{\partial x^\beta_A}{\partial x^\alpha}(x^1, \ldots, x^n)$$

for all α, β and all multi-indices A of order $|A| < k$ respectively $|A| = k$. By a straightforward induction all solutions to these constraints are completely determined by the parameter functions $x^\beta(x^1, \ldots, x^n)$ corresponding to the empty multi-index via the expected formula:

$$x^\beta_A(x^1, \ldots, x^n) \;=\; \frac{\partial^{|A|} x^\beta}{\partial x^A}(x^1, \ldots, x^n) \qquad x^\beta_{A,\alpha} \;=\; \frac{\partial^{|A|+1} x^\beta}{\partial x^{A+\alpha}}(x^1, \ldots, x^n) \tag{51}$$

In consequence every solution submanifold $N \subset \text{Gr}^k_n M$ to the contact system on the generalized Graßmannian $\text{Gr}^k_n M$ is holonomic in the sense that there exists at least locally a submanifold $N_{\text{base}} \subset M$ of dimension n with the property $N = \{ \text{jet}^k_{\pi(p)} N_{\text{base}} \mid p \in N \}$. Concluding our excursion to jet coordinates we recall that the exterior derivative of the scalar component γ^β_A of the contact form γ^{contact} indexed by a multi-index A of order $|A| < k - 1$

$$d\,\gamma_A^\beta \;=\; -\sum_\alpha \Big(\gamma_{A+\alpha}^\beta + \sum_{\tilde\alpha} x_{A+\alpha+\tilde\alpha}^\beta\,dx^{\tilde\alpha}\Big)\wedge dx^\alpha \;\overset{!}{=}\; -\sum_\alpha \gamma_{A+\alpha}^\beta \wedge dx^\alpha$$

lies in the ideal generated by the components of γ^{contact}, because $\sum x_{A+\alpha+\tilde\alpha}^\beta\,dx^{\tilde\alpha}\wedge$ $dx^\alpha = 0$ vanishes due to symmetry. For multi-indices A of order $|A| = k - 1$ on the other hand the exterior derivative $d\gamma_A^\beta = -\sum dx_{A+\alpha}^\beta \wedge dx^\alpha$ of γ_A^β restricts to a non-trivial 2-form

$$(d\gamma_A^\beta)_p\Big(\;\frac{d}{dx^{\tilde\alpha}}\Big|_p + \sum_{\substack{|\tilde A|=k \\ \tilde\beta}} x_{\tilde A,\tilde\alpha}^{\tilde\beta}\,\frac{\partial}{\partial x_{\tilde A}^{\tilde\beta}}\Big|_p\,,\;\; \frac{d}{dx^{\hat\alpha}}\Big|_p + \sum_{\substack{|\hat A|=k \\ \hat\beta}} x_{\hat A,\hat\alpha}^{\hat\beta}\,\frac{\partial}{\partial x_{\hat A}^{\hat\beta}}\Big|_p\;\Big)$$

$$\overset{!}{=}\; x_{A+\tilde\alpha,\hat\alpha}^\beta \;-\; x_{A+\hat\alpha,\tilde\alpha}^\beta \tag{52}$$

on a general linear complement of the vertical in the admissible vectors in a point $p \in \mathrm{Gr}_n^k M$ written in the form (50) with suitably chosen constants $x_{A,\alpha}^\beta \in \mathbb{R}$.

A partial differential equation of order $k \geq 1$ for submanifolds of dimension n of a manifold M is in essence the same as the associated subset $\mathrm{Eq}^k M \subset \mathrm{Gr}_n^k M$ of *algebraic* solutions. In practice $\mathrm{Eq}^k M$ is usually a smooth subbundle of the fiber bundle $\pi : \mathrm{Gr}_n^k M \longrightarrow M$, although in principle it could arbitrarily complicated. Partial differential equations satisfying this regularity assumption can be transformed into an equivalent exterior differential system on the manifold $\mathrm{Eq}^k M$ simply by restricting the differential forms comprising the contact system on $\mathrm{Gr}_n^k M$ to the submanifold $\mathrm{Eq}^k M$. Exterior differential systems of general type for example can be reduced to an exterior differential system in the sense of Definition 5.1, because they are invariably first order partial differential equations for submanifolds.

A peculiar consequence of the observation (52) is that the tangent space $T_p N$ of a solution N to an exterior differential system $CM \subseteq HM \subseteq T^*M$ on a manifold M has to satisfy additional *quadratic* constraints besides being a linear complement to the vertical in the admissible vectors, namely the exterior derivative of every contact form $\gamma \in \Gamma(CM)$ needs to vanish $(d\gamma)_p|_{T_p N \times T_p N} = 0$ when restricted to $T_p N$. More precisely $\gamma|_{T_p N} = 0$ for every contact form $\gamma \in \Gamma(CM)$, because every vector tangent to N is admissible and hence in $C_p^\perp M$, in terms of the inclusion $\iota_N : N \longrightarrow M$ we may write this $\iota_N^* \gamma = 0$ and obtain

$$\iota_N^*(d\gamma) \;=\; d\,(\iota_N^*\gamma) \;=\; 0 \qquad \Longrightarrow \qquad (d\gamma)_p\big|_{T_p N \times T_p N} \;=\; 0 \tag{53}$$

using the naturality of the exterior derivative. Recall now that the linear complements to the vertical in the admissible vectors correspond directly to sections of the short exact sequence

$$0 \;\longrightarrow\; H_p^\perp M \;\overset{\subset}{\longrightarrow}\; C_p^\perp M \;\overset{\text{pr}}{\longrightarrow}\; C_p^\perp M\big/_{H_p^\perp M} \;\longrightarrow\; 0 \tag{54}$$

namely the *image* of a section $s : C_p^\perp M / H_p^\perp M \longrightarrow C_p^\perp M$ is a linear complement and every linear complement T equals the image of a unique section $s_T : C_p^\perp M / H_p^\perp M \longrightarrow C_p^\perp M$. Sections of a short exact sequence like (54) on the other hand form an affine space modelled on the vector space $(H_p M / C_p M) \otimes H_p^\perp M$ of linear maps $\Delta s : C_p^\perp M / H_p^\perp M \longrightarrow H_p^\perp M$. Only the images of those sections $s : C_p^\perp M / H_p^\perp M \longrightarrow C_p^\perp M$ qualify as candidates for the tangent space of a solution N passing through $p \in M$, which satisfy the quadratic constraint

$$(d\gamma)_p(\, s\, X, \, s\, Y\,) \;=\; 0 \tag{55}$$

for every contact form $\gamma \in \Gamma(TM)$ and all $X, \, Y \in C_p^\perp M / H_p^\perp M$. In due course we will analyze this quadratic constraint in more detail, in particular a description of the set of all possible solutions $s : C_p^\perp M / H_p^\perp M \longrightarrow C_p^\perp M$ is given in Corollary 5.6.

Leaving the analytical description of exterior differential systems aside and turning to the associated algebraic theory of comodules we begin by casting the characteristic compatibility condition $d\,\Gamma(CM) \subset \Gamma(HM \wedge T^*M)$ between contact and horizontal forms into more manageable terms. Multilinear algebra tells us that ideal $H \wedge \Lambda^{\circ-1}T^* \subset \Lambda^{\circ}T^*$ in the exterior algebra of alternating forms on a vector space T generated by a subspace $H \subset T^*$ equals the ideal of alternating forms vanishing on all tuples of arguments in $H^\perp \subset T$:

$$H \wedge \Lambda^{\circ-1}T^* \;=\; \{\, \gamma \in \Lambda^{\circ}T^* \mid \gamma(\, V_1, \ldots, V_\circ\,) = 0 \;\text{ for all }\; V_1, \ldots, V_\circ \in H^\perp \,\}$$

More succinctly this statement reads $\Lambda^{\circ}H^\perp = (H \wedge \Lambda^{\circ-1}T^*)^\perp$ in terms of the duality between $\Lambda^{\circ}T$ and $\Lambda^{\circ}T^*$, in particular it can be seen as the supersymmetric analogue of the statement that a polynomial $\gamma \in \mathrm{Sym}\,T^*$ on T lies in the ideal generated by $H \subset T^*$, if and only if it vanishes identically on the subspace $H^\perp \subset T$. In consequence the characteristic compatibility condition imposed on an exterior differential system is equivalent to

$$d\gamma(\, V_1, \, V_2\,) \;=\; 0 \tag{56}$$

for every contact form $\gamma \in \Gamma(CM)$ and all vertical vector fields $V_1, V_2 \in \Gamma(H^\perp M)$. Replacing one of the two vertical vector fields by an admissible vector field $A \in \Gamma(C^\perp M)$ we obtain an expression $\Sigma(\gamma, A, V) := d\gamma(A, V) \in C^\infty(M)$, which does only depend on the class represented by A in the sections of the quotient bundle $C^\perp M / H^\perp M$. Despite first appearance $\Sigma(\gamma, A, V)$ depends $C^\infty(M)$-linearly not only on the vector fields A and V, but on the contact form $\gamma \in \Gamma(CM)$ as well, because $\gamma(A) = 0 = \gamma(V)$ both vanish so that:

$$\Sigma(\, f\gamma, A, V\,) \;=\; (df \wedge \gamma)(A, V) + f\, d\gamma(A, V) \;=\; f\, \Sigma(\gamma, A, V)$$

for every smooth function $f \in C^\infty(M)$ and every contact form $\gamma \in \Gamma(CM)$:

Definition 5.3 (Symbol of an Exterior Differential System). The symbol of an exterior differential system $CM \subseteq HM \subseteq T^*M$ on a manifold M is the $C^\infty(M)$-trilinear map Σ : $\Gamma(CM) \times \Gamma(C^\perp M/H^\perp M) \times \Gamma(H^\perp M) \longrightarrow C^\infty(M)$ defined for a contact form $\gamma \in \Gamma(CM)$ and vector fields $A \in \Gamma(C^\perp M)$ and $V \in \Gamma(H^\perp M)$ by:

$$\Sigma(\gamma, A, V) := d\gamma(A, V)$$

Being $C^\infty(M)$-trilinear the symbol Σ can be thought of as a homomorphism of vector bundles in many different ways, the preferred interpretation for exterior differential system reads:

$$\Sigma_p : \quad H_p^\perp M \longrightarrow (H_p M/C_p M) \otimes C_p^* M,$$

$$V_p \longmapsto \left(A_p \otimes \gamma_p \longmapsto (d\gamma)_p(A, V) \right)$$

Even more important than the symbol of an exterior differential system $CM \subseteq HM \subseteq T^*M$ on a manifold M are the two $\mathrm{Sym}(H_p M/C_p M)$-comodules associated to Σ in a point $p \in M$:

Definition 5.4 (Symbol and Reduced Symbol Comodule). Consider an exterior differential system $CM \subseteq HM \subseteq TM$ on a manifold M. The reduced symbol comodule of this exterior differential system in a point $p \in M$ is the tableau comodule $\mathscr{A}_p^\bullet \subset \mathrm{Sym}^\bullet(H_p M/C_p M) \otimes C_p^* M$ associated to the image of Σ_p considered as a tableau:

$$\mathscr{A}_p^1 := \mathbf{im}\, \Sigma_p \subset (H_p M/C_p M) \otimes C_p^* M$$

The symbol comodule $\mathscr{R}_p^\bullet$ in the point $p \in M$ is the kernel of the composition

$$\mathrm{Sym}^\bullet(H_p M/C_p M) \otimes H_p^\perp M$$

$$\stackrel{\mathrm{id} \otimes \Sigma_p}{\longrightarrow} \mathrm{Sym}^\bullet(H_p M/C_p M) \otimes (H_p M/C_p M) \otimes C_p^* M$$

$$\stackrel{B \otimes \mathrm{id}}{\longrightarrow} \mathrm{Sym}^{\bullet-1}(H_p M/C_p M) \otimes \Lambda^2(H_p M/C_p M) \otimes C_p^* M$$

of comodule homomorphisms involving the Spencer coboundary operator B of Sect. 3.

Interestingly the symbol comodule $\mathscr{R}_p$ of a general exterior differential system is never even mentioned in the otherwise authoritative reference [1] on exterior differential systems. The most important reason for this strange omission seems to be that the symbol comodule $\mathscr{R}_p$ and its reduced counterpart $\mathscr{A}_p$ are related by the very simple short exact sequence

$$0 \longrightarrow \mathrm{Sym}^\bullet(H_p M/C_p M) \otimes \ker \Sigma_p \stackrel{\subset}{\longrightarrow} \mathscr{R}_p^\bullet \stackrel{\mathrm{id} \otimes \Sigma_p}{\longrightarrow} \mathscr{A}_p^\bullet(1) \longrightarrow 0 \quad (57)$$

of comodules and thus have very similar Spencer cohomologies. In addition Σ_p is injective for many interesting examples so that not only the Spencer cohomology of $\mathcal{R}_p$ and $\mathcal{A}_p$, but the comodules themselves are easily confounded. It is symbol comodule $\mathcal{R}_p$ though, which has the direct bearance on the solution space of an exterior differential system erroneously attributed to the reduced symbol comodule in [1]. In any case the family of subspaces $\ker \Sigma_p \subset H_p^\perp M \subset T_p M$ parametrized by $p \in M$ appears in [1] in the guise of the so-called special Cauchy characteristic vector fields $\Gamma(\ker \Sigma) \subset \Gamma(TM)$.

In order to justify the short exact sequence (57) linking the two symbol comodules associated to an exterior differential system we recall from Theorem 4.10 that the twist $\mathcal{A}_p(1)$ of the tableau comodule $\mathcal{A}_p$ is again a tableau comodule, in fact it is the tableau comodule arising from the tableau $\mathcal{A}_p^2 \subset (H_p M / C_p M) \otimes \mathcal{A}_p^1$. In turn this tableau can be written as the kernel of the Spencer coboundary operator $B \otimes \mathrm{id}$ in the exact sequence

$$
0 \longrightarrow \mathcal{A}_p^2 \overset{\subset}{\longrightarrow} (H_p M / C_p M) \otimes \mathcal{A}_p^1 \overset{B \otimes \mathrm{id}}{\longrightarrow} \Lambda^2 (H_p M / C_p M) \otimes C_p^* M
$$

due to the generic property $H^{0,2}(\mathcal{A}_p) = \{0\} = H^{1,1}(\mathcal{A}_p)$ of tableau comodules established in Eq. (30). In consequence the twist $\mathcal{A}_p^\bullet(1)$ of the reduced symbol comodule $\mathcal{A}_p^\bullet$ can be written as the kernel of the following homomorphism of free comodules:

$$
\mathrm{Sym}^\bullet(H_p M / C_p M) \otimes \mathcal{A}_p^1
$$

$$
\overset{B \otimes \mathrm{id}}{\longrightarrow} \mathrm{Sym}^{\bullet - 1}(H_p M / C_p M) \otimes \Lambda^2 (H_p M / C_p M) \otimes C_p^* M
$$

With $\mathcal{A}_p^1 := \mathbf{im}\, \Sigma_p$ the symbol comodule $\mathcal{R}_p^\bullet$ is thus by its very definition the preimage of the subcomodule $\mathcal{A}_p^\bullet(1) \subset \mathrm{Sym}^\bullet(H_p M / C_p M) \otimes (H_p M / C_p M) \otimes C_p^* M$ under the homomorphism $\mathrm{id} \otimes \Sigma_p : \mathrm{Sym}^\bullet(H_p M / C_p M) \otimes H_p^\perp M \longrightarrow \mathrm{Sym}^\bullet(H_p M / C_p M) \otimes (H_p M / C_p M) \otimes C_p^* M$ of free comodules induced by Σ_p so that the sequence (57) is short exact.

With the machinery of symbol and symbol comodules at our disposal let us now come back to the discussion of the quadratic constraint (55) characterizing the set of linear complements to the vertical in the admissible vectors, which are proper candidates for the tangent spaces $T_p N$ of solutions N passing through $p \in M$. Modifying the section $s : C_p^\perp M / H_p^\perp M \longrightarrow C_p^\perp M$ of the short exact sequence (54) corresponding to an arbitrary linear complement by a linear map $\Delta s : C_p^\perp M / H_p^\perp M \longrightarrow H_p^\perp M$ we obtain for all vectors $X, Y \in C_p^\perp M / H_p^\perp M$

$$
(d\gamma)_p((s + \Delta s) X, (s + \Delta s) Y)
$$
$$
= (d\gamma)_p(s X, s Y) + (d\gamma)_p(s X, (\Delta s) Y) - (d\gamma)_p(s Y, (\Delta s) X)
$$
$$
= (d\gamma)_p(s X, s Y) + \Sigma_p(\gamma_p, X, (\Delta s) Y) - \Sigma_p(\gamma_p, Y, (\Delta s) X)
$$

because $(d\gamma)_p$ vanishes on two vertical arguments due to the reformulation (56) of the axiomatic compatibility condition $d\,\Gamma(CM) \subset \Gamma(HM \wedge T^*M)$ between the contact and the horizontal forms. The modified section $s + \Delta s$ is a solution to the quadratic constraint (55), if and only if Δs satisfies the following inhomogeneous linear equation:

$$\Sigma_p(\gamma_p, X, (\Delta s)Y) - \Sigma_p(\gamma_p, Y, (\Delta s)X) = -(d\gamma)_p(sX, sY) \quad (58)$$

Our preferred interpretation $\Sigma_p : H_p^\perp M \longrightarrow (H_p M/C_p M) \otimes C_p^* M$ of the symbol Σ_p on the other hand allows us to write the trilinear form $(X, Y, \gamma_p) \longmapsto \Sigma_p(\gamma_p, Y, (\Delta s)X)$ as the image of Δs considered as an element of $(H_p M/C_p M) \otimes H_p^\perp M$ under the linear map:

$$\mathrm{id} \otimes \Sigma_p : \quad (H_p M/C_p M) \otimes H_p^\perp M \longrightarrow (H_p M/C_p M) \otimes \mathscr{A}_p^1$$

In addition the skew-symmetrization of this trilinear form on the left hand side of the linear equation (58) for Δs implements a special case of the Spencer coboundary operator

$$\left\langle B\left[(\mathrm{id} \otimes \Sigma_p)(\Delta s)\right](X, Y), \gamma_p \right\rangle$$

$$= \Sigma_p(\gamma_p, X, (\Delta s)Y) - \Sigma_p(\gamma_p, Y, (\Delta s)X)$$

namely $B : (H_p M/C_p M) \otimes \mathscr{A}_p^1 \longrightarrow \Lambda^2(H_p M/C_p M) \otimes \mathscr{A}_p^0$ with $\mathscr{A}_p^0 := C_p^* M$ by definition, the most difficult problem here is to convince oneself of the correctness of the sign. Since every section of the short exact sequence (54) can be written in the form $s + \Delta s$ for an arbitrarily chosen base section $s : C_p^\perp M/H_p^\perp M \longrightarrow C_p^\perp M$ and a suitable modification Δs we conclude that the linear complements T to the vertical vectors $H_p^\perp M$ in the admissible vectors $C_p^\perp M$ satisfying the quadratic constraint (55) correspond via $s_T = s + \Delta s$ bijectively to the solutions $\Delta s \in (H_p M/C_p M) \otimes H_p^\perp M$ of the inhomogeneous linear equation

$$B\left((\mathrm{id} \otimes \Sigma_p)(\Delta s)\right) = -\Theta_p(s) \quad (59)$$

with right hand side given by $\langle \Theta_p(s)(X, Y), \gamma_p \rangle := (d\gamma)_p(sX, sY)$, compare the original equation (58). In turn this inhomogeneous linear equation gives rise to the concept of torsion:

Definition 5.5 (Torsion). Consider an exterior differential system $CM \subseteq HM \subseteq TM$ on a manifold M. The torsion of this exterior differential system in a point $p \in M$ is the Spencer cohomology class

$$[\Theta_p(s)] \in H^{0,2}(\mathscr{A}_p)$$

$$:= \left(\Lambda^2(H_p M/C_p M) \otimes C_p^* M\right)\Big/ B\left((H_p M/C_p M) \otimes \mathscr{A}_p^1\right)$$

represented by the 2-form $\Theta_p(s) \in \Lambda^2(H_pM/C_pM) \otimes C_p^*M$ with values in C_p^*M defined for an arbitrary section $s : C_p^\perp M/H_p^\perp M \longrightarrow C_p^\perp M$ of the short exact sequence (54) by:

$$\langle \Theta_p(s)(X, Y), \gamma_p \rangle := (d\gamma)_p(sX, sY)$$

A classical theorem of linear algebra asserts that the inhomogeneous linear equation (59) characterizing the candidates $\mathbf{im}(s + \Delta s) \subset C_p^\perp M$ for the tangent spaces of solution submanifolds $N \subset M$ passing through $p \in M$ has a solution $\Delta s \in (H_pM/C_pM) \otimes H_p^\perp M$, if and only if $-\Theta_p(s)$ lies in the image of the Spencer coboundary operator B, if and only if the torsion vanishes. After all $\mathrm{id} \otimes \Sigma_p : (H_pM/C_pM) \otimes H_p^\perp M \longrightarrow (H_pM/C_pM) \otimes \mathscr{A}_p^1$ is surjective by definition, hence the vanishing $[\Theta_p(s)] = 0$ of the torsion implies that every preimage $\Delta s \in (H_pM/C_pM) \otimes H_p^\perp M$ of an element of $(H_pM/C_pM) \otimes \mathscr{A}_p^1$ making $-\Theta_p(s)$ exact is a solution to the inhomogeneous equation (59). A very similar argument implies that the torsion is actually independent of the section used to define the representative $\Theta_p(s)$ due to the identity $\Theta_p(s + \Delta s) = \Theta_p(s) + B[(\mathrm{id} \otimes \Sigma_p)(\Delta s)]$:

Corollary 5.6 (Significance of Torsion). *No solution submanifold $N \subset M$ to an exterior differential system $CM \subseteq HM \subseteq T^*M$ on a manifold M passes through a point $p \in M$, unless the torsion $[\Theta_p(s)] \in H^{0,2}(\mathscr{A}_p)$ vanishes for one and hence every section $s : C_p^\perp M/H_p^\perp M \longrightarrow C_p^\perp M$ of the short exact sequence (54). In the latter case the linear complements T to the vertical in the admissible vectors satisfying the constraint $(d\gamma)_p(s_TX, s_TY) = 0$ for all contact forms $\gamma \in \Gamma(CM)$ and all $X, Y \in C_p^\perp M/H_p^\perp M$ form an affine space modelled on the vector space $\mathscr{R}_p^1$.*

Somewhat surprisingly it is the homogeneous subspace $\mathscr{R}_p^1$ of the symbol comodule $\mathscr{R}_p$, which parametrizes the possible candidates for the tangent spaces T_pN of solution submanifolds $N \subset M$ in the case of vanishing torsion $[\Theta_p(s)] = 0$ in the point $p \in M$, not a homogeneous subspace of the more prominent reduced symbol comodule $\mathscr{A}_p$. The reason for this is simple: By its very definition $\mathscr{R}_p^1$ equals the kernel of the linear map $B \circ (\mathrm{id} \otimes \Sigma_p)$ and thus acts naturally on the solutions to the inhomogeneous linear equation (59).

The generalization of Corollary 5.6 to higher orders of differentiation forms the cornerstone of the Cartan–Kähler theory of exterior differential systems. Similar to its historic precursor, the theorem of Cauchy–Kovalevskaya the Cartan–Kähler theory tries to reconstruct the solution submanifolds from their infinite order Taylor series in a given point, a notion made precise by the projective limit $\mathrm{Gr}_n^\infty M$ of the tower (16) of Graßmannians:

$$\cdots \xrightarrow{\mathrm{pr}} \mathrm{Gr}_n^3 M \xrightarrow{\mathrm{pr}} \mathrm{Gr}_n^2 M \xrightarrow{\mathrm{pr}} \mathrm{Gr}_n^1 M \xrightarrow{\pi} \mathrm{Gr}_n^0 M = M$$

In such a power series approach we are inevitably led to consider jet solutions of sorts:

Definition 5.7 (Jet Solutions and Semisolutions). A jet solution of order $k \geq 1$ to an exterior differential system $CM \subseteq HM \subseteq T^*M$ on a manifold M is a kth order jet of a submanifold $\mathrm{jet}_p^k N \in \mathrm{Gr}_n^k M$, whose tangent space in p is a linear complement $T_p N \subset C_p^{\perp} M$ to the subspace $H_p^{\perp} M$ of vertical vectors, such that

$$\mathrm{jet}_p^{k-1}(\iota_N^* \gamma) = 0 = \mathrm{jet}_p^{k-1}(\iota_N^* d\gamma)$$

for all contact forms $\gamma \in \Gamma(CM)$. Similarly a jet semisolution of order $k \geq 1$ is an element $\mathrm{jet}_p^k N \in \mathrm{Gr}_n^k M$ represented by a submanifold $N \subset M$ satisfying $C_p^{\perp} M = T_p N \oplus H_p^{\perp} M$ and $\mathrm{jet}_p^{k-1}(\iota_N^* \gamma) = 0$ for all $\gamma \in \Gamma(CM)$. Solutions and semisolutions assemble into the sets:

$$\mathrm{Eq}_p^k M := \{\ \mathrm{jet}_p^k N \ \mid \ \mathrm{jet}_p^k N \in \mathrm{Gr}_n^k M \text{ is a jet solution of order } k \ \}$$

$$\overline{\mathrm{Eq}}_p^k M := \{\ \mathrm{jet}_p^k N \ \mid \ \mathrm{jet}_p^k N \in \mathrm{Gr}_n^k M \text{ is a jet semisolution of order } k \ \}$$

In light of the identification $\mathrm{jet}_p^1 N \leftrightarrow T_p N$ of the generalized Graßmannian $\mathrm{Gr}_n^1 M$ with the Graßmann bundle $\mathrm{Gr}_n(TM)$ of n-dimensional subspaces of TM the preceding definition of jet solutions and jet semisolutions faithfully reflects our considerations above for order $k = 1$. Jet semisolutions of order $k = 1$ say are simply linear complements $T_p N$ to the vertical in the admissible vectors, while jet solutions are linear complements satisfying the quadratic constraint $\mathrm{jet}_p^0(\iota_N^* d\gamma) = (d\gamma)_p|_{T_p N \times T_p N} = 0$. Hence $\overline{\mathrm{Eq}}_p^1 M$ is always the affine space of sections of the short exact sequence (54), while $\mathrm{Eq}_p^1 M$ is described by Corollary 5.6 as an affine space modelled on $\mathscr{R}_p^1$ in the case of vanishing torsion, otherwise it is empty. This classification of jet solutions and jet semisolutions of order $k = 1$ generalizes to the picture at higher orders of differentiation $k \geq 1$, which is best remembered as a tower

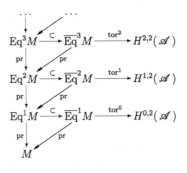

$$\tag{60}$$

we need to climb up one step at a time. The diagonal projections $\mathrm{pr}\ :\ \overline{\mathrm{Eq}}_p^{k+1} M \longrightarrow \mathrm{Eq}_p^k M$ are always surjective with fiber an affine space modelled on $\mathrm{Sym}^{k+1}(H_p M/C_p M) \otimes H_p^{\perp} M$, while the rows in this tower are "exact" in the following sense: There exists a jet solution $\mathrm{jet}_p^{k+1} N \in \mathrm{Eq}_p^{k+1} M$ over a jet solution

$\text{jet}_p^k N \in \text{Eq}_p^k M$, if and only if the higher torsion $\text{tor}^k : \overline{\text{Eq}}_p^{k+1} M \longrightarrow H^{k,2}(\mathscr{A}_p)$ vanishes on some and hence on every $\text{jet}_p^{k+1} \overline{N} \in \overline{\text{Eq}}_p^{k+1} M$ lying over $\text{jet}_p^k N$. Interestingly all obstructions against formal integrability live in the Spencer cohomology $H^{\bullet,2}(\mathscr{A}_p)$ of the reduced symbol comodule of form degree $\circ = 2$.

A more detailed study of the tower (60) has to wait a little bit until we have clarified the subtle interplay between the jets of submanifolds and the jets of differential forms, on which Definition 5.7 is based. Conceptually it is easier in this endeavor to consider the more general case of smooth maps $\varphi : N \longrightarrow M$ between manifolds N and M and specify to canonical inclusions $\iota_N : N \longrightarrow M$ of submanifolds later on. A smooth map $\varphi : N \longrightarrow M$ can be written in local coordinates $(x^1, \ldots, x^n)$ and $(y^1, \ldots, y^m)$ around a point $p \in N$ and its image $\varphi(p) \in M$ as m smooth functions of n variables, namely the pull backs of:

$$\varphi^* y^\mu =: y^\mu(x^1, \ldots, x^n) \qquad \Longrightarrow \qquad \varphi^* dy^\mu = \sum_{\alpha=1}^n \frac{\partial y^\mu}{\partial x^\alpha}(x^1, \ldots, x^n) \, dx^\alpha$$

In this local coordinate description the pull back of a general differential form reads:

$$\varphi^* \left[f(y^1, \ldots, y^m) \, dy^{\mu_1} \wedge \ldots \wedge dy^{\mu_r} \right]$$

$$= \sum_{\alpha_1, \ldots, \alpha_r = 1}^n f(y^1(x), \ldots, y^m(x)) \frac{\partial y^{\mu_1}}{\partial x^{\alpha_1}}(x) \ldots \frac{\partial y^{\mu_r}}{\partial x^{\alpha_r}}(x) \, dx^{\alpha_1} \wedge \ldots \wedge dx^{\alpha_r}$$

In consequence the partial derivatives up to order k of the coefficients of the right hand side with respect to the monomial basis $dx^{\alpha_1} \wedge \ldots \wedge dx^{\alpha_r}$ depend on the partial derivatives of the original coefficient $f(y^1, \ldots, y^m)$ up to order k and on the partial derivatives of the functions y^μ up to order $k + 1$ only. In other words we have a well-defined linear map of vector spaces

$$[\![\text{jet}_p^{k+1} \varphi]\!] : \quad \text{Jet}_{\varphi(p)}^k \Lambda^\circ T^* M \longrightarrow \text{Jet}_p^k \Lambda^\circ T^* N, \qquad \text{jet}_{\varphi(p)}^k \omega \longmapsto \text{jet}_p^k(\varphi^* \omega) \tag{61}$$

which only depends on $\text{jet}_p^{k+1} \varphi \in \text{Jet}_p^{k+1}(N, M)$. In a similar vein we recall that the exterior derivative d is a linear first order differential operator, its composition $\text{jet}^{k-1} \circ d$ is thus a linear differential operator of order k, whose total symbol map in the sense of Definition 2.2 induces for every point $p \in N$ a linear map of the jet fiber vector spaces

$$d^{\text{formal}} : \quad \text{Jet}_p^k \Lambda^\circ T^* N \longrightarrow \text{Jet}_p^{k-1} \Lambda^{\circ+1} T^* N, \qquad \text{jet}_p^k \omega \longmapsto \text{jet}_p^{k-1}(d\omega)$$

as well as its analogue $d^{\text{formal}} : \text{Jet}_{\varphi(p)}^k \Lambda^\circ T^* M \longrightarrow \text{Jet}_{\varphi(p)}^{k-1} \Lambda^{\circ+1} T^* M$, such that the diagram

$$\begin{array}{ccc}
\operatorname{Jet}^k_{\varphi(p)}\Lambda^\circ T^* M & \xrightarrow{\;[\mathrm{jet}^{k+1}_p \varphi]\;} & \operatorname{Jet}^k_p \Lambda^\circ T^* N \\[2pt]
{\scriptstyle d^{\mathrm{formal}}}\Big\downarrow & & {\scriptstyle d^{\mathrm{formal}}}\Big\downarrow \\[2pt]
\operatorname{Jet}^{k-1}_{\varphi(p)}\Lambda^{\circ+1} T^* M & \xrightarrow{\;[\,\mathrm{jet}^k_p \varphi\,]\;} & \operatorname{Jet}^{k-1}_p \Lambda^{\circ+1} T^* N
\end{array}$$

$$(62)$$

commutes due to the naturality $d(\varphi^*\omega) = \varphi^*(d\omega)$ of the exterior derivative. In passing we remark that the *principal* symbol of the differential operator $\mathrm{jet}^{k-1} \circ d$, which is by definition the restriction of its *total* symbol d^{formal} to the symbol subspace defined in (8)

$$\operatorname{Sym}^k T^*_p N \otimes \Lambda^\circ T^*_p N \;=\; \mathbf{ker}\Big(\operatorname{pr}: \operatorname{Jet}^k_p \Lambda^\circ T^* N \longrightarrow \operatorname{Jet}^{k-1}_p \Lambda^\circ T^* N \Big) \quad (63)$$

agrees with the Spencer coboundary operator B defined in Sect. 3. Of course this is the *conditio sine qua non* for the usefulness of Spencer cohomology in the study of exterior differential systems.

A rather surprising aspect of the commutative diagram (62) should not pass by unnoticed, the linear map $[\![\mathrm{jet}^k \varphi]\!]$ does only depend on the jet of φ of order k, whereas $[\![\mathrm{jet}^{k+1}\varphi]\!]$ invariably involves the partial derivatives of φ of order $k + 1$. In order to resolve this apparent contradiction to the commutativity of (62) we observe that the only terms in the partial derivatives of the coefficients of $\varphi^*[f(y^1,\ldots, y^m)\, dy^{\mu_1} \wedge \ldots \wedge dy^{\mu_r}]$ of order up to k, which actually involve partial derivatives of the functions y^μ of order $k + 1$, can be written:

$$f(\varphi(p)) \left(\sum_{\alpha=1}^{n} \frac{\partial^{|A|+1} y^{\mu_1}}{\partial^{A+\alpha} x}\, dx^\alpha \right) \wedge \varphi^*_p dy^{\mu_2} \wedge \ldots \wedge \varphi^*_p dy^{\mu_r}$$

$$+ \; f(\varphi(p))\, \varphi^*_p dy^{\mu_1} \wedge \left(\sum_{\alpha=1}^{n} \frac{\partial^{|A|+1} y^{\mu_2}}{\partial^{A+\alpha} x}\, dx^\alpha \right) \wedge \ldots \wedge \varphi^*_p dy^{\mu_r} \; + \; \ldots$$

Quite remarkably this expression looks like a derivation applied to $f(\varphi(p))\, dy^{\mu_1} \wedge \ldots \wedge dy^{\mu_r}$! The change in the pull back of jets of differential forms resulting from a modification of the highest order partial derivatives of φ by adding $\Delta\varphi \in \operatorname{Sym}^{k+1} T^*_p N \otimes T_{\varphi(p)} M$ thus reads

$$[\![\mathrm{jet}^{k+1}_p \varphi + \Delta \varphi]\!](\mathrm{jet}^k_p \omega) = [\![\mathrm{jet}^{k+1}_p \varphi]\!](\mathrm{jet}^k_p \omega) + B\Big((\mathrm{id} \otimes \varphi^*_p)(\Delta\varphi \lrcorner \, \omega_{\varphi(p)})\Big)$$

$$(64)$$

where the additional term involves the Spencer coboundary operator B and the composition

$$\Lambda^\circ T^*_{\varphi(p)} M \xrightarrow{\;\Delta\varphi \lrcorner\;} \operatorname{Sym}^{k+1} T^*_p N \otimes \Lambda^{\circ-1} T^*_{\varphi(p)} M \xrightarrow{\;\mathrm{id}\otimes\varphi^*_p\;} \operatorname{Sym}^{k+1} T^*_p N \otimes \Lambda^{\circ-1} T^*_p N$$

applied to the value $\omega_{\varphi(p)} \in \Lambda^\circ T^*_{\varphi(p)} M$ of the differential form ω in $\varphi(p)$. Since the Spencer coboundary operator equals d^{formal} on the symbol subspace, the additional term in (64) lies in the kernel of d^{formal} and is thus irrelevant to the commutativity of (62).

Specializing the preceding observations to the inclusion maps $\iota_N : N \longrightarrow M$ of submanifolds of M we remark that $\text{jet}^k_p \iota_N \in \text{Jet}^k(N, M)$ and $\text{jet}^k_p N \in \text{Gr}^k_n M$ encode essentially the same object, the jet of a submanifold, hence we may write in a shorthand notation

$$[\![\text{jet}^k_p N]\!] : \quad \text{Jet}^{k-1}_p \Lambda^\circ T^* M \longrightarrow \text{Jet}^{k-1}_p \Lambda^\circ T^* N, \qquad \text{jet}^{k-1}_p \omega \longmapsto \text{jet}^{k-1}_p (\iota^*_N \omega)$$

for the restriction maps appearing prominently in Definition 5.7 of jet solutions and semisolutions. Feeling somewhat uneasy with the fact that the target vector space depends on the representative submanifold N we sooth our conscience by observing that for every two submanifolds N_1 and N_2 representing $\text{jet}^k_p N_1 = \text{jet}^k_p N_2$ there exists a distinguished class $\text{jet}^k_p \varphi \in \text{Jet}^k_p(N_1, N_2)$ represented by those diffeomorphisms $\varphi : N_1 \longrightarrow N_2$, which satisfy

$$\left. \frac{d^{\leq k}}{dt^{\leq k}} \right|_0 c = \left. \frac{d^{\leq k}}{dt^{\leq k}} \right|_0 (\varphi \circ c) \in T^k_p M$$

for every curve $c : \mathbb{R} \longrightarrow N_1$. For every diffeomorphism φ in this class $\text{jet}^k_p \iota_{N_1} = \text{jet}^k_p (\iota_{N_2} \circ \varphi)$ so that the two realizations of $[\![\text{jet}^k_p N]\!]$ are intertwined by the well-defined isomorphism:

$$[\![\text{jet}^k_p \varphi]\!] : \quad \text{Jet}^{k-1}_p \Lambda^\circ T^* N_2 \xrightarrow{\cong} \text{Jet}^{k-1}_p \Lambda^\circ T^* N_1, \qquad \text{jet}^{k-1}_p \omega \longmapsto \text{jet}^{k-1}_p (\varphi^* \omega)$$

Although these comments may look rather pedantic, they are directly related to a delicate subtlety, which has bothered the author for quite a while. If we modify the highest order partial derivatives of $\text{jet}^{k+1}_p N \in \text{Gr}^{k+1}_n M$ by an element $\Delta N \in \text{Sym}^{k+1} T^*_p N \otimes (T_p M / T_p N)$ using the addition (20), then the modification formula (64) tells us

$$[\![\text{jet}^{k+1}_p N + \Delta N]\!](\text{jet}^k_p \omega)$$
$$= [\![\text{jet}^{k+1}_p N]\!](\text{jet}^k_p \omega) + B \left((\text{id} \otimes \text{res}_{T^*_p N})((\Delta N)^{\text{lift}} \lrcorner \, \omega_p) \right) \qquad (65)$$

the additional term on the right however depends on the lift $(\Delta N)^{\text{lift}} \in \text{Sym}^{k+1} T^*_p N \otimes T_p M$ we need to chose in the addition (20). The resolution to this paradox is that a representative submanifold $\tilde{N}$ for $\text{jet}^{k+1}_p N + \Delta N$ is in contact with N to order k in p only, hence we no longer sport a distinguished vector space isomorphism $[\![\text{jet}^{k+1}_p \varphi]\!] : \text{Jet}^k_p \Lambda^\circ T^* \tilde{N} \xrightarrow{\cong} \text{Jet}^k_p \Lambda^\circ T^* N$. Cum grano salis the modification formula still makes sense: The ambiguity in choosing $(\Delta N)^{\text{lift}}$ given

ΔN is countered exactly by the ambiguity of lifting the distinguished class $\mathrm{jet}_p^k \varphi$ of diffeomorphisms $\varphi : N \longrightarrow \tilde{N}$ to a vector space isomorphism $[\![\ \mathrm{jet}_p^{k+1} \varphi\ ]\!]$.

Let us now put the formulas derived above to the test and study the tower (60) of jet solutions and jet semisolutions in more detail. In a first step we observe that every jet semisolution $\mathrm{jet}_p^{k+1} N \in \overline{\mathrm{Eq}}_p^{k+1} M$ of order $k + 1$ projects under pr : $\mathrm{Gr}_n^{k+1} M \longrightarrow \mathrm{Gr}_n^k N$ to a jet solution $\mathrm{jet}_p^k N$. By assumption $\mathrm{jet}_p^k(\iota_N^* \gamma) = 0$ vanishes for every contact form $\gamma \in \Gamma(CM)$, thus

$$\mathrm{pr}[\,\mathrm{jet}_p^k(\iota_N^* \, d\gamma)\,] \;\;=\;\; \mathrm{jet}_p^{k-1}(\iota_N^* \, d\gamma) \;\;=\;\; d^{\mathrm{formal}}\Big(\mathrm{jet}_p^k(\iota_N^* \gamma)\Big) \;\;=\;\; 0$$

for all $\gamma \in \Gamma(CM)$ as claimed. In this way we have proved the first statement of the lemma:

Lemma 5.8 (Lifting Jet Solutions to Semisolutions). *Every jet semisolution* $\mathrm{jet}_p^{k+1} N$ *of order $k + 1$ of an exterior differential system $CM \subseteq HM$ on a manifold M projects under* pr : $\mathrm{Gr}_n^{k+1} M \longrightarrow \mathrm{Gr}_n^k M$ *to a jet solution* $\mathrm{jet}_p^k N$ *of order k. Conversely the set of all jet semisolutions lying over a given solution* $\mathrm{jet}_p^k N$ *of order $k \geq 1$ is a non-empty affine subspace of* $\mathrm{pr}^{-1}(\mathrm{jet}_p^k N)$ *modelled on the vector subspace:*

$$\mathrm{Sym}^{k+1} T_p^* N \otimes H_p^\perp M \;\;\subset\;\; \mathrm{Sym}^{k+1} T_p^* N \otimes (T_p M / T_p N)$$

Proof. Consider a submanifold $N \subset M$ of dimension n representing a given jet solution $\mathrm{jet}_p^k N$ of order $k \geq 1$. By definition N represents a jet semisolution of order $k+1$, if and only if the $\mathbb{R}$-linear map $\Theta : \Gamma(CM) \longrightarrow \mathrm{Jet}_p^k T^* N$, $\gamma \longmapsto \mathrm{jet}_p^k(\iota_N^* \gamma)$, is trivial. In light of our discussion above Θ depends on the representative submanifold N only through $\mathrm{jet}_p^{k+1} N$:

$$\Theta(\gamma) \;\; := \;\; [\![\ \mathrm{jet}_p^{k+1} N\ ]\!](\mathrm{jet}_p^k \gamma) \;\; := \;\; \mathrm{jet}_p^k(\iota_N^* \gamma)$$

Since N represents a jet solution of order k, we find $\mathrm{pr}[\mathrm{jet}_p^k(\iota_N^* \gamma)] = \mathrm{jet}_p^{k-1}(\iota_N^* \gamma) = 0$, hence $\Gamma(CM)$ gets mapped under Θ into the kernel of the jet projection, the symbol subspace $\mathrm{Sym}^k T_p^* N \otimes T_p^* N$ of observation (63). In particular Θ is $C^\infty(M)$-linear with

$$\Theta(f \gamma) \;\; = \;\; \mathrm{jet}_p^k(\iota_N^* f \wedge \iota_N^* \gamma) \;\; = \;\; \mathrm{jet}_p^k(\iota_N^* f) \wedge \mathrm{jet}_p^k(\iota_N^* \gamma) \;\; = \;\; f(p)\,\mathrm{jet}_p^k(\iota_N^* \gamma)$$

for all $f \in C^\infty(M)$, where we use the natural algebra structure induced on the jet fiber of the algebra bundle $\Lambda^\circ T^* N$ in the second and $\mathrm{jet}_p^{k-1}(\iota_N^* \gamma) = 0$ in the third equality. Due to $C^\infty(M)$-linearity $\Theta(\gamma)$ depends on the value of γ in p only, moreover the composition

$$C_p M \xrightarrow{\Theta} \mathrm{Sym}^k T_p^* N \otimes T_p^* N \xrightarrow{B} \mathrm{Sym}^{k-1} T_p^* N \otimes \Lambda^2 T_p^* N$$

vanishes, because the Spencer coboundary B agrees with d^{formal} on the symbol subspace:

$$B[\,\Theta(\gamma_p)\,] = d^{\mathrm{formal}}[\,\mathrm{jet}_p^k(\iota_N^* \gamma)\,] = \mathrm{jet}_p^{k-1}(d\,\iota_N^* \gamma) = \mathrm{jet}_p^{k-1}(\iota_N^* d\gamma) = 0$$

The calculation of the Spencer cohomology of free comodules on the other hand implies that

$$0 \longrightarrow \mathrm{Sym}^{k+1} T_p^* N \xrightarrow{B} \mathrm{Sym}^k T_p^* N \otimes T_p^* N \xrightarrow{B} \mathrm{Sym}^{k-1} T_p^* N \otimes \Lambda^2 T_p^* N$$

is exact for $k \geq 1$. In consequence there exists a unique $\Theta^{\mathrm{pre}} \in \mathrm{Sym}^{k+1} T_p^* N \otimes C_p^* M$ with

$$B(\langle\, \Theta^{\mathrm{pre}},\, \gamma_p\,\rangle) = \Theta(\gamma_p) = \mathrm{jet}_p^k(\iota_N^* \gamma)$$

What remains to do, now that the existence of Θ^{pre} is established, is to write the standard short exact sequence associated to the 3-step filtration $T_p N \subset C_p^{\perp} M \subset T_p M$ with a view

$$0 \longrightarrow H_p^{\perp} M \longrightarrow T_p M / T_p N \longrightarrow C_p^* M \longrightarrow 0 \qquad (66)$$

on the canonical isomorphisms $T_p M / C_p^{\perp} M \cong C_p^* M$ and $H_p^{\perp} M \cong C_p^{\perp} M / T_p N$. The element $\Theta^{\mathrm{pre}} \in \mathrm{Sym}^{k+1} T_p^* N \otimes C_p^* M$ can thus be lifted to $\Delta N \in \mathrm{Sym}^{k+1} T_p^* N \otimes (T_p M / T_p N)$, although not uniquely, and all lifts satisfy the decisive property $\Delta N \lrcorner\, \gamma_p = \langle\, \Theta^{\mathrm{pre}},\, \gamma_p\,\rangle$. Together with the modification formula (65) this property implies that the modification $\mathrm{jet}_p^{k+1} N - \Delta N \in \mathrm{Gr}_n^{k+1} M$ is a jet semisolution of order $k+1$ lying over $\mathrm{jet}_p^k N$ due to:

$$[\![\, \mathrm{jet}_p^{k+1} N - \Delta N\,]\!](\mathrm{jet}_p^k \gamma) = \mathrm{jet}_p^k(\iota_N^* \gamma) - B(\Delta N \lrcorner\, \gamma_p) = 0$$

Last but not least the exactness of the sequence (66) tells us that the difference of two lifts of Θ^{pre} corresponds exactly to an element of $\mathrm{Sym}^{k+1} T_p^* N \otimes H_p^{\perp} M$. ☐

Lemma 5.9 (Obstructions against Formal Integrability). *Consider a jet semisolution* $\mathrm{jet}_p^{k+1} N \in \overline{\mathrm{Eq}}_p^{k+1} M$ *of order* $k+1$ *to an exterior differential system on a manifold* M. *There exists a jet solution* $\mathrm{jet}_p^{k+1} \tilde{N} \in \mathrm{Eq}_p^{k+1} M$ *of order* $k+1$ *lifting the jet solution* $\mathrm{jet}_p^k N = \mathrm{jet}_p^k \tilde{N}$, *if and only if* $k+1$ *recursively defined obstructions*

$$B_{r+1}\left[\, \Theta^{\mathrm{pre}}(\,\mathrm{jet}_p^{k+1} N\,)\,\right] \in \mathrm{Sym}^{k-r} T_p^* N \otimes H^{r,2}(\mathscr{A}_p)$$

vanish for all $r = 0, \ldots, k$. *In the latter case the set of all possible jet solutions* $\mathrm{jet}_p^{k+1} N$ *over the jet solution* $\mathrm{jet}_p^k N$ *form an affine space modelled on the vector space* $\mathscr{R}_p^{k+1}$.

In a very precise sense the statement of this lemma reflects the standard spectral sequence of Lemma 3.7, the operators B_{r+1} say are exactly the higher order coboundary operators of this spectral sequence. In particular the obstructions are defined strictly recursively in the sense that $B_{r+1} [\Theta^{\mathrm{pre}}(\mathrm{jet}_p^{k+1} N)]$ is only defined, if the preceding obstructions $B_1 [\Theta^{\mathrm{pre}}(\mathrm{jet}_p^{k+1} N)], \ldots, B_r [\Theta^{\mathrm{pre}}(\mathrm{jet}^{k+1} N)] = 0$ all vanish. For this reason in particular it is rather difficult to calculate these integrability obstructions explicitly.

Proof. In order to begin we choose a representative submanifold $N \subset M$ for the given jet semisolution $\mathrm{jet}_p^{k+1} N \in \overline{\mathrm{Eq}}_p^{k+1} M$ of order $k \geq 0$ and consider the associated the $\mathbb{R}$-linear map $\Theta : \Gamma(CM) \longrightarrow \mathrm{Jet}_p^k \Lambda^2 T^* N, \gamma \longmapsto \mathrm{jet}_p^k(\iota_N^* d\gamma)$, whose triviality characterizes $\mathrm{jet}_p^{k+1} N$ as a jet solution of order $k + 1$. We recall that $\mathrm{jet}_p^{k+1} N$ projects to a jet solution $\mathrm{jet}_p^k N$

$$\mathrm{pr}[\mathrm{jet}_p^k(\iota_N^* d\gamma)] = \mathrm{jet}_p^{k-1}(d\,\iota_N^* \gamma) = d^{\mathrm{formal}}(\mathrm{jet}_p^k(\iota_N^* d\gamma)) = 0$$

so that the image of Θ lies in the symbol subspace $\mathrm{Sym}^k T_p^* N \otimes \Lambda^2 T_p^* N \subset \mathrm{Jet}_p^k \Lambda^2 T^* N$ defined in (63). In consequence the $\mathbb{R}$-linear map Θ is actually $C^\infty(M)$-linear with

$$\Theta(f\,\gamma) = \mathrm{jet}_p^k(\iota_N^*(df \wedge \gamma + f\,d\gamma))$$
$$= \mathrm{jet}_p^k(\iota_N^* df) \wedge \mathrm{jet}_p^k(\iota_N^* \gamma) + \mathrm{jet}_p^k(\iota_N^* f) \wedge \mathrm{jet}_p^k(\iota_N^* d\gamma)$$
$$= f(p)\,\mathrm{jet}_p^k(\iota_N^* d\gamma)$$

for all $f \in C^\infty(M)$, because $\mathrm{jet}_p^k(\iota_N^* \gamma) = 0$ as well as $\mathrm{jet}_p^{k-1}(\iota_N^* d\gamma) = 0$. We may thus think of Θ as a linear map $C_p M \longrightarrow \mathrm{Sym}^k T_p^* N \otimes \Lambda^2 T_p^* N$ with the additional property that

$$C_p M \xrightarrow{\Theta} \mathrm{Sym}^k T_p^* N \otimes \Lambda^2 T_p^* N \xrightarrow{B} \mathrm{Sym}^{k-1} T_p^* N \otimes \Lambda^3 T_p^* N$$

vanishes as a linear map, after all B agrees with d^{formal} on the symbol subspace (63) and so:

$$B[\Theta(\gamma_p)] = d^{\mathrm{formal}}[\mathrm{jet}_p^k(\iota_N^* d\gamma)] = \mathrm{jet}_p^{k-1}(\iota_N^* d^2\gamma) = 0$$

Up to this point we have followed the proof of Lemma 5.8 closely with only minute changes in the argument, but now we have to deviate from the path laid out above.

Although we may still choose a preimage $\Theta^{\mathrm{pre}} \in \mathrm{Sym}^{k+1} T_p^* N \otimes T_p^* N \otimes C_p^* M$ of Θ with the property

$$B(\langle\, \Theta^{\mathrm{pre}},\ \gamma_p \,\rangle) \;=\; \Theta(\gamma_p) \;=\; \mathrm{jet}_p^k(\iota_N^* d\gamma)$$

this preimage is no longer unique, because $B \;:\; \mathrm{Sym}^{k+1} T_p^* N \otimes T_p^* N \longrightarrow \mathrm{Sym}^k T_p^* N \otimes \Lambda^2 T_p^* N$ is no longer injective, to wit its kernel equals the image of $\mathrm{Sym}^{k+2} T_p^* N$ under B.

Keeping an eye on this non-uniqueness problem of the chosen preimage Θ^{pre} we observe that the restriction $\mathrm{res}_{T_p N} \;:\; H_p M \longrightarrow T_p^* N$ is surjective due to $T_p N \cap H_p^\perp M = \{0\}$ with kernel equal to $C_p M$ by $T_p N \subset C_p^\perp M$, in other words it induces a canonical isomorphism

$$H_p M / C_p M \;\overset{\mathrm{res}_{T_p N}}{\longrightarrow}\; T_p^* N \tag{67}$$

equivalently $T_p N \subset C_p^\perp M$ is a complete set of representatives for the quotient $C_p^\perp M / H_p^\perp M$. This canonical isomorphism by restriction allows us to interpret the tableau $\mathscr{A}_p^1$ as a subspace of $T_p^* N \otimes C_p^* M \cong (H_p M / C_p M) \otimes C_p^* M$, in turn we will consider the class

$$[\,\Theta^{\mathrm{pre}}\,] \;\in\; \mathrm{Sym}^{k+1} T_p^* N \otimes \Big[\, T_p^* N \otimes C_p^* M \big/ {\mathscr{A}_p^1} \,\Big] \;=\; \mathrm{Sym}^{k+1} T_p^* N \otimes H^{0,1}(\,\mathscr{A}_p\,)$$

represented by Θ^{pre}. The vector space on the right is one the trihomogeneous subspaces of the E^1-term of the standard spectral sequence for the reduced symbol comodule $\mathscr{A}_p$

$$\mathrm{Sym}^\bullet T_p^* N \otimes H^{\bullet,\circ}(\,\mathscr{A}_p\,) \;\Longrightarrow\; \delta_{\circ=0=\bullet}\,\mathscr{A}_p^\bullet$$

constructed in Lemma 3.7. More precisely the total degree $k+2$ part of the E^1-term reads

$$\mathrm{Sym}^{k+2} T_p^* N \otimes C_p^* M \overset{B_1}{\longrightarrow} \mathrm{Sym}^{k+1} T_p^* N \otimes \Big[T_p^* N \otimes C_p^* M \big/ {\mathscr{A}_p^1}\Big] \overset{B_1}{\longrightarrow} \mathrm{Sym}^k T_p^* N \otimes H^{0,2}(\mathscr{A}_p)$$

0	0	$\mathrm{Sym}^{k-1} T_p^* N \otimes H^{1,2}(\mathscr{A}_p)$
$\vdots$	$\vdots$	$\vdots$
0	0	$\mathrm{Sym}^1 T_p^* N \otimes H^{k-1,2}(\mathscr{A}_p)$
0	0	$\mathrm{Sym}^0 T_p^* N \otimes H^{k,2}(\mathscr{A}_p)$

in form degree 0, 1 and 2, where B_1 is the coboundary operator for the E^1-term and the higher order coboundary operators $B_2, \dots, B_{k+1}$ relevant for our argument have been indicated, although they are defined only on the kernel of all preceding coboundary operators. In consequence the standard spectral sequence results in $k+1$ recursively defined obstructions

$$B_{r+1}[\,\Theta^{\mathrm{pre}}\,] \;\in\; \mathrm{Sym}^{k-r} T_p^* N \otimes H^{r,2}(\,\mathscr{A}_p\,)$$

which are independent of the preimage Θ^{pre} chosen for Θ, because the resulting ambiguity of the class $[\,\Theta^{\mathrm{pre}}\,]$ lies in the image of the left coboundary operator B_1.

Because the standard spectral sequence converges to $\{0\}$ in all positive form degrees, the class $[\,\Theta^{\mathrm{pre}}\,]$ lies in the image of the left B_1 coboundary operator, if and only if all the recursively defined obstructions $B_{r+1}[\,\Theta^{\mathrm{pre}}\,] = 0$ vanish for all $r = 0, \ldots, k$. Under this assumption we can modify our chosen preimage to a possibly different preimage of Θ with:

$$\Theta^{\mathrm{pre}} \;\in\; \mathrm{Sym}^{k+1} T_p^* N \otimes \mathscr{A}_p^1 \;\subset\; \mathrm{Sym}^{k+1} T_p^* N \otimes T_p^* N \otimes C_p^* M \qquad (68)$$

Recall now that the set of jet semisolutions of order $k + 1$ lying over $\mathrm{jet}_p^k N$ is an affine space modelled on $\mathrm{Sym}^{k+1} T_p^* N \otimes H_p^\perp M$ according to Lemma 5.8, where $H_p^\perp M$ serves as a set of representatives for the subspace $C_p^\perp M / T_p N \subset T_p M / T_p N$.

In case that we can chose a preimage Θ^{pre} of Θ of the special form (68), equivalently in case that all recursively defined obstructions vanish, we can lift such a Θ^{pre} to a preimage $\Delta N \in \mathrm{Sym}^{k+1} T_p^* N \otimes H_p^\perp M$ under the surjective symbol map $\mathrm{id} \otimes \Sigma_p$. Since the symbol map Σ_p is based on the idea of inserting vertical vectors $V \in H_p^\perp M$ into the exterior derivatives

$$\mathrm{res}_{T_p N}\Big(V \,\lrcorner\, (d\gamma)_p \Big) \;=\; -\,\mathrm{res}_{T_p N}\Big(\Sigma_p(\gamma_p, \cdot\,, V) \Big) \;=\; -\Sigma_p(\gamma_p, \cdot\,, V)$$

of contact forms, every ΔN chosen in this way satisfies the decisive equation:

$$(\mathrm{id} \otimes \mathrm{res}_{T_p N})(\Delta N \,\lrcorner\, (d\gamma)_p) \;=\; -\Sigma_p(\gamma_p, \cdot\,, \Delta N) \;=\; -\langle\, \Theta^{\mathrm{pre}}, \gamma_p \rangle$$

Note that the restriction $\mathrm{res}_{T_p N}$ implements the canonical isomorphism $H_p M / C_p M \xrightarrow{\cong} T_p^* N$ only and can be dropped from notation. For all contact forms $\gamma \in \Gamma(CM)$ we thus find:

$$[\![\, \mathrm{jet}_p^{k+1} N + \Delta N \,]\!](\mathrm{jet}_p^k(d\gamma))$$

$$= \mathrm{jet}_p^k(\iota_N^* d\gamma) + B\Big((\mathrm{id} \otimes \mathrm{res}_{T_p N})(\Delta N \,\lrcorner\, (d\gamma)_p) \Big)$$

$$= \mathrm{jet}_p^k(\iota_N^* d\gamma) - B\langle\, \Theta^{\mathrm{pre}}, \gamma_p \rangle \;=\; 0$$

In consequence the modification $\mathrm{jet}_p^k N + \Delta N \in \mathrm{Eq}_p^{k+1} M$ is a jet solution of order $k + 1$ lying over $\mathrm{jet}_p^k N$. Being the kernel of $(B \otimes \mathrm{id}) \circ (\mathrm{id} \otimes \Sigma_p)$ in $\mathrm{Sym}^{k+1} T_p^* N \otimes$

$H_p^\perp M$ the homogeneous subspace $\mathscr{R}_p^{k+1}$ of the symbol comodule $\mathscr{R}_p$ parametrizes the possible choices for the difference element $\Delta N \in \mathrm{Sym}^{k+1} T_p^* N \otimes H_p^\perp M$. □

Unluckily Lemma 5.9 does not exclude the possibility of an infinite number of obstructions against formal integrability occurring at arbitrarily high orders of differentiation. Only the last of the recursively defined obstructions $B_{k+1}[\Theta^{\mathrm{pre}}(\mathrm{jet}_p^{k+1} N)]$ however appears to convey genuine information, the preceding obstructions are simply partial derivatives of the obstructions at lower order of differentiation. This intuitive idea gives rise to the conjecture:

Conjecture 5.10 (Vanishing Criterion for Obstructions). Consider an exterior differential system $CM \subseteq HM \subseteq T^*M$ on a manifold M. If the sets $\mathrm{Eq}^k M$ and $\mathrm{Eq}^{k-1} M$ of jet solutions of order k and $k - 1$ form a smooth subtower

$$
\begin{array}{ccccc}
\mathrm{Eq}^k M & \xrightarrow{\ \mathrm{pr}\ } & \mathrm{Eq}^{k-1} M & \xrightarrow{\ \pi\ } & M \\
{\scriptstyle\subset}\big\downarrow & & {\scriptstyle\subset}\big\downarrow & & {\scriptstyle=}\big\downarrow \\
\mathrm{Gr}_n^k M & \xrightarrow{\ \mathrm{pr}\ } & \mathrm{Gr}_n^{k-1} M & \xrightarrow{\ \pi\ } & M
\end{array}
$$

of the tower (16) of Graßmannians in a neighborhood of a point $p \in M$, then all the recursively defined obstructions on $\overline{\mathrm{Eq}}_p^{k+1} M$ vanish except possibly the last, the higher torsion:

$$
\mathrm{tor}^k : \quad \overline{\mathrm{Eq}}_p^{k+1} M \ \longrightarrow \ H^{k,2}(\mathscr{A}_p)
$$

For the time being the author has been unable to prove this conjecture, nevertheless he is quite convinced of its validity. The point is that the conjecture is definitely true in an essentially dual formulation of the formal theory of partial differential equations, however this proof appears to require the use of so-called semiholonomic jets and is thus not easily translated into the language of exterior differential systems. Assuming the validity of this conjecture and climbing up the tower (60) one step at a time using Lemmas 5.8 and 5.9 alternatingly the reader will find no difficulties to prove the following version of the Theorem of Cartan–Kähler inductively starting with the fact that $\mathrm{Eq}^1 M \longrightarrow M$ is a smooth subbundle of the tower of generalized Graßmannians in the case of vanishing torsion:

Theorem 5.11 (Formal Version of Cartan–Kähler). *Consider an exterior differential system $CM \subseteq HM \subseteq T^*M$ on a manifold M. In a given point $p \in M$ we choose $d_{\max}$ so that the Spencer cohomology of the reduced symbol comodule vanishes $H^{d,2}(\mathscr{A}_p) = \{0\}$ in form degree $\circ = 2$ for all $d > d_{\max}$. If the torsion maps*

$$
\mathrm{tor}^k : \quad \overline{\mathrm{Eq}}_p^{k+1} M \ \longrightarrow \ H^{k,2}(\mathscr{A}_p)
$$

all vanish for $k = 0, \ldots, d_{max}$, *then* $\mathrm{Eq}_p^{k+1} M$ *is an affine fiber bundle over* $\mathrm{Eq}_p^k M$ *with fiber modelled on* $\mathscr{R}_p^{k+1}$ *for all* $k \geq 0$. *In particular there exist as many formal submanifold solutions in* $\mathrm{Gr}_n^\infty M$ *as predicted by the dimensions of the homogeneous subspaces of* $\mathscr{R}_p$.

In its original formulation the theorem of Cartan–Kähler treats involutive reduced symbol comodules only, for which we may choose $d_{max} = 0$ according to Lemma 4.8, the theorem of Cauchy–Kovalevskaya for underdetermined partial can be seen as the case, where $d_{max} = -1$ is already sufficient. In general a formal solution to a given exterior differential system need not correspond to a real submanifold solution, under the additional assumption that M is an analytical manifold and both CM and HM are analytical subbundles of the cotangent bundle T^*M however, every formal solution defines an actual submanifold solution within its radius of convergence. Under this analyticity assumption the theorem of Cartan–Kähler extends to the statement that there exist as many formal solutions with *positive* radius of convergence as predicted by the dimensions of the homogeneous subspaces of $\mathscr{R}_p$.

References

1. R.L. Bryant, S.S. Chern, R.B. Gardner, H.L. Goldshmidt, P.A. Griffiths, *Exterior Differential Systems*. MSRI Lecture Notes, vol. 18 (Springer, New York, 1990)
2. D. Eisenbud, *Commutative Algebra with a View Toward Algebraic Geometry*. Graduate Texts in Mathematics, vol. 150 (Springer, Berlin, 2004)
3. W. Fulton, J. Harris, *Representation Theory: A First Course*. Readings in Mathematics, vol. 131 (Springer, New York, 1990)
4. M. Gromov, *Partial Differential Relations*. Ergebnisse der Mathematik und ihrer Grenzgebiete, vol. 9 (Springer, Berlin, 1986)
5. R. Hartshorne, *Algebraic Geometry*. Graduate Texts in Mathematics, vol. 52 (Springer, New York, 1977)
6. P.J. Olver, *Applications of Lie Groups to Differential Equations*. Graduate Texts in Mathematics, vol. 107 (Springer, New York, 1986)
7. J. Pommaret, *Differential Galois Theory*. Mathematics & its Applications (Routledge, New York, 1983)
8. F.W. Warner, *Foundations of Differentiable Manifolds and Lie Groups*. Graduate Texts in Mathematics, vol. 94 (Springer, New York, 1983)

LECTURE NOTES IN MATHEMATICS

 **Springer**

Edited by J.-M. Morel, B. Teissier; P.K. Maini

Editorial Policy (for Multi-Author Publications: Summer Schools / Intensive Courses)

1. Lecture Notes aim to report new developments in all areas of mathematics and their applications - quickly, informally and at a high level. Mathematical texts analysing new developments in modelling and numerical simulation are welcome. Manuscripts should be reasonably selfcontained and rounded off. Thus they may, and often will, present not only results of the author but also related work by other people. They should provide sufficient motivation, examples and applications. There should also be an introduction making the text comprehensible to a wider audience. This clearly distinguishes Lecture Notes from journal articles or technical reports which normally are very concise. Articles intended for a journal but too long to be accepted by most journals, usually do not have this "lecture notes" character.

2. In general SUMMER SCHOOLS and other similar INTENSIVE COURSES are held to present mathematical topics that are close to the frontiers of recent research to an audience at the beginning or intermediate graduate level, who may want to continue with this area of work, for a thesis or later. This makes demands on the didactic aspects of the presentation. Because the subjects of such schools are advanced, there often exists no textbook, and so ideally, the publication resulting from such a school could be a first approximation to such a textbook. Usually several authors are involved in the writing, so it is not always simple to obtain a unified approach to the presentation.

   For prospective publication in LNM, the resulting manuscript should not be just a collection of course notes, each of which has been developed by an individual author with little or no coordination with the others, and with little or no common concept. The subject matter should dictate the structure of the book, and the authorship of each part or chapter should take secondary importance. Of course the choice of authors is crucial to the quality of the material at the school and in the book, and the intention here is not to belittle their impact, but simply to say that the book should be planned to be written by these authors jointly, and not just assembled as a result of what these authors happen to submit.

   This represents considerable preparatory work (as it is imperative to ensure that the authors know these criteria before they invest work on a manuscript), and also considerable editing work afterwards, to get the book into final shape. Still it is the form that holds the most promise of a successful book that will be used by its intended audience, rather than yet another volume of proceedings for the library shelf.

3. Manuscripts should be submitted either online at www.editorialmanager.com/lnm/ to Springer's mathematics editorial, or to one of the series editors. Volume editors are expected to arrange for the refereeing, to the usual scientific standards, of the individual contributions. If the resulting reports can be forwarded to us (series editors or Springer) this is very helpful. If no reports are forwarded or if other questions remain unclear in respect of homogeneity etc, the series editors may wish to consult external referees for an overall evaluation of the volume. A final decision to publish can be made only on the basis of the complete manuscript; however a preliminary decision can be based on a pre-final or incomplete manuscript. The strict minimum amount of material that will be considered should include a detailed outline describing the planned contents of each chapter.

   Volume editors and authors should be aware that incomplete or insufficiently close to final manuscripts almost always result in longer evaluation times. They should also be aware that parallel submission of their manuscript to another publisher while under consideration for LNM will in general lead to immediate rejection.

4. Manuscripts should in general be submitted in English. Final manuscripts should contain at least 100 pages of mathematical text and should always include

   - a general table of contents;
   - an informative introduction, with adequate motivation and perhaps some historical remarks: it should be accessible to a reader not intimately familiar with the topic treated;
   - a global subject index: as a rule this is genuinely helpful for the reader.

   Lecture Notes volumes are, as a rule, printed digitally from the authors' files. We strongly recommend that all contributions in a volume be written in the same LaTeX version, preferably LaTeX2e. To ensure best results, authors are asked to use the LaTeX2e style files available from Springer's web-server at
   ftp://ftp.springer.de/pub/tex/latex/svmonot1/ (for monographs) and
   ftp://ftp.springer.de/pub/tex/latex/svmultt1/ (for summer schools/tutorials).
   Additional technical instructions, if necessary, are available on request from:
   lnm@springer.com.

5. Careful preparation of the manuscripts will help keep production time short besides ensuring satisfactory appearance of the finished book in print and online. After acceptance of the manuscript authors will be asked to prepare the final LaTeX source files and also the corresponding dvi-, pdf- or zipped ps-file. The LaTeX source files are essential for producing the full-text online version of the book. For the existing online volumes of LNM see:
   http://www.springerlink.com/openurl.asp?genre=journal&issn=0075-8434.
   The actual production of a Lecture Notes volume takes approximately 12 weeks.

6. Volume editors receive a total of 50 free copies of their volume to be shared with the authors, but no royalties. They and the authors are entitled to a discount of 33.3 % on the price of Springer books purchased for their personal use, if ordering directly from Springer.

7. Commitment to publish is made by letter of intent rather than by signing a formal contract. Springer-Verlag secures the copyright for each volume. Authors are free to reuse material contained in their LNM volumes in later publications: a brief written (or e-mail) request for formal permission is sufficient.

Addresses:
Professor J.-M. Morel, CMLA,
École Normale Supérieure de Cachan,
61 Avenue du Président Wilson, 94235 Cachan Cedex, France
E-mail: morel@cmla.ens-cachan.fr

Professor B. Teissier, Institut Mathématique de Jussieu,
UMR 7586 du CNRS, Équipe "Géométrie et Dynamique",
175 rue du Chevaleret, 75013 Paris, France
E-mail: teissier@math.jussieu.fr

For the "Mathematical Biosciences Subseries" of LNM:

Professor P. K. Maini, Center for Mathematical Biology,
Mathematical Institute, 24-29 St Giles,
Oxford OX1 3LP, UK
E-mail: maini@maths.ox.ac.uk

Springer, Mathematics Editorial I,
Tiergartenstr. 17,
69121 Heidelberg, Germany,
Tel.: +49 (6221) 4876-8259
Fax: +49 (6221) 4876-8259
E-mail: lnm@springer.com